Vectors and Tensors in Engineering and Physics,

SECOND EDITION

D. A. Danielson

Naval Postgraduate School

Advanced Book Program

CRC Press
Taylor & Francis Group
Boca Raton London New York

CRC Press is an imprint of the
Taylor & Francis Group, an **informa** business
A CHAPMAN & HALL BOOK

In memory of my mother and father

First published 2003 by Westview Press

Published 2018 by CRC Press
Taylor & Francis Group
6000 Broken Sound Parkway NW, Suite 300
Boca Raton, FL 33487-2742

CRC Press is an imprint of the Taylor & Francis Group, an informa business

Visit the Taylor & Francis Web site at
http://www.taylorandfrancis.com

and the CRC Press Web site at
http://www.crcpress.com

A Cataloging-in-Publication data record for this book is available from the Library of Congress.

ISBN 13: 978-0-8133-4080-7 (pbk)

Text designed and typeset by the author

CONTENTS

PREFACE

I have written this text for those who want to understand the close correspondence between mathematics and the physical world. The emphasis herein on physical interpretation means that certain aspects of mathematical rigor are neglected. "Rigor" to a mathematician can be "rigor mortis" to an engineer.

An advanced undergraduate student majoring in a physical science, engineering, or mathematics would have the necessary prerequisites for this book. The reader is expected to know only some elementary calculus and linear algebra. In order to make the subject accessible to a wide audience, certain valuable mathematical tools such as the calculus of variations have not been used.

The focus of revision for this second edition is the insertion of additional steps in the derivations of mathematical formulas and problem solutions. The changes will make it easier for the reader to learn the subject matter solely from this text.

Whereas my own students were the proving ground for the material in the first edition, other readers have suggested many of the improvements in this second edition. My heartfelt thanks goes to each of these individuals who seek the truth.

The universe stands continually open to our gaze, but it cannot be understood unless one first learns to comprehend the language and interpret the characters in which it is written. It is written in the language of mathematics, and its characters are triangles, circles, and other geometrical figures, without which it is humanly impossible to understand a single word of it; without these, one is wandering about in a dark labyrinth.

– Galileo Galilei, 1623

Chapter 1

VECTOR ALGEBRA

1.1 INTRODUCTION

Tensor mathematics is a beautiful, simple, and useful language for the description of natural phenomena. *Tensor fields* are the abstract symbols of this language. Each tensor field represents a single physical quantity that is associated with certain *places* in three-dimensional space and *instants* of time.

Quantities such as the mass of a satellite, the temperature at points in a body, and the charge of an electron have a definite magnitude. They can be represented adequately by pure numbers or *scalars* (tensors of order *zero*). Properties such as the position or velocity of a satellite, the flow of heat in a body, and the electromagnetic force on an electron have both magnitude and direction. They are better represented by directed line segments or *vectors* (tensors of order *one*). Other quantities such as the stress inside a solid or fluid may be characterized by tensors of order *two* or higher.

Tensor equations express the relationship between physical quantities. We will study the basic equations governing the trajectories of point masses in a gravitational field, motion of finite rigid bodies, transfer of heat by conduction, deformation of solids, flow of fluids, and electromagnetic phenomena. By solving these equations, we will be able to predict the course of future events.

You will gain from this book a manipulative skill with tensors and an ability to apply them to model our physical world. Then when you encounter a tensor in your studies or research, you will recognize it as such and have the correct mathematical tools to work with.

1.2 HISTORICAL NOTES

Great scientific breakthroughs result from the accumulated efforts of many researchers working over a long period of time. Nevertheless, later generations inaccurately attribute these achievements solely to just one or two individuals who first assembled the ideas of their predecessors into a unifying theory or book. We list here some of the individuals who are now credited with the development of our field. For an exciting experience in the world of mathematics, read the classic works under historical references at the back of this book.

Euclidean geometry is based on the *Elements* by the Greek Euclid (300 B.C.). Cartesian coordinates are named after the French scientist Descartes (1596–1650). Newton (1642–1727) presented the motion of the planets and other bodies in strictly geometrical terms. The great mathematician Euler (1707–1783) used Cartesian components for force and moment of inertia. The French mathematician

Cauchy (1789–1857) correctly described the general state of stress in an elastic body. The words "scalar," "vector," and "tensor"[1] were used by the Irish mathematician Hamilton (1805–1865). The German mathematicians Gauss (1777–1855) and Riemann (1826–1866) were major contributors to the metrical geometry of non-Euclidean manifolds. The great physicist Maxwell (1831–1879) was aware that the dielectric permittivity, the magnetic permeability, and the conductivity for electric currents may be linear vector operators. Our present vector and dyadic notation is almost the same as that used in the old book by the Americans Gibbs and Wilson (1901). Tensor calculus was perfected by the Italian mathematicians Ricci (1853–1925) and Levi-Civita (1873–1941). Einstein (1879–1955) gets credit for first applying the generalized calculus of tensors to gravitation.

1.3 VECTOR BASICS

In the first eight chapters of this book, we model physical space by three-dimensional Euclidean geometry. In a Euclidean space we may construct straight line segments. If we ascribe a direction to a line segment, it becomes a *vector*. A vector may be represented geometrically by an arrow (see Fig. 1.1). The *direction* of a vector is the direction of its arrow, and the *magnitude* of a vector is the length of its arrow. We represent a vector algebraically by a *lowercase* boldface letter, such as $\mathbf{v}$, e, or $\mathbf{f}$.

A *scalar* is simply a pure number. In the text of this book every number is *real*; that is, every number can be represented by a terminating or nonterminating decimal. Scalars are designated by italic letters, such as c, k, or W. The magnitude (length) of a vector $\mathbf{v}$ is denoted by $|\mathbf{v}|$ or v. Of course, the magnitude (absolute value) of a scalar c is also denoted by $|c|$. (Both $|\mathbf{v}|$ and $|c|$ are nonnegative

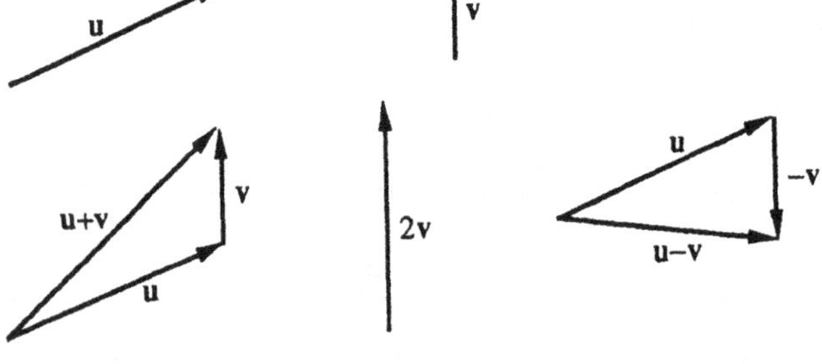

Figure 1.1. The sum and difference of vectors $\mathbf{u}$ and $\mathbf{v}$; multiplication of $\mathbf{v}$ by the scalars 2 and 0.

[1]The word "scalar" stems from the Latin "scalae" meaning "stairs" (scale of numbers from negative to positive infinity). The word "vector" stems from the Latin "vectus" meaning "carried" (denoting a direction). The word "tensor" stems from the Latin "tensus" meaning "stretched" (tension).

numbers.) Any vector whose length is one is called a *unit* vector. *Throughout this book the symbol* $\mathbf{e}$ *will always denote a unit vector.*

We say that two vectors are *equal* if they have the same magnitude and direction. Thus vectors in vector algebra are *free*, meaning that a vector is free to move about under parallel displacements.

Two vectors $\mathbf{u}$ and $\mathbf{v}$ may be *added* by placing the tail of one on the head of the other. The sum $\mathbf{u}+\mathbf{v}$ is then defined to be the vector joining the unattached tail to the unattached head. A vector $\mathbf{v}$ may be *multiplied* by a scalar c. The product $c\mathbf{v}$ is defined to be a new vector parallel to $\mathbf{v}$ whose magnitude has been multiplied by $|c|$ and whose direction is the same as $\mathbf{v}$ if $c > 0$ or opposite to $\mathbf{v}$ if $c < 0$. Any vector whose magnitude is zero is called the *zero vector* and given the symbol $\mathbf{0}$ or 0. These definitions are illustrated in Fig. 1.1.

The laws that govern addition and scalar multiplication of vectors are identical with those governing these operations in the ordinary arithmetic of real numbers:

$$\begin{aligned}
\mathbf{u}+\mathbf{v} &= \mathbf{v}+\mathbf{u}, \\
(\mathbf{u}+\mathbf{v})+\mathbf{w} &= \mathbf{u}+(\mathbf{v}+\mathbf{w}) = \mathbf{u}+\mathbf{v}+\mathbf{w}, \\
c(d\mathbf{v}) &= d(c\mathbf{v}) = (cd)\mathbf{v} = cd\mathbf{v}, \\
(c+d)\mathbf{v} &= c\mathbf{v}+d\mathbf{v}, \\
c(\mathbf{u}+\mathbf{v}) &= c\mathbf{u}+c\mathbf{v}.
\end{aligned}$$

It is common practice to omit parentheses in vector expressions when they are not needed to prevent ambiguity. These rules are equivalent to propositions in solid geometry dealing with the properties of triangles and other polygons.

Vector algebra can be a powerful tool in geometry.

Intersection of the medians of a triangle. A median of a triangle is a line connecting a vertex with the midpoint of the opposite side. Construct two medians, and label vectors emanating from the point P of intersection to the vertices by $\mathbf{a}, \mathbf{b}, \mathbf{c}$ and to the midpoints of the sides by $\mathbf{a}', \mathbf{b}'$ (Fig. 1.2). From vector addition, we have

$$\begin{aligned}
\mathbf{a}' &= \mathbf{c} + \frac{\mathbf{b}-\mathbf{c}}{2} = \frac{\mathbf{b}}{2} + \frac{\mathbf{c}}{2}, \\
\mathbf{b}' &= \mathbf{c} + \frac{\mathbf{a}-\mathbf{c}}{2} = \frac{\mathbf{a}}{2} + \frac{\mathbf{c}}{2}.
\end{aligned}$$

Eliminating $\mathbf{c}$ from these equations yields

$$2\mathbf{a}' + \mathbf{a} = 2\mathbf{b}' + \mathbf{b}. \tag{1.1}$$

But the vector on the left of (1.1) is not parallel to the vector on the right of (1.1), and hence each must be zero:

$$\mathbf{a} = -2\mathbf{a}', \qquad \mathbf{b} = -2\mathbf{b}'.$$

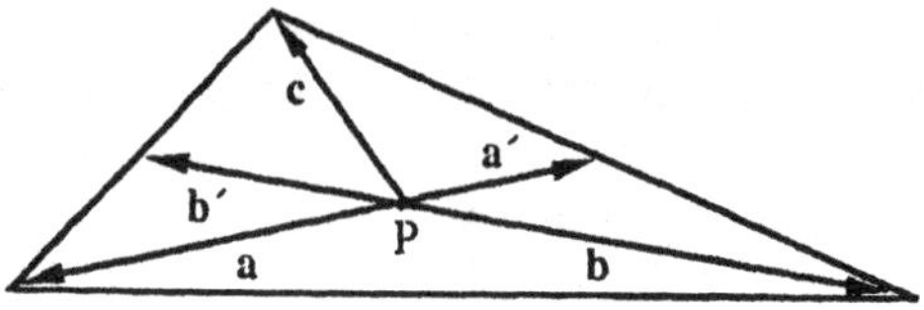

Figure 1.2. The medians of a triangle meet at a point P which is $\frac{2}{3}$ of the distance from the vertices.

Thus P is $\frac{2}{3}$ of the distance along the medians from the vertices. A similar argument applied to the third median and one of the already constructed medians implies that all three medians meet at the single point P, which divides each median in the ratio 2 to 1.

PROBLEMS 1.3

Work as many of the homework problems at the end of each section as you have time for. Solving the homework problems will deepen your understanding of the text. Answers are at the back of the book.

1. Interpret $\frac{\mathbf{v}}{|\mathbf{v}|}$ for $\mathbf{v} \neq 0$.
2. If the vector $\mathbf{u}$ is 1 unit long and points north and the vector $\mathbf{v}$ is 2 units long and points west, determine $2\mathbf{u} - \mathbf{v}$.
3. Find the magnitude of the sum of three unit vectors drawn from a common vertex of a cube along three of its sides.
4. Show that to add many vectors we can join all the vectors together in the head-to-tail method and then the sum will be the vector joining the unattached tail to the unattached head.
5. Consecutive sides of a quadrilateral are labeled $\mathbf{a}$, $\mathbf{b}$, $\mathbf{c}$, and $\mathbf{d}$. Show that the figure is a parallelogram if and only if $\mathbf{a}+\mathbf{c} = 0$, in which case $\mathbf{b}+\mathbf{d} = 0$.
6. Show that if two vectors add up to zero, they must be parallel to a straight line, and if three vectors add up to zero, they must be parallel to a plane.
7. Prove that the diagonals of a parallelogram bisect each other.
8. Prove the inequalities: $|\mathbf{u}| - |\mathbf{v}| \leq |\mathbf{u}+\mathbf{v}| \leq |\mathbf{u}| + |\mathbf{v}|$.
9. Show that the lines connecting a vertex of a tetrahedron with the point of intersection of the medians on the opposite face meet at a point that divides each line in the ratio 3 to 1.

1.4 DOT AND CROSS PRODUCT

The *dot* or scalar product of two vectors $\mathbf{u}$ and $\mathbf{v}$ is a scalar quantity defined by[2]

$$\boxed{\mathbf{u} \cdot \mathbf{v} = uv \cos\theta .}$$

Here θ denotes the angle between $\mathbf{u}$ and $\mathbf{v}$ when their origins coincide (Fig. 1.3a). To investigate the geometrical significance of the dot product, define a unit vector $\mathbf{e} = \frac{\mathbf{u}}{u}$ in the direction of $\mathbf{u}$. The vector $\mathbf{e}(\mathbf{e} \cdot \mathbf{v})$ that we get by orthogonally projecting $\mathbf{v}$ onto a line parallel to $\mathbf{e}$ is called the *vector projection* of $\mathbf{v}$ onto $\mathbf{e}$ and denoted by

$$\mathbf{P} \cdot \mathbf{v} = \mathbf{e}(\mathbf{e} \cdot \mathbf{v}). \tag{1.2}$$

Since $\mathbf{v}$ can be written as the sum of vector projections in two perpendicular directions (Fig. 1.3b), we call $\mathbf{e} \cdot \mathbf{v}$ the *component* of $\mathbf{v}$ in the direction of $\mathbf{e}$. Thus $\mathbf{u} \cdot \mathbf{v}$ is the magnitude of $\mathbf{u}$ times the component of $\mathbf{v}$ in the direction of $\mathbf{u}$.

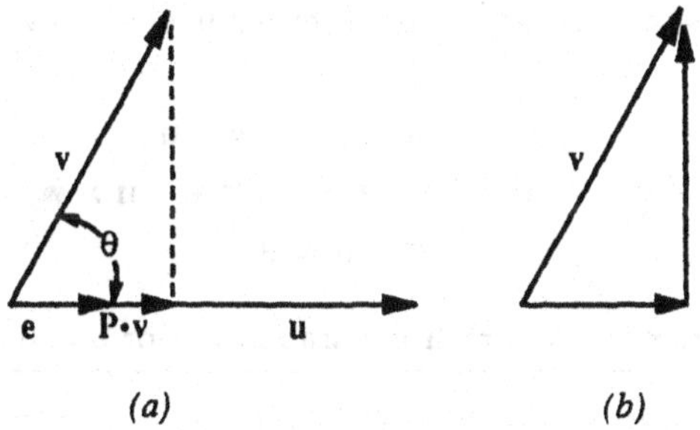

Figure 1.3. (a) Vector projection $\mathbf{P} \cdot \mathbf{v}$ of $\mathbf{v}$ onto a unit vector $\mathbf{e}$ in the direction of $\mathbf{u}$. (b) Decomposition of $\mathbf{v}$ into components in the direction of $\mathbf{e}$ and perpendicular to $\mathbf{e}$.

The *cross* or vector product of two vectors $\mathbf{u}$ and $\mathbf{v}$, denoted by $\mathbf{u} \times \mathbf{v}$, is a vector quantity. The direction of $\mathbf{u} \times \mathbf{v}$ is the direction of the extended thumb when the fingers of the right hand are closed from $\mathbf{u}$ to $\mathbf{v}$ (origins coinciding) through the smallest possible angle θ $(0 \leq \theta \leq \pi)$. The magnitude of $\mathbf{u} \times \mathbf{v}$ is defined by

$$\boxed{|\mathbf{u} \times \mathbf{v}| = uv \sin\theta .}$$

The magnitude of $\mathbf{u} \times \mathbf{v}$ is the area of the parallelogram determined by $\mathbf{u}$ and $\mathbf{v}$, as shown in Fig. 1.4.

[2] Formulas in this text which are enclosed in a box should be *memorized*, if you do not already know them.

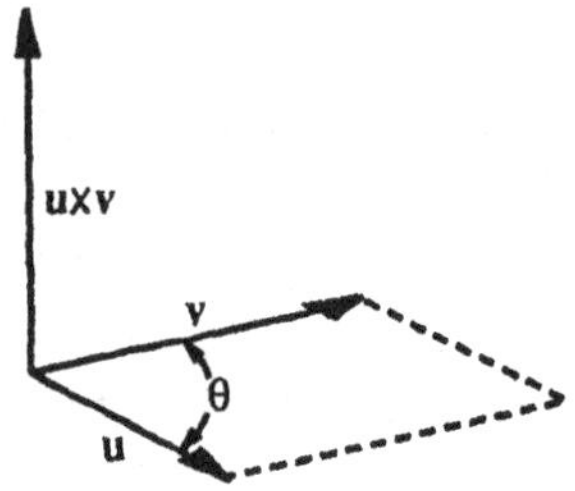

Figure 1.4. The cross product of vectors **u** and **v** is in the indicated direction and has magnitude equal to the area of the parallelogram.

You should easily be able to verify the following rules governing these products:

$$\mathbf{u}\cdot\mathbf{v}=\mathbf{v}\cdot\mathbf{u},$$
$$\mathbf{u}\cdot(\mathbf{v}+\mathbf{w})=\mathbf{u}\cdot\mathbf{v}+\mathbf{u}\cdot\mathbf{w},$$
$$\mathbf{u}\cdot\mathbf{u}=u^2,$$

$\mathbf{u}\cdot\mathbf{v}=0 \iff \mathbf{u}$ is orthogonal (perpendicular) to $\mathbf{v}$ or $\mathbf{u}=0$ or $\mathbf{v}=0$;

$$\mathbf{u}\times\mathbf{v}=-\mathbf{v}\times\mathbf{u},$$
$$\mathbf{u}\times(\mathbf{v}+\mathbf{w})=\mathbf{u}\times\mathbf{v}+\mathbf{u}\times\mathbf{w},$$
$$\mathbf{u}\times\mathbf{u}=0,$$

$\mathbf{u}\times\mathbf{v}=0 \iff \mathbf{u}$ is parallel to $\mathbf{v}$ or $\mathbf{u}=0$ or $\mathbf{v}=0$.

The product $(\mathbf{u}\times\mathbf{v})\cdot\mathbf{w}$ is called the box or *triple scalar product.* Its geometrical interpretation depends upon its sign:

(a) If $(\mathbf{u}\times\mathbf{v})\cdot\mathbf{w}>0$, then $(\mathbf{u}\times\mathbf{v})\cdot\mathbf{w}$ is the volume of the parallelepiped determined by $\mathbf{u},\mathbf{v},\mathbf{w}$ (Fig. 1.5a). In this case we say that $\mathbf{u},\mathbf{v},\mathbf{w}$ form a *right-handed* triad of vectors.

(b) If $(\mathbf{u}\times\mathbf{v})\cdot\mathbf{w}<0$, then $(\mathbf{u}\times\mathbf{v})\cdot\mathbf{w}$ is the *negative* of the volume of the parallelepiped whose sides are $\mathbf{u},\mathbf{v},\mathbf{w}$ (Fig. 1.5b). In this case $\mathbf{u},\mathbf{v},\mathbf{w}$ form a *left-handed* triad.

(c) If $(\mathbf{u}\times\mathbf{v})\cdot\mathbf{w}=0$, the vectors $\mathbf{u},\mathbf{v},\mathbf{w}$ are coplanar (parallel to the same plane).

It follows from this geometrical interpretation that the dot and cross in the triple scalar product may be interchanged:

$$\boxed{(\mathbf{u}\times\mathbf{v})\cdot\mathbf{w}=\mathbf{u}\cdot(\mathbf{v}\times\mathbf{w}).} \tag{1.3}$$

Note that the parentheses in (1.3) can be eliminated because the products can be interpreted in only one way.

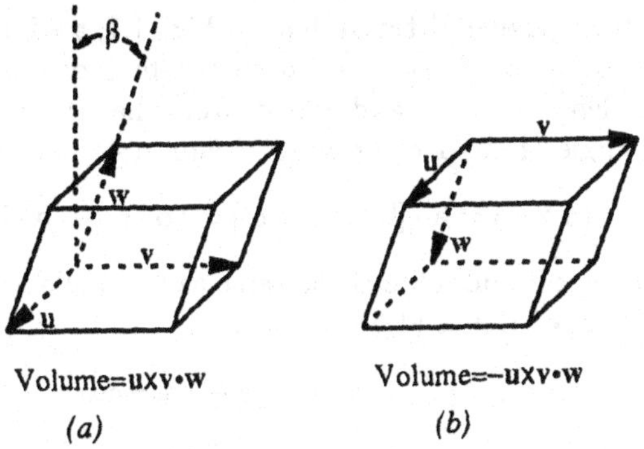

Figure 1.5. The triple scalar product of $\mathbf{u}$, $\mathbf{v}$, and $\mathbf{w}$ is plus or minus the volume of these parallelepipeds.

The product $(\mathbf{u} \times \mathbf{v}) \times \mathbf{w}$ is called a *triple vector product*. In general, the triple vector product is not associative: $(\mathbf{u} \times \mathbf{v}) \times \mathbf{w} \neq \mathbf{u} \times (\mathbf{v} \times \mathbf{w})$. Vector products of three or more vectors can be reduced to simpler products with the aid of the identity

$$\boxed{\mathbf{a} \times (\mathbf{b} \times \mathbf{c}) = \mathbf{b}(\mathbf{a} \cdot \mathbf{c}) - \mathbf{c}(\mathbf{a} \cdot \mathbf{b}).} \tag{1.4}$$

This is known as the "back cab" rule.

We can use vector products to solve geometrical problems.

Law of cosines for plane triangles. Let $\mathbf{a}, \mathbf{b}$, and $\mathbf{c} = \mathbf{a} + \mathbf{b}$ be vectors along the sides of a triangle, as shown in Fig. 1.6. Using the properties of the dot products, we obtain

$$\begin{aligned} c^2 &= (\mathbf{a} + \mathbf{b}) \cdot (\mathbf{a} + \mathbf{b}) = \mathbf{a} \cdot \mathbf{a} + 2\mathbf{a} \cdot \mathbf{b} + \mathbf{b} \cdot \mathbf{b} \\ &= a^2 + 2ab \cos(\pi - C) + b^2 \\ &= a^2 + b^2 - 2ab \cos C, \end{aligned}$$

where C denotes the angle opposite the side c.

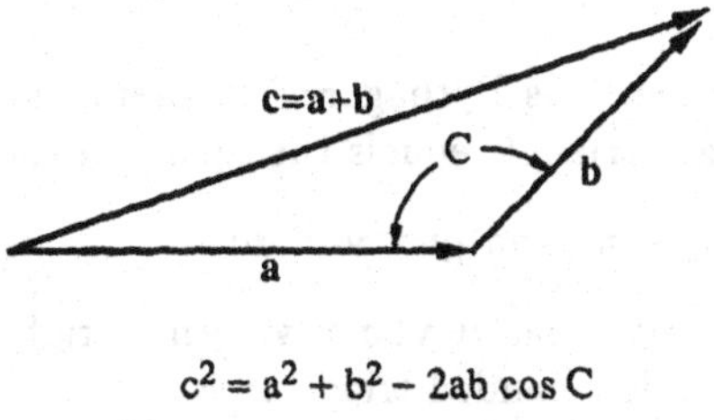

Figure 1.6. Law of cosines for plane triangles.

Law of cosines for spherical triangles. In Fig. 1.7, a triangle whose sides are arcs of great circles has been constructed on a sphere of radius 1. The sides of the triangle are of length α, β, γ, and vectors from the center of the sphere to the vertices of the triangle are $\mathbf{a}, \mathbf{b}, \mathbf{c}$. From (1.3) and (1.4), we have

$$(\mathbf{a} \times \mathbf{b}) \cdot (\mathbf{a} \times \mathbf{c}) = \mathbf{a} \cdot [\mathbf{b} \times (\mathbf{a} \times \mathbf{c})] = (\mathbf{b} \cdot \mathbf{c}) - (\mathbf{a} \cdot \mathbf{b})(\mathbf{a} \cdot \mathbf{c}). \qquad (1.5)$$

The angle between $\mathbf{a} \times \mathbf{b}$ and $\mathbf{a} \times \mathbf{c}$ is the same as the angle A between the planes formed by $\mathbf{a}, \mathbf{b}$ and $\mathbf{a}, \mathbf{c}$. Hence (1.5) becomes

$$\sin\gamma \sin\beta \cos A = \cos\alpha - \cos\gamma \cos\beta.$$

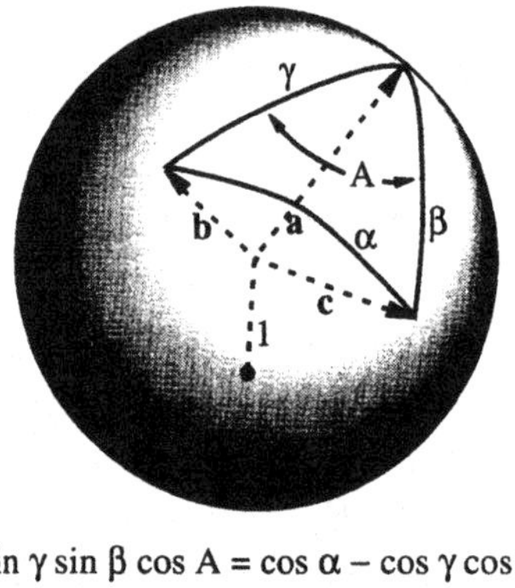

Figure 1.7. Law of cosines for spherical triangles.

PROBLEMS 1.4

1. Can an arbitrary vector be added, dotted, or crossed into each side of a vector equation?

2. The four chemical bonds of the carbon atom are directed to the four corners of a regular tetrahedron surrounding the atom. Find the angle between the bonds.

3. Prove by vector methods Pythagoras' theorem: The sum of the squares of the sides of a right triangle equals the square of the hypotenuse.

4. Prove that $|\mathbf{u} \cdot \mathbf{v}| \le uv$ and $|\mathbf{u} \times \mathbf{v}| \le uv$.

5. If $\mathbf{u} \times \mathbf{v} \ne 0$, verify that $\mathbf{u}, \mathbf{v}, \mathbf{u} \times \mathbf{v}$ form a right-handed triad and that $\mathbf{u}, \mathbf{v}, \mathbf{v} \times \mathbf{u}$ form a left-handed triad.

6. Verify the geometrical interpretation of $(\mathbf{u} \times \mathbf{v}) \cdot \mathbf{w}$. Then show that every switch of the vectors in the triple scalar product causes a change in sign: $\mathbf{u} \times \mathbf{v} \cdot \mathbf{w} = -\mathbf{u} \times \mathbf{w} \cdot \mathbf{v} = \mathbf{w} \times \mathbf{u} \cdot \mathbf{v} = -\mathbf{w} \times \mathbf{v} \cdot \mathbf{u} = \mathbf{v} \times \mathbf{w} \cdot \mathbf{u} = -\mathbf{v} \times \mathbf{u} \cdot \mathbf{w}$.

7. Express an arbitrary vector $\mathbf{v}$ in terms of three noncoplanar vectors $\mathbf{g}_1, \mathbf{g}_2, \mathbf{g}_3$.

8. Prove that $\mathbf{u} \times (\mathbf{v} \times \mathbf{w}) - (\mathbf{u} \times \mathbf{v}) \times \mathbf{w} = \mathbf{v} \times (\mathbf{u} \times \mathbf{w})$.

9. Prove that

$$(\mathbf{a} \times \mathbf{b}) \times (\mathbf{c} \times \mathbf{d}) = \mathbf{c}(\mathbf{a} \times \mathbf{b} \cdot \mathbf{d}) - \mathbf{d}(\mathbf{a} \times \mathbf{b} \cdot \mathbf{c})$$

and

$$(\mathbf{a} \times \mathbf{b}) \cdot (\mathbf{c} \times \mathbf{d}) = (\mathbf{a} \cdot \mathbf{c})(\mathbf{b} \cdot \mathbf{d}) - (\mathbf{a} \cdot \mathbf{d})(\mathbf{b} \cdot \mathbf{c}).$$

10. Derive the law of sines for plane triangles:

$$\frac{\sin A}{a} = \frac{\sin B}{b} = \frac{\sin C}{c}.$$

11. Derive the law of sines for spherical triangles:

$$\frac{\sin A}{\sin \alpha} = \frac{\sin B}{\sin \beta} = \frac{\sin C}{\sin \gamma}.$$

1.5 PHYSICAL SCALARS AND VECTORS

Vector analysis was originally developed to satisfy the needs of scientists who needed a mathematical symbolism in which to express physical laws. In this section we discuss some physical quantities that can be represented by scalars or vectors and some equations from mechanics relating these quantities. For more information on the various applications of vectors and tensors, consult the references at the back of this book.

Temperature, mass, humidity, speed, and voltage. Each of these has magnitude that can be represented by a pure number. A temperature of 8 degrees, a mass of 8 grams, and a mass of 8 kilograms are all represented by the same scalar 8. But it would make no sense to add a scalar representing temperature to a scalar representing mass. We can add only tensors representing different values of the same object expressed in the same units.

Position. Suppose you walk along a curved path, as sketched in Fig. 1.8. We can represent the distance from a fixed point O to your point P by the *position* vector $\mathbf{r} = \vec{OP}$. We can represent your change of position from point P_1 to P_2 by the *displacement* vector $\vec{P_1P_2} = \mathbf{r}_2 - \mathbf{r}_1$, and from point P_2 to P_3 by another arrow $\vec{P_2P_3} = \mathbf{r}_3 - \mathbf{r}_2$. The displacement vector represents only the net effect of the motion and does not depend upon the actual path taken; the net effect of the two displacements is the same as a displacement from P_1 to P_3. Thus actual displacements obey the head-to-tail rule of vector addition: $\vec{P_1P_2} + \vec{P_2P_3} = \vec{P_1P_3} = \mathbf{r}_3 - \mathbf{r}_1$. Displacement is a vector rather than a scalar quantity because the magnitude of displacements do not add, in general: $|\vec{P_1P_2}| + |\vec{P_2P_3}| \neq |\vec{P_1P_3}|$.

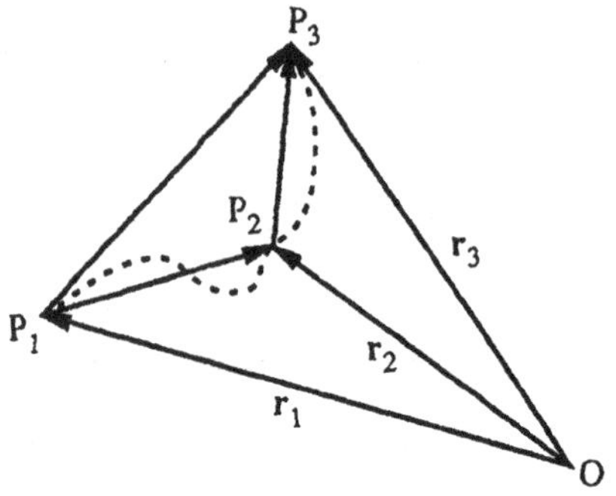

Figure 1.8. Displacement vectors $\overrightarrow{P_1P_2}$, $\overrightarrow{P_1P_3}$, $\overrightarrow{P_2P_3}$ and position vectors $\vec{\mathbf{r}}_1$, $\vec{\mathbf{r}}_2$, $\vec{\mathbf{r}}_3$.

Velocity. The average *velocity* vector $\mathbf{v}$ of your walk from P_1 to P_2 is the displacement vector $\overrightarrow{P_1P_2}$ divided by the time it takes you to walk from P_1 to P_2. The magnitude of $\mathbf{v}$ is your average speed. To convince yourself that velocities usually add like vectors, imagine your walk to occur on a people conveyer moving with average velocity $\mathbf{v}$ relative to the ground (Fig. 1.9). Suppose your velocity relative to the moving conveyer is $\bar{\mathbf{v}}$. Since a people conveyer does not move too fast, your velocity relative to the ground can be taken to be $\mathbf{v} + \bar{\mathbf{v}}$. However, if the speed of the conveyer were near the speed of light, you would have to calculate your velocity relative to the ground by a more complicated formula based

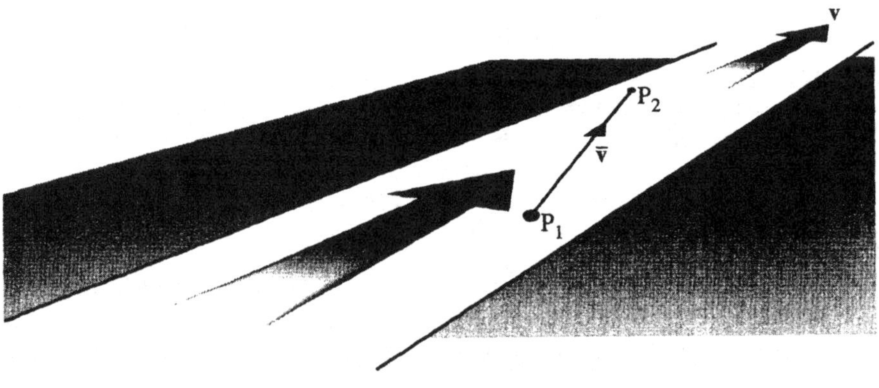

Figure 1.9. If a conveyer moves with velocity $\mathbf{v}$ relative to the ground and you move with velocity $\bar{\mathbf{v}}$ relative to the conveyer, your velocity relative to the ground is $\mathbf{v} + \bar{\mathbf{v}}$.

on relativity theory (see Section 9.1). Thus velocities add like three-dimensional vectors at normal speeds but not at speeds encountered in particle accelerators.

Gravity assist. The increase in speed of a spacecraft which occurs as a result of its passage close by a planet is called *gravity assist.* When a spacecraft passes close to a planet, it will be temporarily captured by the gravitational field of the planet. Neglecting the gravitational attraction of the sun and other perturbations during the flyby, we can show (see Section 7.4) that the spacecraft will follow a hyperbolic trajectory about the planet. For simplicity, let us assume that the spacecraft's orbit and planet's orbit lie in the same plane. Let $\bar{\mathbf{v}}_1$ be the velocity of the spacecraft *relative to the planet* at the time it is captured by the gravitational field of the planet, and let δ_1 be the angle $\bar{\mathbf{v}}_1$ makes with the planetary orbit, as sketched in Fig. 1.10. The velocity $\bar{\mathbf{v}}_2$ of the spacecraft *relative to the planet* at the time it is released by the gravitational field of the planet will have the same magnitude $\bar{v}$ but a different direction δ_2 (the turning angle $\delta_2 - \delta_1$ is determined by the closeness of approach of the spacecraft to the planet). If $\mathbf{v}$ denotes the velocity of the planet *relative to the sun*, then the velocity of the spacecraft *relative to the sun* is $\mathbf{v} + \bar{\mathbf{v}}_1$ at the time it is captured by the gravitational field of the planet and $\mathbf{v} + \bar{\mathbf{v}}_2$ at the time it is released by the gravitational field of the planet. From the vector addition diagram in Fig. 1.10 and the law of cosines, the initial and final speeds of the spacecraft *relative to the sun* are, respectively,

$$\sqrt{v^2 + \bar{v}^2 - 2v\bar{v}\cos\delta_1} \quad \text{and} \quad \sqrt{v^2 + \bar{v}^2 - 2v\bar{v}\cos\delta_2}.$$

For the Voyager 1 flyby of the planet Jupiter (which occurred on March 5, 1979) $v = 12.8\ \frac{\text{km}}{\text{s}}\ \left(\frac{\text{kilometers}}{\text{second}}\right)$, $\bar{v} = 10.8\ \frac{\text{km}}{\text{s}}$, $\delta_1 = 63.8°$, $\delta_2 = 162.4°$, so the speed of Voyager 1 relative to the sun increased from 12.6 $\frac{\text{km}}{\text{s}}$ to 23.3 $\frac{\text{km}}{\text{s}}$.

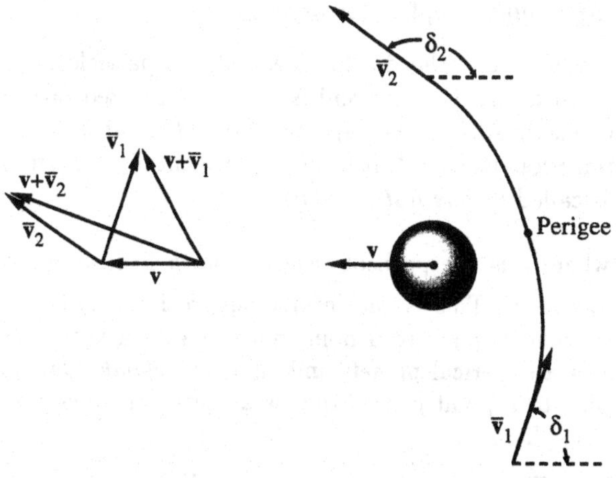

Figure 1.10. Vector addition diagram for figuring the increase in speed of a spacecraft passing by a planet.

Force. A *force* vector $\mathbf{f}$ represents a push or a pull exerted in a given direction. Whether or not forces add like vectors can be tested experimentally. Pull on a spring with two different known forces $\mathbf{f}_1$ and $\mathbf{f}_2$, as depicted in Fig. 1.11a. Then release these two forces and measure the new force $\mathbf{f}$ required to extend the spring to the same position (Fig. 1.11b). It will be found that $\mathbf{f}$ is the resultant vector obtained by adding the original force vectors: $\mathbf{f} = \mathbf{f}_1 + \mathbf{f}_2$. Forces acting at any point of a solid body may be replaced by their *resultant* obtained from the parallelogram law of addition. However, two parallel forces applied at different points of a body may not be equal in their mechanical effects. When determining physical effects, we usually must associate a given tensor with only a single place on the physical object.

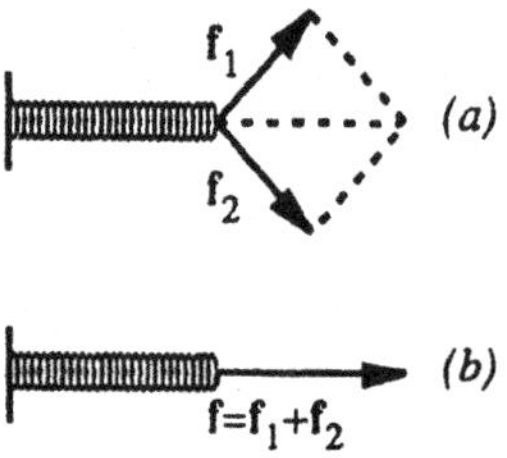

Figure 1.11. The forces $\mathbf{f}_1$ and $\mathbf{f}_2$ add like vectors.

Mechanics of a particle. The motion of a particle of constant mass m is governed by *Newton's law*,

$$\mathbf{f} = m\mathbf{a}, \tag{1.6}$$

where $\mathbf{f}$ is the force acting on the particle and $\mathbf{a}$ is its acceleration. The vector $\mathbf{f}$ is the sum of the *applied* forces and the *reactive* forces due to any constraints on the motion. Some common applied forces are:

- $\mathbf{f} = -k\mathbf{u}$, where $\mathbf{u}$ is the displacement of the particle and k is a scalar. This is known as *Hooke's law* and is the force exerted on a mass attached to a linear elastic spring. The *spring constant* or *stiffness* k depends upon the material properties and must be experimentally determined. Thus this equation is called a *constitutive relation*.

- $\mathbf{f} = \dfrac{k\mathbf{r}}{r^3}$, where $\mathbf{r}$ is the position vector from a fixed origin to the particle and k is a scalar. This is the inverse square law, called *Newton's law of gravitation* when applied to a point mass that is a vector distance $\mathbf{r}$ from the center of a spherical planet, and called *Coulomb's law of electrostatics* when applied to a point charge that is a vector distance $\mathbf{r}$ from the center of a charged sphere.

- $\mathbf{f} = q(\boldsymbol{\mathcal{E}} + \mathbf{v} \times \boldsymbol{\mathcal{B}})$, where $\mathbf{v}$ is the velocity of the particle, q is a scalar, and $\boldsymbol{\mathcal{E}}$ and $\boldsymbol{\mathcal{B}}$ are vectors. This is the *Lorentz force* on a particle of charge q in an *electric field* $\boldsymbol{\mathcal{E}}$ and a *magnetic field* $\boldsymbol{\mathcal{B}}$.

Certain combinations of the above scalars and vectors are given special names. A particle's *linear momentum* is the product $m\mathbf{v}$. Its *angular momentum* about an origin is the combination $\mathbf{h} = m\mathbf{r} \times \mathbf{v}$. Its *kinetic energy* is the scalar $\frac{1}{2}mv^2$.

Statics of a motionless body. Suppose a body of finite size is completely motionless. Imagine that any portion of the body is cut away or isolated from the rest. Then the vector sums of all the forces and moments acting on that portion must vanish (this follows from Newton's law — see Section 5.5):

$$\mathbf{f}_1 + \mathbf{f}_2 + \mathbf{f}_3 + \cdots = 0, \tag{1.7}$$

$$\mathbf{r}_1 \times \mathbf{f}_1 + \mathbf{r}_2 \times \mathbf{f}_2 + \mathbf{r}_3 \times \mathbf{f}_3 + \cdots = 0. \tag{1.8}$$

Here $\mathbf{f}_1$ denotes a force acting on the isolated portion, $\mathbf{r}_1$ denotes the position vector $\vec{OP_1}$ from a fixed point O to the point P_1 of application of $\mathbf{f}_1$, $\mathbf{r}_1 \times \mathbf{f}_1$ is the *moment* or *torque* of $\mathbf{f}_1$ about O, and so on (see Fig. 1.12). *Statics* is the study of the forces in bodies when the force and moment equilibrium equations (1.7) and (1.8) are sufficient to determine them.

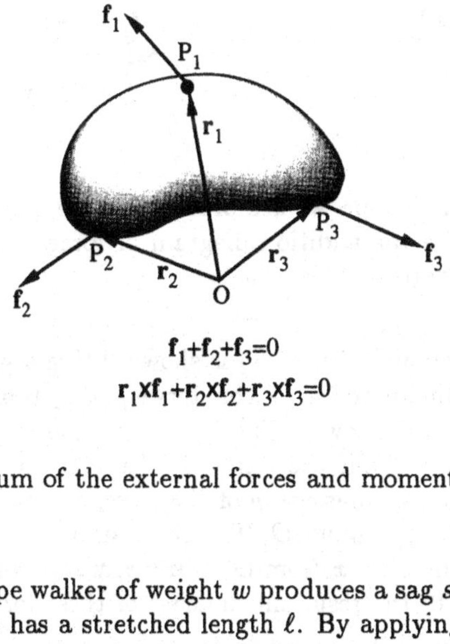

Figure 1.12. The sum of the external forces and moments acting on a body at rest must be zero.

Example: A tightrope walker of weight w produces a sag s when standing on the middle of a wire that has a stretched length ℓ. By applying (1.7) to an arbitrary length of wire, and neglecting the weight of the wire, we conclude that the tension or reactive forces $\mathbf{f}_1$ and $\mathbf{f}_2$ in each portion of the constraining wire not under the walker have constant direction and magnitude f. Hence, if the person occupies a very small portion of the wire, the wire has the shape of the V sketched in Fig. 1.13a and

$$\sin\theta = \frac{2s}{\ell},$$

where θ is the angle of sag. By applying (1.7) to the central bit of wire on which the person stands, we conclude that $\mathbf{f}_1 + \mathbf{f}_2 + \mathbf{w} = 0$. Hence the vector addition diagram of these forces is the closed triangle in Fig. 1.13b and

$$\sin\theta = \frac{w}{2f}.$$

It follows that the force in the wire under the walker is

$$f = \frac{w\ell}{4s}.$$

Suppose $w = 500$ N (1 newton = 0.2247 pounds), $\ell = 20$ m, and $s = 0.2$ m. Then $\theta \approx 1.15°$ and $f = 12{,}500$ N.

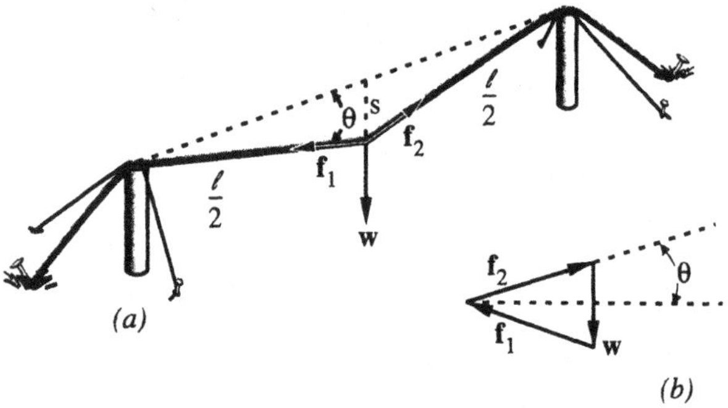

Figure 1.13. (a) A tightrope walker produces a sag s when standing on the center of a wire. (b) Vector addition diagram of forces $\mathbf{f}_1$, $\mathbf{f}_2$, $\mathbf{w}$ acting on an isolated portion of the wire.

Example: The tower crane in Fig. 1.14 is shown lifting a weight w_1 that is at a distance ℓ_1 from the climbing tower. A counterweight w_2 is placed at a distance ℓ_2 from the opposing side of the tower. The forces acting upon the horizontal boom, which produce a moment about the center O of the tower, have been shown. If we assume that the reactive pressure p of the tower upon the boom is constant, it will produce no net torque about O. The gravitational torque of a small length Δx of the boom at a distance x from O is $\rho x \Delta x$, where ρ is the boom's weight per unit length. By (1.8) the resultant moment of these forces about O must be zero:

$$\ell_1 w_1 - \ell_2 w_2 + \int_0^{\ell_1} \rho x dx - \int_0^{\ell_2} \rho x dx = 0.$$

Approximating ρ by a constant, we obtain a formula for the distance at which to place the counterweight:

$$\ell_2 = \frac{\sqrt{w_2^2 + 2\rho\ell_1 w_1 + \rho^2\ell_1^2} - w_2}{\rho}.$$

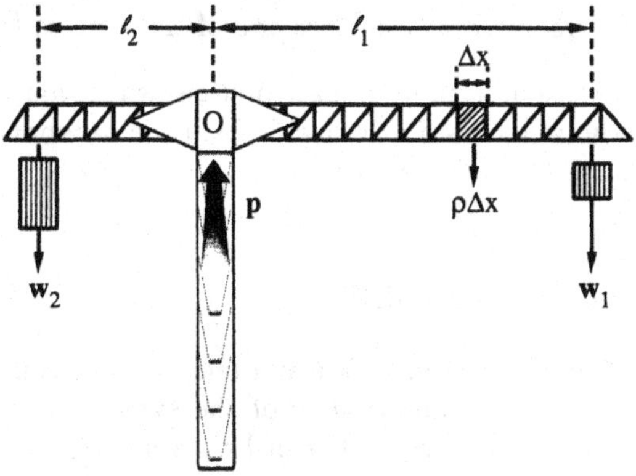

Figure 1.14. The resultant moment of the weights w_1, w_2, and $\rho\Delta x$ about the point O on the tower crane must be zero.

PROBLEMS 1.5

1. What is the maximum possible increase in speed of a spacecraft relative to the sun as a result of a planetary flyby?

2. A plank rests on two logs, each having a circumference of one meter. If each log is simultaneously turned over one complete revolution, how far forward will the plank move? What is the velocity, as seen by an observer on the ground, of each point on the outer surface of the logs at the instant it touches the ground?

3. The rain is falling straight down on your head with speed c. You start walking forward with speed $v \ll c$ and notice the rain is now hitting your face. What is the angle of *aberration* that the rain now appears to make with the vertical?

4. A magnetic field points horizontally from south to north. In what direction would an electron moving from east to west be deflected?

5. Calculate the torque about O on a particle under the action of a *central force* $\mathbf{f} = k\mathbf{r}$, where $\mathbf{r}$ is the position vector from O to the particle.

6. Show that the moment $\mathbf{r} \times \mathbf{f}$ about a point O is not altered by moving the force vector $\mathbf{f}$ along its line of action but does depend upon the location of O.

7. Suppose the vector sum of the forces acting on a solid body vanish (Eq. 1.7), and the vector sum of the moments about a given point O vanish (Eq. 1.8). Show that the vector sum of the moments about *any* point vanish.

8. Why does a tightrope walker carry a pole? Find the forces in the wire if the walker is at an arbitrary distance d_1 along the wire from one end.

9. A block and tackle has n pulleys at both its upper and lower ends. Wh the force f required at the loose end of the rope to lift a weight w?

10. Find the tension f in your back muscle when you lift a heavy weigh Assume the weight is a distance ℓ_1 from the backbone and the back mu is a distance ℓ_2 from the backbone.

1.6 NOTATION OF OTHERS

Unfortunately the notation in vector and tensor analysis is not standardized. list at the end of each chapter *some* of the symbols and terminology used other writers. Although every author utilizes a different notation, the underly mathematical ideas are the same.

Mathematicians often use the word "vector" in a more general sense t in this book, as an element of an *abstract vector space*. In that sense, n-tu (sequences of n numbers) and $n \times m$ matrices are also vectors. The directed segments (arrows) used here are sometimes called *Gibbsian vectors* to disting them from the more abstract vectors.

Table 1

OUR NOTATION	OUR NAME	OTHER NOTATIONS	OTHER NAMES
$\mathbf{v}$	Vector.	v, V, $\mathbf{V}$, $\vec{V}$, $\hat{\mathbf{v}}$	
$\lvert\mathbf{v}\rvert$	Magnitude.	$\lVert\mathbf{v}\rVert$	Norm.
$\mathbf{u}\cdot\mathbf{v}$	Dot product.	$(\mathbf{u}, \mathbf{v})$, $\langle\mathbf{u}, \mathbf{v}\rangle$	Inner product.
$\mathbf{u}\times\mathbf{v}$	Cross product.	$\mathbf{u}\wedge\mathbf{v}$	Outer product.
$\mathbf{f}$	Force vector.	$\mathbf{F}$	
$\mathcal{E}$	Electric field vector.	$\mathbf{E}$	

Chapter 2

TENSOR ALGEBRA

2.1 TENSORS DEFINED

Tensors may be defined using the mathematical objects introduced in Chapter 1. A *tensor of order zero* is simply another name for a *scalar.*

We have seen that a given vector $\mathbf{a}$ may be *dotted* into any vector $\mathbf{v}$ to produce a scalar $\mathbf{a} \cdot \mathbf{v}$. The combination $\mathbf{a}\cdot$ may be regarded as a *function* or rule that associates a vector $\mathbf{v}$ with a scalar $\mathbf{a} \cdot \mathbf{v}$. It is in fact a *linear* function because

$$\mathbf{a} \cdot (c\mathbf{u} + \mathbf{v}) = c\mathbf{a} \cdot \mathbf{u} + \mathbf{a} \cdot \mathbf{v}$$

for all vectors $\mathbf{u}$ and $\mathbf{v}$ and scalars c.

A tensor of order one is a linear function that maps every vector into a scalar. Thus any vector $\mathbf{a}$ is a tensor of order one, if it is understood that $\mathbf{a}$ is to be *dotted* into vectors. It will be shown in Section 3.2 that *every* tensor of order one can be represented by a vector.

We also have seen that a given vector $\mathbf{a}$ may be *crossed* into any vector $\mathbf{v}$ to produce a vector $\mathbf{a} \times \mathbf{v}$. The combination $\mathbf{a}\times$ is a function that associates a vector $\mathbf{v}$ with a vector $\mathbf{a} \times \mathbf{v}$. It is a linear function because

$$\mathbf{a} \times (c\mathbf{u} + \mathbf{v}) = c\mathbf{a} \times \mathbf{u} + \mathbf{a} \times \mathbf{v}$$

for all vectors $\mathbf{u}$ and $\mathbf{v}$ and scalars c.

More generally, consider a pair of vectors like $\mathbf{ab}$ written without the intervention of a dot or cross. The pair $\mathbf{ab}$ is called a *dyad* or *direct product* with *antecedent* $\mathbf{a}$ and *consequent* $\mathbf{b}$. A dyad maps any vector $\mathbf{v}$ into a vector parallel to $\mathbf{a}$ according to the definition

$$\boxed{\mathbf{ab} \cdot \mathbf{v} = \mathbf{a}(\mathbf{b} \cdot \mathbf{v}).} \tag{2.1}$$

It follows from (2.1) that a dyad has the linearity property:

$$\mathbf{ab} \cdot (c\mathbf{u} + \mathbf{v}) = c\mathbf{ab} \cdot \mathbf{u} + \mathbf{ab} \cdot \mathbf{v}.$$

A linear combination of dyads with scalar coefficients, for example, $m\mathbf{ab} + n\mathbf{cd}$, is termed a *dyadic.*

A tensor of order two is a linear function that maps every vector into a vector. Thus $\mathbf{a}\times$ is a tensor of order two. Also, any dyad $\mathbf{ab}$ is a tensor of order two, if it is understood that $\mathbf{ab}$ is to be dotted into vectors. It will be shown in Section 3.2 that *every* tensor of order two can be represented by a dyadic.

We denote tensors of order two or higher by boldface *capital* symbols, such as $\mathbf{R}$, $\mathbf{S}$, $\mathbf{T}$. When a tensor is denoted by a boldface symbol, its action on vectors is denoted by a dot. Thus, a tensor $\mathbf{T}$ of order two maps any vector $\mathbf{v}$ into a vector $\mathbf{T}\cdot\mathbf{v}$. Since a tensor $\mathbf{T}$ is linear,

$$\boxed{\mathbf{T}\cdot(c\mathbf{u}+\mathbf{v}) = c\mathbf{T}\cdot\mathbf{u}+\mathbf{T}\cdot\mathbf{v}}$$

for all vectors $\mathbf{u}$ and $\mathbf{v}$ and scalars c.

Tensors of higher order may be defined in a similar way. For instance, *a tensor of order three is a linear function that maps every vector into a dyadic.* If $\mathbf{B}$ is a tensor of order three, $\mathbf{B}\cdot\mathbf{u}$ is a dyadic. Three adjacent vectors, such as $\mathbf{uvw}$, form a third-order tensor called a *triad*. Any tensor of order three can be represented as a *triadic* (linear combination of triads).

In other language, a tensor $\mathbf{T}$ is a *linear machine* or *linear operator*. If you put any vector $\mathbf{v}$ into the machine, it outputs another tensor $\mathbf{T}\cdot\mathbf{v}$ of one lower order. The function can be defined as the set of all pairs $(\mathbf{v}, \mathbf{T}\cdot\mathbf{v})$. The *domain* of the function is all vectors $\mathbf{v}$; the *range* of the function is all *image* tensors $\mathbf{T}\cdot\mathbf{v}$.

Example: The wind exerts a force $\mathbf{f}$ normal to a sail on a boat heading in the direction of a unit vector $\mathbf{e}$ (Fig. 2.1). Only the component $\mathbf{e}\cdot\mathbf{f}$ propels the boat in the forward direction. The vector $\mathbf{e}$ together with the dot product is a tensor of order one that maps every sail force $\mathbf{f}$ into the component pushing the boat in the forward direction. The sailboat will move faster when $\mathbf{e}\cdot\mathbf{f}$ is larger.

Example: A force $\mathbf{f}$ is applied to a wrench at a vector distance $\mathbf{r}$ from the center O of a nut. The moment (torque) of $\mathbf{f}$ about the point O is $\mathbf{r}\times\mathbf{f}$. The combination $\mathbf{r}\times$ is notation for the tensor of order two that maps every force $\mathbf{f}$ into the moment $\mathbf{r}\times\mathbf{f}$. The forces $\mathbf{f}_1, \mathbf{f}_2, \mathbf{f}_3$ applied to the wrench in Fig. 2.2 produce respective moments $\mathbf{r}\times\mathbf{f}_1$, $\mathbf{r}\times\mathbf{f}_2$, $\mathbf{r}\times\mathbf{f}_3$ about O. The nut will turn easier when $\mathbf{r}\times\mathbf{f}$ is larger.

Example: Given an arbitrary unit vector $\mathbf{e}$, we can form the dyad $\mathbf{ee}$. The operator $\mathbf{ee}$ maps any vector $\mathbf{v}$ into its vector projection $\mathbf{e}(\mathbf{e}\cdot\mathbf{v})$ on $\mathbf{e}$. To agree with our previous notation (see Eq. 1.2), we let

$$\mathbf{P} = \mathbf{ee}.$$

$\mathbf{P}$ is a second-order tensor that we can call the *projection tensor*. The effect of $\mathbf{P}$ on several vectors is illustrated in Fig. 2.3.

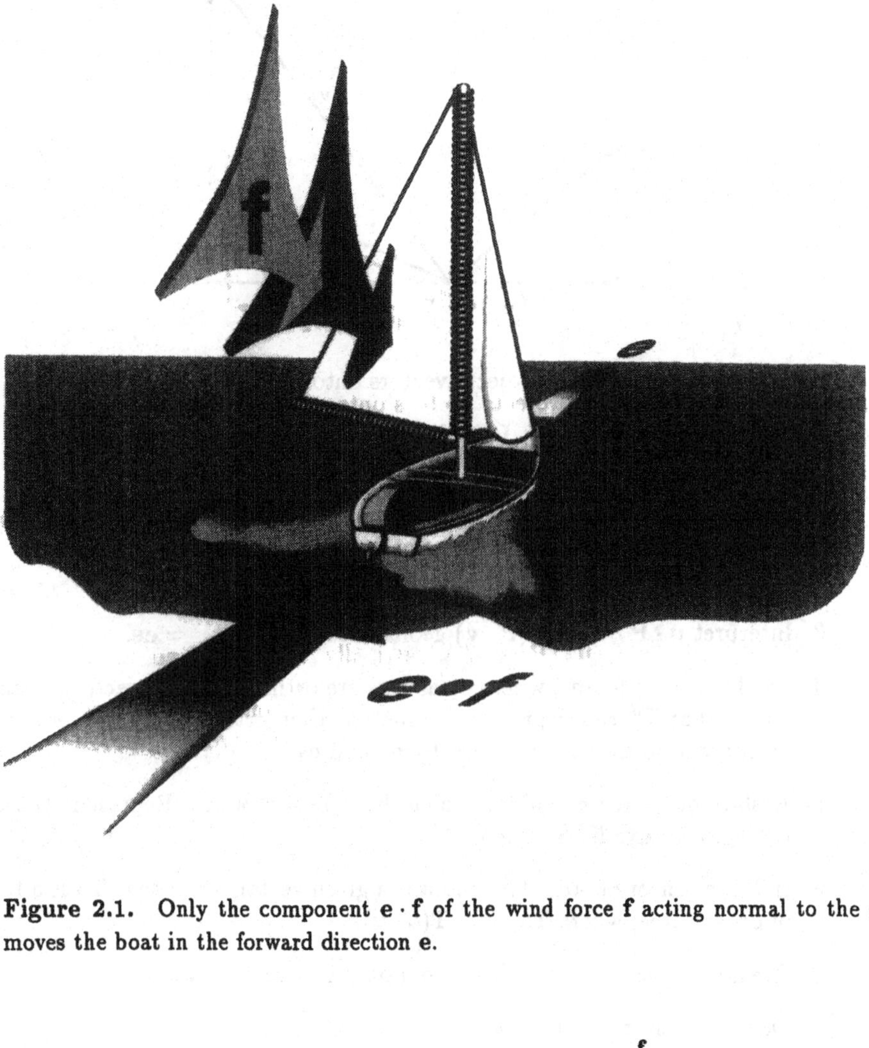

Figure 2.1. Only the component $\mathbf{e}\cdot\mathbf{f}$ of the wind force $\mathbf{f}$ acting normal to the sail moves the boat in the forward direction $\mathbf{e}$.

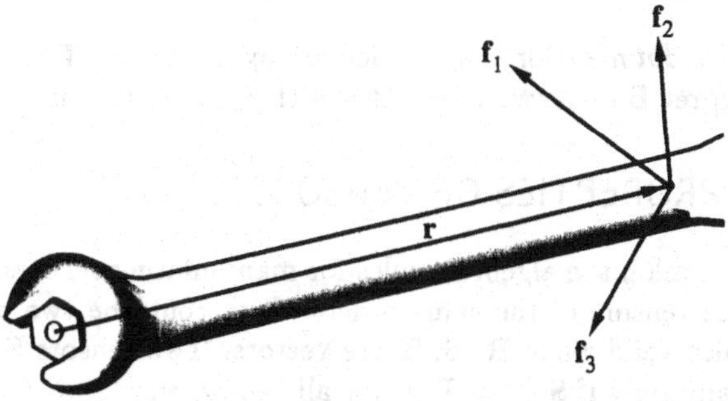

Figure 2.2. The forces exert differing torques $\mathbf{r}\times\mathbf{f}_1$, $\mathbf{r}\times\mathbf{f}_2$, $\mathbf{r}\times\mathbf{f}_3$ on the nut.

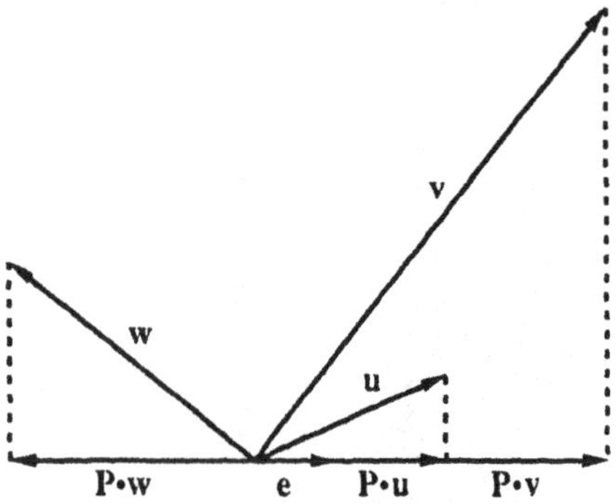

Figure 2.3. The tensor **P** projects vectors onto a line through **e**.

PROBLEMS 2.1

1. Consider the function **f**, where $\mathbf{f}(\mathbf{v}) = \mathbf{a} \cdot \mathbf{v} + d$. Is **f** a tensor? Think of some other functions of vectors that are not tensors.

2. Sketch the effect of the tensors $\mathbf{a}\times$ and **ab** on several vectors.

3. Interpret $\mathbf{u} \cdot \mathbf{P} \cdot \mathbf{v} = \mathbf{u} \cdot (\mathbf{P} \cdot \mathbf{v})$ geometrically, where $\mathbf{P} = \mathbf{ee}$.

4. Let $\mathbf{P}' = \mathbf{e}_1\mathbf{e}_1 + \mathbf{e}_2\mathbf{e}_2$, where $\mathbf{e}_1$ and $\mathbf{e}_2$ are orthogonal unit vectors. Demonstrate that $\mathbf{P}'$ can represent a function that orthogonally projects every vector **v** onto the *plane* formed by $\mathbf{e}_1$ and $\mathbf{e}_2$.

5. A slide projector magnifies a slide by a factor of K. Represent the slide projector in dyadic notation.

6. If **T** is a tensor of order two and **a** is a given vector, show that **Ta** is a tensor of order three, where $\mathbf{Ta} \cdot \mathbf{v} = \mathbf{T}(\mathbf{a} \cdot \mathbf{v})$.

7. Create a tensor of order three from only two vectors **a** and **b**.

8. Define a tensor of order four and a *polyad*.

9. The *double dot* notation may be defined by $\mathbf{T} \cdot \cdot \mathbf{uv} = (\mathbf{T} \cdot \mathbf{u}) \cdot \mathbf{v}$. How would you interpret $\mathbf{B} \cdots \mathbf{uvw}$, where **B** is a third-order tensor?

2.2 BASIC PROPERTIES OF TENSORS

We must now develop the algebraic rules for manipulating tensors. In this section **R**, **S**, **T** denote tensors of the same order, which could be two or higher. Most formulas are also valid when **R**, **S**, **T** are vectors. Two tensors **S** and **T** are said to be *equal* if and only if $\mathbf{S} \cdot \mathbf{v} = \mathbf{T} \cdot \mathbf{v}$ for all vectors **v**:

$$\boxed{\mathbf{S} = \mathbf{T} \Longleftrightarrow \mathbf{S} \cdot \mathbf{v} = \mathbf{T} \cdot \mathbf{v}.} \tag{2.2}$$

Since the defining equations will be required to hold for *all* vectors $\mathbf{v}$, it follows from (2.2) that the quantities to be defined will be *unique.*

The *sum* $\mathbf{S}+\mathbf{T}$ of two tensors is the tensor that obeys the rule

$$(\mathbf{S}+\mathbf{T})\cdot\mathbf{v}=\mathbf{S}\cdot\mathbf{v}+\mathbf{T}\cdot\mathbf{v}.$$

Multiplication by a scalar c is defined by

$$(c\mathbf{T})\cdot\mathbf{v}=c(\mathbf{T}\cdot\mathbf{v}).$$

The product $0\mathbf{T}$ is called a *zero tensor* and given the symbol $\mathbf{0}$ or 0.

The *dot product* $\mathbf{S}\cdot\mathbf{T}$ of two tensors is defined by the requirement

$$(\mathbf{S}\cdot\mathbf{T})\cdot\mathbf{v}=\mathbf{S}\cdot(\mathbf{T}\cdot\mathbf{v})=\mathbf{S}\cdot\mathbf{T}\cdot\mathbf{v}. \tag{2.3}$$

In (2.3) $\mathbf{S}$ and $\mathbf{T}$ need not be of the same order; a tensor may be pre- or post-multiplied by another tensor of any order. The order of $\mathbf{S}\cdot\mathbf{T}$ is two less than the sum of the orders of $\mathbf{S}$ and $\mathbf{T}$.

The function that maps every tensor $\mathbf{T}$ into itself is called the *identity tensor* or *metric tensor* and given the special symbol $\mathbf{1}$:

$$\boxed{\mathbf{1}\cdot\mathbf{T}=\mathbf{T}\cdot\mathbf{1}=\mathbf{T}.}$$

$\mathbf{1}$ is a second-order tensor. The function that maps every vector $\mathbf{v}$ into the negative of the second-order tensor $\mathbf{v}\times$ is called the *permutation tensor* and given the special symbol $\mathbf{E}$:

$$\boxed{\mathbf{E}\cdot\mathbf{v}=-\mathbf{v}\times.}$$

$\mathbf{E}$ is a third-order tensor.

Integral powers of a tensor $\mathbf{T}$ are defined inductively by

$$\begin{aligned}\mathbf{T}^0&=\mathbf{1},\\ \mathbf{T}^n&=\mathbf{T}^{n-1}\cdot\mathbf{T}\end{aligned}$$

where n is a positive integer.

The summation, multiplication by scalars, and dot product of tensors are governed by many of the same rules as in the arithmetic of real numbers. For example,

$$\begin{aligned}\mathbf{S}+\mathbf{T}&=\mathbf{T}+\mathbf{S},\\ c(d\mathbf{T})&=(cd)\mathbf{T}=cd\mathbf{T},\\ (\mathbf{R}\cdot\mathbf{S})\cdot\mathbf{T}&=\mathbf{R}\cdot(\mathbf{S}\cdot\mathbf{T})=\mathbf{R}\cdot\mathbf{S}\cdot\mathbf{T},\\ (\mathbf{R}+\mathbf{S})\cdot\mathbf{T}&=\mathbf{R}\cdot\mathbf{T}+\mathbf{S}\cdot\mathbf{T},\\ \mathbf{T}^m\cdot\mathbf{T}^n&=\mathbf{T}^{m+n},\\ (\mathbf{T}^m)^n&=\mathbf{T}^{mn},\end{aligned}$$

where m and n are nonnegative integers. These rules can be verified by dotting each side of an equation with an arbitrary vector and showing that the resulting expressions are equal.

It should be noted that tensors do not obey *all* of the rules of ordinary arithmetic. The dot product is generally not commutative: $\mathbf{S}\cdot\mathbf{T} \neq \mathbf{T}\cdot\mathbf{S}$ and $\mathbf{v}\cdot\mathbf{T} \neq \mathbf{T}\cdot\mathbf{v}$. The equation $\mathbf{S}\cdot\mathbf{T} = 0$ does not necessarily imply that $\mathbf{S}$ or $\mathbf{T}$ is zero, and $\mathbf{T}\cdot\mathbf{v} = 0$ does not necessarily imply that $\mathbf{T}$ or $\mathbf{v}$ is zero.

Following are some of the laws of dyad algebra:

$$\boxed{\begin{aligned} \mathbf{v}\cdot(\mathbf{ab}) &= (\mathbf{v}\cdot\mathbf{a})\mathbf{b}, \\ \mathbf{ab}\cdot\mathbf{cd} &= (\mathbf{b}\cdot\mathbf{c})\mathbf{ad}. \end{aligned}}$$

Note that in general the direct product is not commutative:

$$\mathbf{ab} \neq \mathbf{ba} \quad \text{and} \quad \mathbf{ab}\cdot\mathbf{cd} \neq \mathbf{cd}\cdot\mathbf{ab}.$$

PROBLEMS 2.2

1. Represent symbolically the tensor that maps any vector $\mathbf{v}$ into its negative $-\mathbf{v}$.
2. Show that $\mathbf{1}^n = \mathbf{1}$ and $\mathbf{P}^n = \mathbf{P}$, where $\mathbf{P} = \mathbf{ee}$ is the projection tensor and n is a positive integer. Interpret geometrically.
3. If $\mathbf{1}\cdot\mathbf{v} = \mathbf{v}$ for all vectors $\mathbf{v}$ and $\bar{\mathbf{1}}\cdot\mathbf{T} = \mathbf{T}$ for all tensors $\mathbf{T}$ of order two, prove that $\mathbf{1} = \bar{\mathbf{1}}$, that is, that $\mathbf{1}$ is unique.
4. Find a tensor $\mathbf{T} \neq 0$ that satisfies $\mathbf{T}\cdot\mathbf{v} = -\mathbf{v}\cdot\mathbf{T}$ and $\mathbf{T}\cdot\mathbf{a} = 0$.
5. Prove that $\mathbf{a}\cdot\mathbf{bc}\cdot\mathbf{d} = \mathbf{b}\cdot\mathbf{ad}\cdot\mathbf{c}$.
6. Prove that $\mathbf{ab} - \mathbf{ba} = (\mathbf{b}\times\mathbf{a})\times$.
7. When is $\mathbf{ab} = \mathbf{ba}$?
8. Prove that $(\mathbf{e}\times)^2 = \mathbf{ee} - \mathbf{1}$ and $(\mathbf{e}\times)^3 = -\mathbf{e}\times$, where $\mathbf{e}$ is an arbitrary unit vector.

2.3 MORE PROPERTIES OF SECOND-ORDER TENSORS

In this section $\mathbf{S}$, $\mathbf{T}$ denote tensors of *order two only*. The definitions and properties to be stated are equivalent to well-known results of matrix theory (see Section 3.2 and the mathematical references). They can be extended to tensors of higher order.

Every second-order tensor $\mathbf{T}$ has a unique *transpose* $\mathbf{T}^t$, which obeys the equation

$$\boxed{\mathbf{v}\cdot\mathbf{T}^t = \mathbf{T}\cdot\mathbf{v}}$$

for all vectors $\mathbf{v}$. The transpose operation has the properties

$$\boxed{\begin{aligned}(\mathbf{T}^t)^t &= \mathbf{T},\\ (c\mathbf{S}+\mathbf{T})^t &= c\mathbf{S}^t+\mathbf{T}^t,\\ (\mathbf{S}\cdot\mathbf{T})^t &= \mathbf{T}^t\cdot\mathbf{S}^t,\\ (\mathbf{ab})^t &= \mathbf{ba}.\end{aligned}}$$

In a three-dimensional space every second-order tensor $\mathbf{T}$ maps certain vectors into scalar multiples of themselves; that is,

$$\boxed{\mathbf{T}\cdot\mathbf{v}=\lambda\mathbf{v}.} \tag{2.4}$$

The scalar λ in (2.4) is called an *eigenvalue* or *principal value* of $\mathbf{T}$. The vector $\mathbf{v}$ in (2.4) is said to be an *eigenvector* corresponding to λ, and the direction of $\mathbf{v}$ is said to be a *principal direction* of $\mathbf{T}$. Every scalar multiple of an eigenvector is also an eigenvector; a *normalized* eigenvector is an eigenvector of unit length. If a nonzero eigenvector corresponds to an eigenvalue $\lambda = 0$, $\mathbf{T}$ is said to be *singular*.

Only *nonsingular* tensors $\mathbf{T}$ have an *inverse* $\mathbf{T}^{-1}$ that satisfies the equations

$$\boxed{\mathbf{T}\cdot\mathbf{T}^{-1}=\mathbf{T}^{-1}\cdot\mathbf{T}=\mathbf{1}.}$$

If $\mathbf{S}$ and $\mathbf{T}$ are invertible tensors, then

$$\boxed{\begin{aligned}(\mathbf{T}^{-1})^{-1} &= \mathbf{T},\\ (c\mathbf{S})^{-1} &= \mathbf{S}^{-1}/c,\\ (\mathbf{S}\cdot\mathbf{T})^{-1} &= \mathbf{T}^{-1}\cdot\mathbf{S}^{-1}.\end{aligned}}$$

A tensor $\mathbf{S}$ is *symmetric* if $\mathbf{S}^t = \mathbf{S}$. A tensor $\mathbf{A}$ is *antisymmetric* or *skew* if $\mathbf{A}^t = -\mathbf{A}$. A tensor $\mathbf{Q}$ is *orthogonal* if $\mathbf{Q}^t = \mathbf{Q}^{-1}$.

Example: Consider the second-order tensor $\mathbf{a}\times$, where $\mathbf{a}$ is a given vector. Since $\mathbf{u}\cdot\mathbf{a}\times\mathbf{v} = -\mathbf{v}\cdot\mathbf{a}\times\mathbf{u}$, $\mathbf{a}\times$ is antisymmetric. Since $\mathbf{a}\times\mathbf{a} = 0$, the eigenvectors of $\mathbf{a}\times$ are multiples of $\mathbf{a}$ with corresponding eigenvalue $\lambda = 0$. Hence $\mathbf{a}\times$ is singular and doesn't have an inverse.

Example: Consider the projection tensor $\mathbf{P} = \mathbf{ee}$, where $\mathbf{e}$ is a unit vector. Since $\mathbf{P}^t = \mathbf{ee} = \mathbf{P}$, $\mathbf{P}$ is symmetric. Setting $\mathbf{P}\cdot\mathbf{v} = \mathbf{e}(\mathbf{e}\cdot\mathbf{v}) = \lambda\mathbf{v}$, we see that the eigenvectors are vectors parallel to $\mathbf{e}$ with corresponding eigenvalue $\lambda = 1$, and vectors perpendicular to $\mathbf{e}$ with corresponding eigenvalue $\lambda = 0$. Hence $\mathbf{P}$ is singular and doesn't have an inverse.

Example: Consider the identity tensor $\mathbf{1}$. Since $\mathbf{u}\cdot\mathbf{1}\cdot\mathbf{v} = \mathbf{u}\cdot\mathbf{v} = \mathbf{v}\cdot\mathbf{1}\cdot\mathbf{u}$, $\mathbf{1}$ is symmetric. Since $\mathbf{1}\cdot\mathbf{v} = \mathbf{v}$, any vector is an eigenvector with corresponding eigenvalue $\lambda = 1$. Since $\mathbf{1}\cdot\mathbf{1} = \mathbf{1}$, we see that $\mathbf{1}^{-1} = \mathbf{1}$. Since $\mathbf{1}^t = \mathbf{1}^{-1} = \mathbf{1}$, $\mathbf{1}$ is also orthogonal.

PROBLEMS 2.3

1. Show that $\mathbf{v} \cdot \mathbf{T}^t \cdot \mathbf{u} = \mathbf{u} \cdot \mathbf{T} \cdot \mathbf{v}$ for all second-order tensors $\mathbf{T}$ and vectors $\mathbf{u}$ and $\mathbf{v}$.

2. Prove that $(\mathbf{T}^{-1})^t = (\mathbf{T}^t)^{-1}$.

3. Define $\mathbf{T}^{-n}$, where $\mathbf{T}$ is an invertible tensor and n is an arbitrary positive integer.

4. Find the eigenvectors and eigenvalues of the dyad $\mathbf{ab}$.

5. Deduce that the (three-dimensional) tensor $\mathbf{P}'$ in Problem 4 of Section 2.1 is symmetric and singular.

6. If $\mathbf{A}$ is antisymmetric, prove that $\mathbf{v} \cdot \mathbf{A} \cdot \mathbf{v} = 0$ for all vectors $\mathbf{v}$.

7. If $\mathbf{Q}$ is orthogonal, show that $(\mathbf{Q} \cdot \mathbf{u}) \cdot (\mathbf{Q} \cdot \mathbf{v}) = \mathbf{u} \cdot \mathbf{v}$, and hence deduce that orthogonal tensors preserve the lengths of and angles between vectors.

8. Suppose $\mathbf{T}$ is symmetric, antisymmetric, or orthogonal. Is $\mathbf{T}^n$ (where n is any integer) symmetric, antisymmetric, or orthogonal?

2.4 PHYSICAL TENSORS

We can now discuss some of the physical quantities characterized by tensors of order two and higher. These tensors enter into physical laws that will be presented in later chapters when we have developed tensor calculus.

Moment of inertia. Suppose a rigid body is rotating so that every particle in the body is instantaneously moving in a circle about some axis fixed in space (Fig. 2.4). The body's *angular velocity* $\boldsymbol{\omega}$ is defined as the vector whose magnitude is the angular speed ω and whose direction is along the axis of rotation. Then a particle's *linear velocity* is

$$\mathbf{v} = \boldsymbol{\omega} \times \mathbf{r}, \tag{2.5}$$

where $v = \omega d$ is its linear speed, d is the distance between the axis and the particle, and $\mathbf{r}$ is the position vector from a fixed point O on the axis to the particle. The particle's *angular momentum* $\mathbf{h}$ about the point O is

$$\mathbf{h} = m\mathbf{r} \times \mathbf{v} = \mathbf{I} \cdot \boldsymbol{\omega}, \tag{2.6}$$

where m is the mass of the particle and

$$\mathbf{I} = m\left(r^2\mathbf{1} - \mathbf{rr}\right). \tag{2.7}$$

$\mathbf{I}$ is called the *moment of inertia tensor of the particle about the point* O and maps any vector $\boldsymbol{\omega}$ into the angular momentum $\mathbf{h}$ the particle would have if the rigid

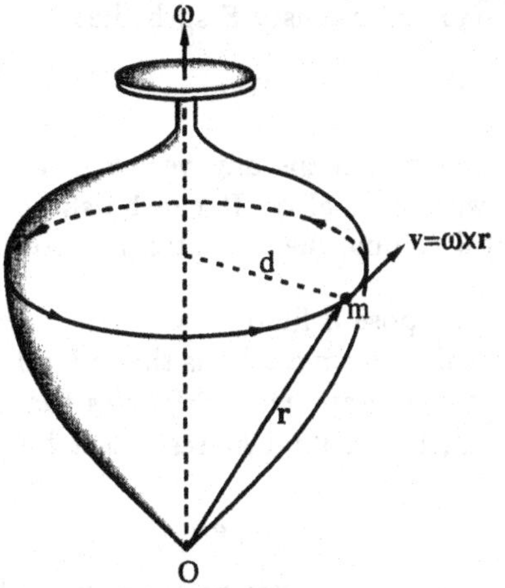

Figure 2.4. The linear velocity **v** of a particle on a rotating body is the angular velocity $\boldsymbol{\omega}$ crossed into the position vector **r**.

body were rotating about an axis through O with angular velocity $\boldsymbol{\omega}$. Note that **I** is a symmetric second-order tensor.

Stress. Suppose a continuous solid or fluid (liquid or gas) is subjected to a system of applied forces. The reactive force per unit area interacting between elemental internal portions of the continuum is called the *traction* or *stress vector* **t**. To visualize **t**, imagine that within the continuum a small cut or slice is made (Fig. 2.5). This creates a small surface specified by its surface area ΔA and a unit normal vector **e** perpendicular to the cut. Then $\mathbf{t}\Delta A$ is the force that is exerted on the surface element $\mathbf{e}\Delta A$ by the matter toward which **e** points. The traction **t** depends linearly on the orientation **e** of the surface element but not on its shape;

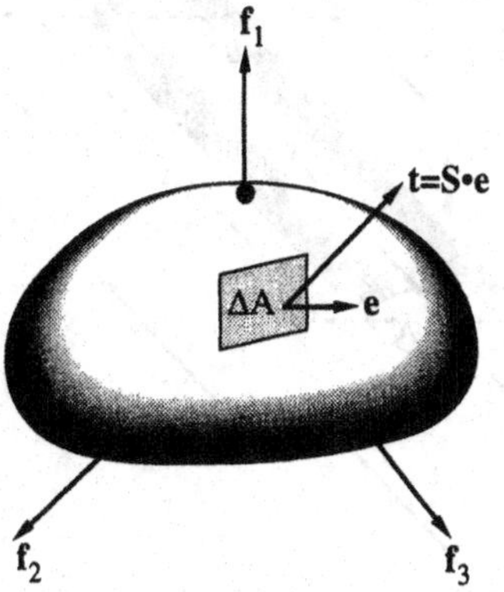

Figure 2.5. The traction **t** acting on an elemental cut with unit normal **e** is $\mathbf{S}\cdot\mathbf{e}$.

that is, there exists a second-order tensor $\mathbf{S}$ such that

$$\mathbf{t} = \mathbf{S} \cdot \mathbf{e}. \tag{2.8}$$

$\mathbf{S}$ is called the *stress tensor* and maps any vector $\mathbf{e}$ into the traction $\mathbf{t}$ acting on an elemental surface with normal $\mathbf{e}$. It can be shown by enforcing moment equilibrium of an elemental volume that the stress tensor is symmetric.

Hydrostatic pressure. Suppose a fluid is completely motionless. The characteristic that distinguishes a fluid from a solid is that a fluid will continue to deform when tangential or *shear* force is applied. It follows that the force $\mathbf{t}\Delta A$ on any internal surface element $\mathbf{e}\Delta A$ of a fluid at rest must be *normal* to the surface; that is,

$$\mathbf{t} = -p\mathbf{e}, \tag{2.9}$$

where the scalar $p \geq 0$ is called the *static pressure* in the fluid. Comparing (2.8) and (2.9), we see that the stress tensor $\mathbf{S}$ in a fluid at rest is given simply by

$$\mathbf{S} = -p\mathbf{1}. \tag{2.10}$$

Equations (2.9) and (2.10) also describe the state of stress in *inviscid* (nonviscous) fluids in motion, and in a solid subjected only to a normal pressure p on its entire surface.

Momentum flux density. Consider now a continuous fluid in motion. Imagine a small surface at a fixed place within the continuum with surface area ΔA and

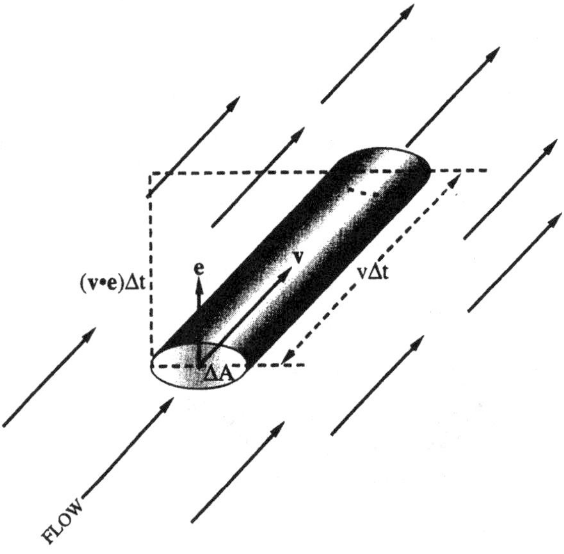

Figure 2.6. The mass per unit time and area passing through a small surface with unit normal $\mathbf{e}$ is $\rho\mathbf{v} \cdot \mathbf{e}$. The momentum per unit time and area passing through the small surface is $\rho\mathbf{v}\mathbf{v} \cdot \mathbf{e}$.

unit normal vector $\mathbf{e}$ making an acute angle with the direction of flow (Fig 2.6). Let us calculate the *momentum* of the matter passing through ΔA in time Δt. Its momentum is the mass passing through ΔA times its velocity $\mathbf{v}$. Let ρ denote the *density* (mass per unit volume) of the continuum. The volume of matter flowing through ΔA in time Δt is equal to the volume of the cylinder with base ΔA and slant height $v\Delta t$ parallel to $\mathbf{v}$. Since the altitude of the cylinder is $(\mathbf{v}\cdot\mathbf{e})\Delta t$, its volume is $(\mathbf{v}\cdot\mathbf{e})\Delta t\Delta A$. Hence the mass of the matter passing through ΔA in time Δt is $(\rho\mathbf{v})\cdot\mathbf{e}\Delta t\Delta A$, and the momentum of this matter is $(\rho\mathbf{v}\mathbf{v})\cdot\mathbf{e}\Delta t\Delta A$. The vector $\rho\mathbf{v}$ is called the *mass flux density* (mass per unit time and area), and the dyad $\rho\mathbf{v}\mathbf{v}$ is called the *momentum flux density*. In particular, taking $\mathbf{e}'$ to be directed perpendicular to the velocity $\mathbf{v}$, we find that $(\rho\mathbf{v})\cdot\mathbf{e}' = 0$ and $(\rho\mathbf{v}\mathbf{v})\cdot\mathbf{e}' = 0$; that is, no matter moves in a direction perpendicular to $\mathbf{v}$. Note that $\rho\mathbf{v}\mathbf{v}$ is a singular symmetric second-order tensor.

Piezoelectricity or pressure electricity is a physical phenomenon that can be modeled with higher-order tensors. Piezoelectric crystals, used for example in sonar transducers and quartz wristwatches, transform electrical force into mechanical force and vice versa. If an electric field $\boldsymbol{\mathcal{E}}$ exists at a point of space, then a force $\boldsymbol{\mathcal{E}}$ would be exerted on a unit test charge placed in the field. When a piezoelectric crystal is placed in the electric field and not allowed to deform, it will develop an internal stress $\mathbf{S}$, which is a linear function of the applied field $\boldsymbol{\mathcal{E}}$:

$$\mathbf{S} = \mathbf{B}\cdot\boldsymbol{\mathcal{E}}. \tag{2.11}$$

Since $\boldsymbol{\mathcal{E}}$ is a vector and $\mathbf{S}$ is a second-order tensor, $\mathbf{B}$ is a *third-order tensor* called a *piezoelectric tensor*. From (2.8) and (2.11) the traction $\mathbf{t}$ acting on an elemental surface in the crystal with unit normal $\mathbf{e}$ is

$$\mathbf{t} = \mathbf{e}\cdot\mathbf{B}\cdot\boldsymbol{\mathcal{E}}.$$

PROBLEMS 2.4

1. Use (2.5) and (2.7) to verify that the two formulas for $\mathbf{h}$ given in (2.6) are equivalent.

2. Show that the moment of inertia tensor $\mathbf{I} = -m(\mathbf{r}\times)^2$.

3. The *moment of inertia of a particle about the axis* of a rotating rigid body is the scalar defined by $I = \mathbf{e}\cdot\mathbf{I}\cdot\mathbf{e}$, where $\mathbf{e} = \frac{\boldsymbol{\omega}}{\omega}$. Show that $I = md^2$, no matter where the point O is on the axis. Show that the particle's rotational kinetic energy is $\frac{1}{2}I\omega^2 = \frac{1}{2}\boldsymbol{\omega}\cdot\mathbf{I}\cdot\boldsymbol{\omega}$. Show that $\mathbf{h} = I\boldsymbol{\omega}$ if the point O is at the center of the circle traced by the particle.

4. Determine the normal component of the traction $\mathbf{t}$ acting on an elemental surface with orientation $\mathbf{e}$.

5. Using the tightrope example of Section 1.5, find the stress tensor $\mathbf{S}$ and the traction vector $\mathbf{t}$ on a surface whose normal makes an angle ϕ with the wire.

6. Interpret physically the symmetry of the moment flux tensor: $\mathbf{e}_1 \cdot \rho\mathbf{vv} \cdot \mathbf{e}_2 = \mathbf{e}_2 \cdot \rho\mathbf{vv} \cdot \mathbf{e}_1$.

7. Explain why a stream of water from a faucet becomes narrower as it falls.

8. A gas of density ρ is impacting upon a wall with a velocity $\mathbf{v}$ at an angle ϕ to the surface normal. Assuming the angle of reflection equals the angle of incidence and the speed remains the same, determine the pressure exerted by the gas on the wall.

9. Prove that $\mathbf{e}_2 \cdot (\mathbf{e}_1 \cdot \mathbf{B}) = \mathbf{e}_1 \cdot (\mathbf{e}_2 \cdot \mathbf{B})$, where $\mathbf{B}$ is the piezoelectric tensor.

2.5 NOTATION OF OTHERS

Sometimes we use a symbol like **a** to denote a *directed line segment*, and at other times **a** denotes a *function* or rule for associating vectors with scalars (using the dot product). The situation is similar to that in elementary calculus, where $f(t)$ can denote either the *image* number associated with the variable t, or the actual *rule* for associating numbers with numbers. In this text we shall be content with such ambiguity. Which meaning to give a symbol should be obvious from the context.

The basic method of defining tensors may be different in other works. Some authors' definitions are based on components, which are numbers that depend upon the coordinate system. However, physical field theories must be *independent* of the coordinate systems used to describe the objects. For instance, the predicted motion of a satellite will be the same whether we solve Newton's law in vector form or in a spherical coordinate system. Tensors, as the symbols we use to represent the physical quantities, are *invariant under coordinate transformations.* To emphasize this invariance our development so far has been based on solid geometry without reference to a specific coordinate system.

Table 2

OUR NOTATION	OUR NAME	OTHER NOTATIONS	OTHER NAMES
$\mathbf{a}\times$		$\tilde{\mathbf{a}}$	
$\mathbf{ab}$	Dyad.	$\mathbf{a}\otimes\mathbf{b}$	Tensor product.
$\mathbf{T}$	Tensor of order two.	T, t, $\mathbf{t}$, $\vec{T}$, $\mathcal{T}$, $\overset{\sqcup}{\mathrm{T}}$, $\underline{\underline{\mathrm{T}}}$, $\overleftrightarrow{\mathbf{T}}$	Tensor of rank 2, linear transformation.
$\mathbf{T}\cdot\mathbf{v}$		$\mathbf{T}\,(\mathbf{v})$, $\mathbf{T}\,\mathbf{v}$	
$\mathbf{u}\cdot\mathbf{T}\cdot\mathbf{v}$		$\mathbf{T}\!:\mathbf{uv}$, $\mathbf{T} \cdot\cdot\, \mathbf{uv}$, $T(u,v)$	
$\mathbf{P}$	Projection tensor.	Proj.	
$\mathbf{T}^t$	Transpose.	$\mathbf{T}^T$, $\tilde{\mathbf{T}}$, $\mathbf{T}$	Same name and symbol may be transpose of ours.
$\mathbf{S}$, p	Stress tensor, pressure.	$-\mathbf{S}$, $-p$	Same name and symbol may be negative of ours.

Chapter 3

CARTESIAN COMPONENTS

3.1 COMPONENTS OF A VECTOR AND COORDINATES OF A POINT

In this chapter we resolve vectors and tensors along a triad of base vectors $(\mathbf{e}_1, \mathbf{e}_2, \mathbf{e}_3)$ that are *right-handed* and *orthonormal*, that is, mutually orthogonal and of unit length. We define the three *rectangular* or *Cartesian components* of a vector $\mathbf{v}$ by

$$\mathbf{v}\cdot\mathbf{e}_1 = v_1, \qquad \mathbf{v}\cdot\mathbf{e}_2 = v_2, \qquad \mathbf{v}\cdot\mathbf{e}_3 = v_3.$$

Then a vector $\mathbf{v}$ may be written as the sum of three vector components parallel to the base vectors (see Fig. 3.1):

$$\mathbf{v} = v_1\mathbf{e}_1 + v_2\mathbf{e}_2 + v_3\mathbf{e}_3.$$

Thus, given a set of base vectors, there is a one-to-one correspondence between vectors $\mathbf{v}$ and *ordered triples* of numbers (v_1, v_2, v_3). Instead of having to draw geometrical pictures, we can do vector algebra with numbers. All of the definitions and rules in Chapter 1 can be written in component form. You can easily show that

$$\begin{aligned}
|\mathbf{v}| &= \sqrt{v_1^2 + v_2^2 + v_3^2},\\
\mathbf{u} = \mathbf{v} &\iff u_1 = v_1, u_2 = v_2, u_3 = v_3,\\
\mathbf{u} + \mathbf{v} &= (u_1 + v_1)\mathbf{e}_1 + (u_2 + v_2)\mathbf{e}_2 + (u_3 + v_3)\mathbf{e}_3,\\
\mathbf{u}\cdot\mathbf{v} &= u_1v_1 + u_2v_2 + u_3v_3,
\end{aligned}$$

and so forth.

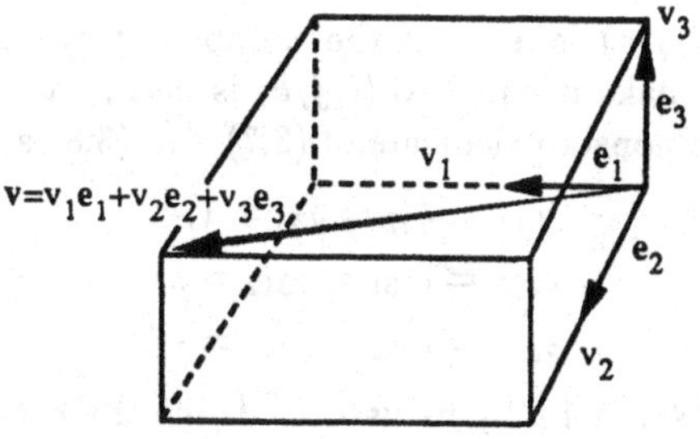

Figure 3.1. The vector $\mathbf{v}$ is resolved into Cartesian components (v_1, v_2, v_3) along the orthonormal triad $(\mathbf{e}_1, \mathbf{e}_2, \mathbf{e}_3)$.

Whenever possible in this book, we will shorten our formulas by using *index notation*. Here is how the previous equations look in index notation:

$$\mathbf{v} \cdot \mathbf{e}_i = v_i, \tag{3.1}$$

$$\mathbf{v} = v_i \mathbf{e}_i = v_k \mathbf{e}_k, \tag{3.2}$$

$$|\mathbf{v}| = \sqrt{v_i v_i}, \tag{3.3}$$

$$\mathbf{u} = \mathbf{v} \iff u_i = v_i, \tag{3.4}$$

$$\mathbf{u} + \mathbf{v} = (u_i + v_i)\mathbf{e}_i, \tag{3.5}$$

$$\mathbf{u} \cdot \mathbf{v} = u_i v_i. \tag{3.6}$$

The repeated appearance of the index i in (3.2), (3.3), (3.5), (3.6) indicates that the right side of each of these equations is a summation with the index i having values 1, 2, and 3 (the usual summation sign is left out). Whenever an index occurs two times in a term, it is implied that the term is to be summed over values of the index from 1 to 3. This shorthand notation for a sum is called the *summation convention*. A repeated index is said to be a *dummy* index, because we can replace it by another letter without changing the sum; for example, in (3.2) we can replace the dummy index i with the dummy index k. An unrepeated index is said to be *free*, meaning that it may have the value 1, 2, or 3; for example, in (3.1) and (3.4) i is a free index.

Other formulas may be shortened by introducing the *Kronecker delta* δ_{ij} and the *permutation symbol* ϵ_{ijk}:

$$\delta_{ij} = \begin{cases} 1 & \text{if } i = j, \\ 0 & \text{if } i \neq j, \end{cases} \tag{3.7}$$

$$\epsilon_{ijk} = \begin{cases} +1 & \text{if } (i,j,k) \text{ is an } \textit{even} \text{ permutation} \\ & \text{of } (1,2,3), \\ -1 & \text{if } (i,j,k) \text{ is an } \textit{odd} \text{ permutation} \\ & \text{of } (1,2,3), \\ 0 & \text{if two or more indices are equal.} \end{cases} \tag{3.8}$$

A permutation of (i,j,k) is *even* if the number of switches of the entries of (1, 2, 3) required to make it equal to (i,j,k) is *even*, and similarly for odd permutations. Thus, the nonzero elements of (3.7) and (3.8) are

$$\delta_{11} = \delta_{22} = \delta_{33} = 1,$$
$$\epsilon_{123} = \epsilon_{231} = \epsilon_{312} = 1,$$
$$\epsilon_{132} = \epsilon_{321} = \epsilon_{213} = -1.$$

The permutation symbol may be used to define the *determinant* of a matrix:

$$\det \begin{bmatrix} u_1 & u_2 & u_3 \\ v_1 & v_2 & v_3 \\ w_1 & w_2 & w_3 \end{bmatrix} = \begin{vmatrix} u_1 & u_2 & u_3 \\ v_1 & v_2 & v_3 \\ w_1 & w_2 & w_3 \end{vmatrix} = \epsilon_{ijk} u_i v_j w_k$$
$$= u_1 v_2 w_3 + u_2 v_3 w_1 + u_3 v_1 w_2 - u_3 v_2 w_1 - u_1 v_3 w_2 - u_2 v_1 w_3 \,. \tag{3.9}$$

'he determinant of a 3 × 3 matrix can be evaluated by repeating the first two ɔlumns of the determinant to the right of it, and then adding the products on ıe downward diagonals and subtracting the products on the upward diagonals.

The usefulness of these symbols is indicated by the following concise formulas:

$$\mathbf{u} \times \mathbf{v} = \epsilon_{ijk} u_i v_j \mathbf{e}_k, \tag{3.10}$$

$$\mathbf{u} \times \mathbf{v} \cdot \mathbf{w} = \epsilon_{ijk} u_i v_j w_k, \tag{3.11}$$

$$(\mathbf{a} \times \mathbf{b} \cdot \mathbf{c})(\mathbf{u} \times \mathbf{v} \cdot \mathbf{w}) = \begin{vmatrix} \mathbf{a}\cdot\mathbf{u} & \mathbf{a}\cdot\mathbf{v} & \mathbf{a}\cdot\mathbf{w} \\ \mathbf{b}\cdot\mathbf{u} & \mathbf{b}\cdot\mathbf{v} & \mathbf{b}\cdot\mathbf{w} \\ \mathbf{c}\cdot\mathbf{u} & \mathbf{c}\cdot\mathbf{v} & \mathbf{c}\cdot\mathbf{w} \end{vmatrix}. \tag{3.12}$$

pplying (3.12) to the base vectors, we obtain the following succinct identities ːlating the permutation symbol and Kronecker delta:

$$\epsilon_{ijk}\epsilon_{pqr} = \begin{vmatrix} \delta_{ip} & \delta_{iq} & \delta_{ir} \\ \delta_{jp} & \delta_{jq} & \delta_{jr} \\ \delta_{kp} & \delta_{kq} & \delta_{kr} \end{vmatrix}. \tag{3.13}$$

$$\epsilon_{ijk}\epsilon_{pqk} = \delta_{ip}\delta_{jq} - \delta_{iq}\delta_{jp}. \tag{3.14}$$

'ote that the 6 indices in (3.13) are free, so (3.13) is equivalent to $3^6 = 729$ ɋuations!

With index notation the verification of vector identities becomes easy:

roof of (1.4):

$$\begin{aligned} \mathbf{a} \times (\mathbf{b} \times \mathbf{c}) &= \mathbf{a} \times (\epsilon_{ijk} b_i c_j \mathbf{e}_k) = \epsilon_{mkn} a_m (\epsilon_{ijk} b_i c_j) \mathbf{e}_n \\ &= \epsilon_{mkn}\epsilon_{ijk} a_m b_i c_j \mathbf{e}_n = \epsilon_{nmk}\epsilon_{ijk} a_m b_i c_j \mathbf{e}_n \\ &= (\delta_{in}\delta_{jm} - \delta_{jn}\delta_{im}) a_m b_i c_j \mathbf{e}_n = \delta_{in}\delta_{jm} a_m b_i c_j \mathbf{e}_n - \delta_{jn}\delta_{im} a_m b_i c_j \mathbf{e}_n \\ &= a_j b_i c_j \mathbf{e}_i - a_i b_i c_j \mathbf{e}_j = (b_i \mathbf{e}_i)(a_j c_j) - (c_j \mathbf{e}_j)(a_i b_i) \\ &= \mathbf{b}(\mathbf{a}\cdot\mathbf{c}) - \mathbf{c}(\mathbf{a}\cdot\mathbf{b}). \end{aligned}$$

Analytic geometry is the numerical representation of geometric objects. The asic idea is the establishment of a one-to-one correspondence between ordered ːiples of numbers and *points* or locations in three-dimensional space. Given an rigin O, we can associate with each point P a position vector $\mathbf{r} = \vec{OP}$ drawn from ɔ to P (see Fig. 3.2). Then, given a basis $(\mathbf{e}_1, \mathbf{e}_2, \mathbf{e}_3)$, we call the components $x_1, x_2, x_3)$ of $\mathbf{r} = x_i \mathbf{e}_i$ the *rectangular* or *Cartesian coordinates* of P. We measure ıe distances (x_1, x_2, x_3) from O along lines called the (x_1, x_2, x_3) axes drawn ırough the base vectors. Thus, using vector methods, we can derive equations ɔr geometric figures.

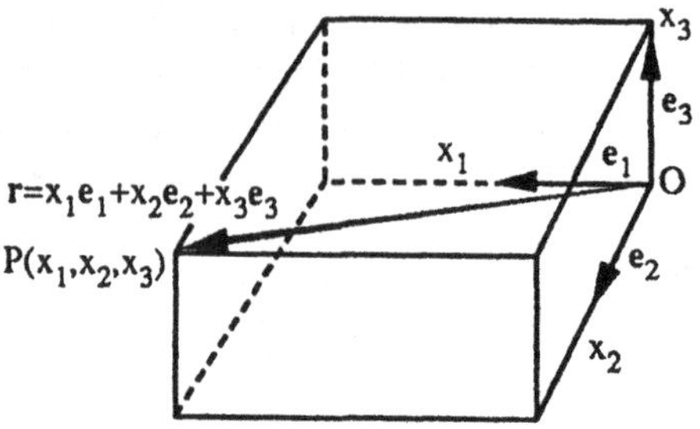

Figure 3.2. The point P is associated with the Cartesian coordinates (x_1, x_2, x_3).

Example: We will derive the equation of the line that passes through the point $P(a_1, a_2, a_3)$ and is parallel to the vector $\mathbf{v} = v_i\mathbf{e}_i$. Let $Q(x_1, x_2, x_3)$ be an arbitrary point on the line (Fig. 3.3). Since the vector $\vec{PQ} = \mathbf{r} - \mathbf{a}$ is parallel to $\mathbf{v}$, there must be a scalar t such that

$$\mathbf{r} - \mathbf{a} = t\mathbf{v}. \tag{3.15a}$$

Separating components in (3.15) with $\mathbf{r} = x_i\mathbf{e}_i$ and $\mathbf{a} = a_i\mathbf{e}_i$, and allowing t to vary from $-\infty$ to $+\infty$, we obtain the *parametric* equations of the line:

$$x_i - a_i = tv_i. \tag{3.15b}$$

We can obtain the *Cartesian* equations of the line by eliminating t from (3.15b):

$$\frac{x_1 - a_1}{v_1} = \frac{x_2 - a_2}{v_2} = \frac{x_3 - a_3}{v_3}.$$

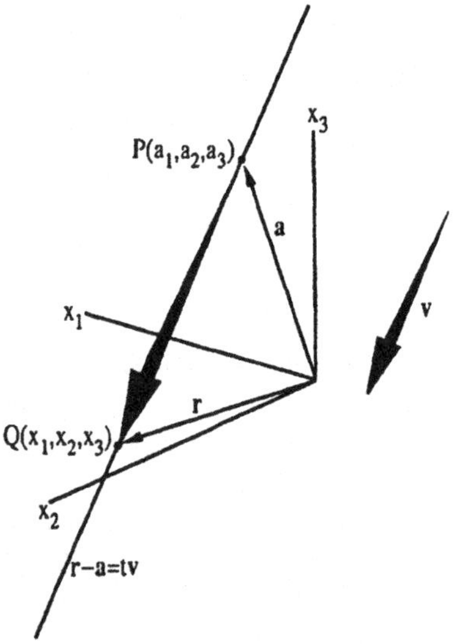

Figure 3.3. Equation of a straight line.

Example: We will derive the equation of the plane that passes through the point $P(a_1, a_2, a_3)$ and is perpendicular to the unit vector $\mathbf{n} = n_i \mathbf{e}_i$. Let $Q(x_1, x_2, x_3)$ be an arbitrary point on the plane (Fig. 3.4). Since the vector $\vec{PQ} = \mathbf{r} - \mathbf{a}$ is perpendicular to $\mathbf{n}$, we must have $(\mathbf{r} - \mathbf{a}) \cdot \mathbf{n} = 0$, or, with $c = \mathbf{a} \cdot \mathbf{n}$,

$$\mathbf{n} \cdot \mathbf{r} = c. \tag{3.16a}$$

The equation of this plane in Cartesian coordinates is

$$n_i x_i = c. \tag{3.16b}$$

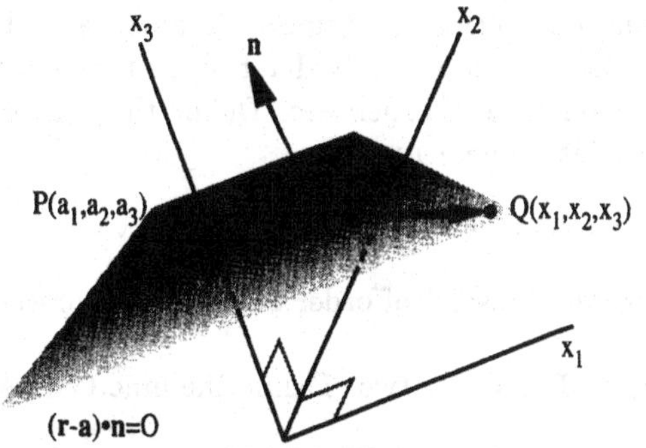

Figure 3.4. Equation of a plane.

PROBLEMS 3.1

1. Show that $\frac{\mathbf{v}}{v} = \mathbf{e}_i \cos\theta_i$, where θ_i is the angle between $\mathbf{v}$ and the x_i axis. These cosines are called the *direction cosines* of $\mathbf{v}$.

2. Find the angle between the diagonal of a cube and one of its edges.

3. Evaluate δ_{ii} and $\epsilon_{ijk}\epsilon_{ijk}$.

4. Write $(\mathbf{a} \times \mathbf{b}) \cdot (\mathbf{c} \times \mathbf{d})$ in component form.

5. Find the volume of the parallelepiped determined by $\mathbf{u} = \mathbf{e}_1 + \mathbf{e}_2 + \mathbf{e}_3$, $\mathbf{v} = \mathbf{e}_2 + 2\mathbf{e}_3$, $\mathbf{w} = -\mathbf{e}_1 + 2\mathbf{e}_3$.

6. Use (3.9), (3.11), and the facts that $\det A = \det A^t$ and $\det A \det B = \det AB$, where A and B are arbitrary 3×3 matrices, to prove (3.12).

7. Directly verify (3.14) for all possible values of the free indices. How many equations are there to check?

8. Find the equation of the line passing through $P(1,2,3)$ and parallel to $\mathbf{v} = 3\mathbf{e}_1 + 2\mathbf{e}_2 + \mathbf{e}_3$. Find the equation of the plane passing through P and perpendicular to $\mathbf{v}$.

9. Find the equation of the line passing through the points (0, 0, 0) and (1, 2, 3). Find the equation of the plane passing through the points (0, 0, 0), (1, 1, 1), and (1, 2, 3).

10. Find the equations of the lines normal to the plane (3.16).

11. Find the minimum distances from the origin to the line (3.15) and to the plane (3.16).

3.2 TENSOR COMPONENTS

We can express any tensor in terms of components along an orthonormal basis. The components of a tensor are completely determined by its action on the base vectors. Consider first a tensor $\mathbf{f}$ of order one. Define the Cartesian components of $\mathbf{f}$ by $a_i = \mathbf{f}(\mathbf{e}_i)$. Since $\mathbf{f}$ is a linear function,

$$\mathbf{f}(\mathbf{v}) = \mathbf{f}(v_i\mathbf{e}_i) = v_i\mathbf{f}(\mathbf{e}_i) = a_i v_i = (a_i\mathbf{e}_i) \cdot \mathbf{v} = \mathbf{a} \cdot \mathbf{v}$$

for all vectors $\mathbf{v}$; that is, any tensor $\mathbf{f}$ of order one can be represented by a vector $\mathbf{a}$.

Next consider a tensor $\mathbf{T}$ of order two. Define the nine Cartesian components of $\mathbf{T}$ by

$$\boxed{T_{ij} = \mathbf{e}_i \cdot \mathbf{T} \cdot \mathbf{e}_j\,.}$$

Then

$$\mathbf{u} \cdot \mathbf{T} \cdot \mathbf{v} = u_i T_{ij} v_j = \mathbf{u} \cdot (T_{ij}\mathbf{e}_i\mathbf{e}_j) \cdot \mathbf{v}$$

for all vectors $\mathbf{u}$ and $\mathbf{v}$, so

$$\boxed{\mathbf{T} = T_{ij}\mathbf{e}_i\mathbf{e}_j\,;}$$

that is, any tensor of order two can be decomposed into a linear combination of the dyads formed from the base vectors.

Tensors of higher order may be decomposed in a similar fashion. For instance, any tensor $\mathbf{B}$ of order three may be represented by the triadic

$$\mathbf{B} = B_{ijk}\mathbf{e}_i\mathbf{e}_j\mathbf{e}_k.$$

The tensors defined in Chapter 2 can all be written in component form. If $\mathbf{S}$ and $\mathbf{T}$ are second-order tensors

$$\boxed{\begin{aligned} \mathbf{S} = \mathbf{T} \Longleftrightarrow S_{ij} &= T_{ij}, \\ \mathbf{T} \cdot \mathbf{v} &= T_{ij} v_j \mathbf{e}_i, \\ \mathbf{u} \cdot \mathbf{T} \cdot \mathbf{v} &= u_i T_{ij} v_j, \\ \mathbf{S} \cdot \mathbf{T} &= S_{ik} T_{kj} \mathbf{e}_i \mathbf{e}_j, \\ \mathbf{T}^t &= T_{ji} \mathbf{e}_i \mathbf{e}_j. \end{aligned}}$$

The Cartesian components of the identity and permutation tensors are the Kronecker delta and permutation symbol, respectively:

$$\boxed{\begin{aligned} \mathbf{1} &= \delta_{ij}\mathbf{e}_i\mathbf{e}_j = \mathbf{e}_i\mathbf{e}_i, \\ \mathbf{E} &= \epsilon_{ijk}\mathbf{e}_i\mathbf{e}_j\mathbf{e}_k. \end{aligned}}$$ (3.17) (3.18)

It follows from this that the components of symmetric, antisymmetric, and orthogonal tensors are related by the formulas

$$\boxed{\begin{aligned} \mathbf{S} \text{ symmetric} &\iff S_{ij} = S_{ji}, \\ \mathbf{A} \text{ antisymmetric} &\iff A_{ij} = -A_{ji}, \\ \mathbf{Q} \text{ orthogonal} &\iff Q_{ik}Q_{jk} = \delta_{ij}. \end{aligned}}$$

The Cartesian components T_{ij} of a second-order tensor $\mathbf{T}$ can be used to create a square matrix T:

$$\boxed{T = \begin{bmatrix} T_{11} & T_{12} & T_{13} \\ T_{21} & T_{22} & T_{23} \\ T_{31} & T_{32} & T_{33} \end{bmatrix}.}$$

We call T the *matrix of the tensor* $\mathbf{T}$ (relative to the given basis). The components can then be manipulated by standard techniques of matrix algebra. And all of the properties of second-order tensors given in Chapter 2 are equivalent to well-known results of matrix theory.

Example: Let's find the inverse of the tensor $\mathbf{T} = \mathbf{1} + \mathbf{e}_2\mathbf{e}_1$ by row reduction. The matrix of $\mathbf{1}$ is the 3×3 *identity matrix* I:

$$I = \begin{bmatrix} 1 & 0 & 0 \\ 0 & 1 & 0 \\ 0 & 0 & 1 \end{bmatrix}.$$

The matrix of $\mathbf{T}$ is

$$T = \begin{bmatrix} 1 & 0 & 0 \\ 1 & 1 & 0 \\ 0 & 0 & 1 \end{bmatrix}. \tag{3.19}$$

Since $\det T = 1$ (not zero), T is nonsingular and invertible. Adjoin the identity matrix to the right side of (3.19):

$$[T|I] = \left[\begin{array}{ccc|ccc} 1 & 0 & 0 & 1 & 0 & 0 \\ 1 & 1 & 0 & 0 & 1 & 0 \\ 0 & 0 & 1 & 0 & 0 & 1 \end{array}\right].$$

Now reduce T to the identity matrix by row operations and simultaneously apply these operations to I to produce A^{-1}. In this case simply multiply the first row by (-1) and add it to the second row to get

$$\left[\begin{array}{ccc|ccc} 1 & 0 & 0 & 1 & 0 & 0 \\ 0 & 1 & 0 & -1 & 1 & 0 \\ 0 & 0 & 1 & 0 & 0 & 1 \end{array}\right] = [I|T^{-1}].$$

Thus

$$\mathbf{T}^{-1} = \mathbf{1} - \mathbf{e}_2\mathbf{e}_1 .$$

The components of tensors of order higher than two may be arranged to form matrices of higher dimensions. For instance, the 27 components B_{ijk} of a third-order tensor **B** form a cubical array in three dimensions. It is difficult to work with such matrices, so index notation is usually used when dealing with higher-order tensors.

PROBLEMS 3.2

1. Determine the Cartesian components of the projection tensor $\mathbf{P} = \mathbf{ee}$.
2. Given an orthonormal basis, how many arbitrary numbers are needed to specify a general second-order tensor? A symmetric tensor? An antisymmetric tensor? An orthogonal tensor?
3. How many components does a tensor of order n have (in three dimensions)?
4. Show that $\det T = \frac{1}{6}\epsilon_{ijk}\epsilon_{pqr}T_{ip}T_{jq}T_{kr}$.
5. Find all vectors $\mathbf{v}$ that get mapped into the vector $\mathbf{T}\cdot\mathbf{v} = \mathbf{e}_1 + \mathbf{e}_2 + \mathbf{e}_3$ by the tensor $\mathbf{T} = \mathbf{1} + \mathbf{e}_1\mathbf{e}_2 + \mathbf{e}_2\mathbf{e}_1$.
6. Find the transpose and inverse of the tensor $\mathbf{T} = \mathbf{e}_2\mathbf{e}_1 + \mathbf{e}_3\mathbf{e}_2 + \mathbf{e}_1\mathbf{e}_3$.
7. A *completely antisymmetric* tensor of order three is a tensor **B** whose Cartesian components are related by $B_{ijk} = -B_{ikj} = B_{kij} = -B_{kji} = B_{jki} = -B_{jik}$. Show that every completely antisymmetric tensor of order three is a scalar multiple of **E**.

3.3 EIGENVECTOR REPRESENTATIONS

Certain second-order tensors have simple representations involving their eigenvalues and eigenvectors. The vector equation (2.4) for the eigenvalues and eigenvectors is equivalent to the following three component scalar equations:

$$T_{ij}v_j = \lambda v_i . \tag{3.20}$$

The set of homogeneous equations (3.20) has nontrivial solutions for v_i if and only if

$$\boxed{\det(T - \lambda I) = 0.} \tag{3.21}$$

Expansion of the determinant leads to the *characteristic polynomial* of **T**:

$$\lambda^3 - J_1(\mathbf{T})\lambda^2 + J_2(\mathbf{T})\lambda - J_3(\mathbf{T}) = 0, \tag{3.22}$$

where

$$\left.\begin{aligned} J_1(\mathbf{T}) &= \text{tr } \mathbf{T} = T_{ii} = \lambda_1 + \lambda_2 + \lambda_3 \\ J_2(\mathbf{T}) &= \tfrac{1}{2}\left[T_{ii}T_{jj} - T_{ij}T_{ji}\right] = \lambda_1\lambda_2 + \lambda_1\lambda_3 + \lambda_2\lambda_3 \\ J_3(\mathbf{T}) &= \det \mathbf{T} = \det T = \lambda_1\lambda_2\lambda_3 \end{aligned}\right\} . \tag{3.23}$$

Here tr $\mathbf{T}$ or the *trace* or *contraction* of $\mathbf{T}$ is the sum of the diagonal elements of the matrix T, $\det \mathbf{T}$ is the determinant of T, and (J_1, J_2, J_3) are the *principal invariants* of $\mathbf{T}$ (to be discussed in Section 4.4). The cubic equation (3.22) has either three real roots $(\lambda_1, \lambda_2, \lambda_3)$ or one real and two complex roots.

Example: Let's find the real eigenvalues and eigenvectors of $\mathbf{T} = \mathbf{1} + \mathbf{e}_2\mathbf{e}_1$. From (3.21),

$$\begin{vmatrix} 1-\lambda & 0 & 0 \\ 1 & 1-\lambda & 0 \\ 0 & 0 & 1-\lambda \end{vmatrix} = (1-\lambda)^3 = 0.$$

The characteristic polynomial is thus

$$\lambda^3 - 3\lambda^2 + 3\lambda - 1 = 0,$$

which agrees with (3.22) and (3.23). The only real eigenvalue is $\lambda = 1$. The eigenvectors corresponding to $\lambda = 1$ are obtained from the linear system

$$\begin{bmatrix} 1 & 0 & 0 \\ 1 & 1 & 0 \\ 0 & 0 & 1 \end{bmatrix} \begin{bmatrix} v_1 \\ v_2 \\ v_3 \end{bmatrix} = \begin{bmatrix} v_1 \\ v_2 \\ v_3 \end{bmatrix} .$$

This system has the solution $v_1 = 0$, v_2 and v_3 arbitrary, so any vector of the form $v_2\mathbf{e}_2 + v_3\mathbf{e}_3$ is an eigenvector.

An *antisymmetric* tensor $\mathbf{A}$ has a single real eigenvalue 0 with a corresponding eigenvector $\boldsymbol{\omega}$, which is called the *axis* of $\mathbf{A}$, and has the simple representation

$$\boxed{\mathbf{A} = \boldsymbol{\omega} \times .}$$

The component form of this is

$$A_{ij} = -\epsilon_{ijk}\omega_k , \tag{3.24a}$$

and the matrix form is

$$A = \begin{bmatrix} 0 & -\omega_3 & \omega_2 \\ \omega_3 & 0 & -\omega_1 \\ -\omega_2 & \omega_1 & 0 \end{bmatrix} . \tag{3.24b}$$

The introduction of an antisymmetric tensor is often avoided by use of the associated axis vector.

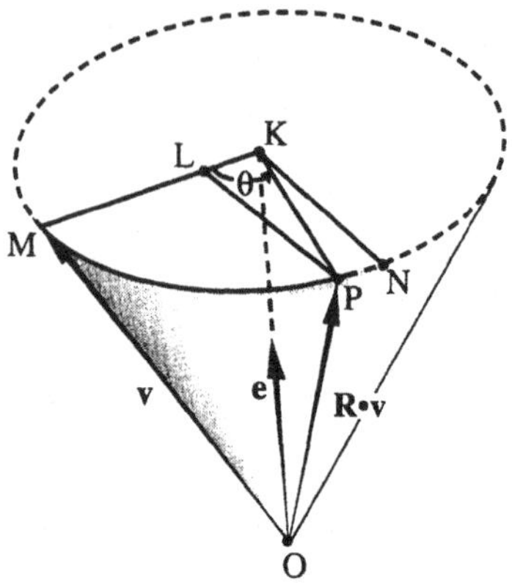

Figure 3.5. The tensor **R** rotates vectors about an axis parallel to e through an angle θ.

Every *orthogonal* tensor **Q** is either a *rotation* tensor **R** or the product of a rotation tensor and a *reflection* tensor **H**. A rotation tensor **R** has a single real eigenvalue + 1 with normalized eigenvector e, and has the representation

$$\boxed{\mathbf{R} = \cos\theta\,\mathbf{1} + (1 - \cos\theta)\mathbf{ee} + \sin\theta\,\mathbf{e}\times .} \tag{3.25}$$

R rotates vectors about an axis parallel to **e** through an angle θ, which is reckoned positive ($0 \leq \theta < 2\pi$) by the closing of the fingers of the right hand with the extended thumb in the direction of **e** (Fig. 3.5). A reflection tensor **H** has an eigenvalue -1 corresponding to e, and has the representation

$$\mathbf{H} = \mathbf{1} - 2\mathbf{ee}. \tag{3.26}$$

H reflects vectors across a plane with normal **e** (Fig. 3.6). One rotation or reflection after another corresponds to the dot product of the corresponding rotation or reflection tensors.

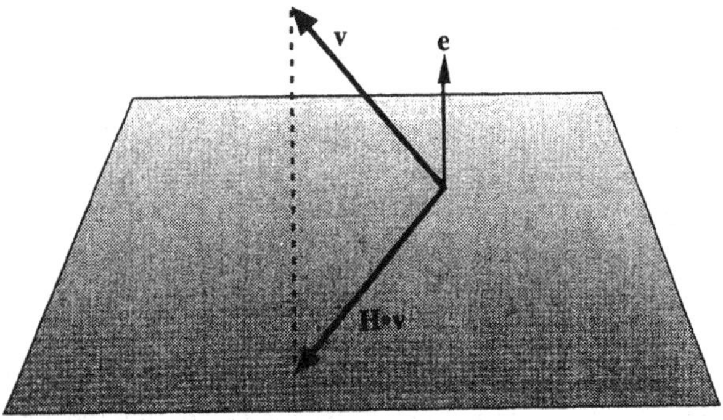

Figure 3.6. The tensor **H** reflects vectors across a plane with unit normal e.

Verification of the properties of R. In Fig. 3.5 the vector $\mathbf{v}$ is rotated about the axis $\mathbf{e}$ through an angle θ to produce the vector $\vec{OP}$. From the construction,

$$\vec{OP} = \vec{OK} + \vec{KL} + \vec{LP}$$

and

$$\begin{aligned} &\vec{OM} = \mathbf{v}, \qquad \vec{OK} = \mathbf{e}(\mathbf{e}\cdot\mathbf{v}), \\ &\vec{KM} = \vec{OM} - \vec{OK} = \mathbf{v} - \mathbf{e}(\mathbf{e}\cdot\mathbf{v}), \qquad \vec{KL} = \cos\theta\ \vec{KM} = \cos\theta[\mathbf{v} - \mathbf{e}(\mathbf{e}\cdot\mathbf{v})], \\ &\vec{KN} = \mathbf{e}\times\mathbf{v}, \qquad \vec{LP} = \sin\theta\ \vec{KN} = \sin\theta\ \mathbf{e}\times\mathbf{v}. \end{aligned}$$

Hence

$$\begin{aligned} \vec{OP} &= \mathbf{e}(\mathbf{e}\cdot\mathbf{v}) + \cos\theta[\mathbf{v} - \mathbf{e}(\mathbf{e}\cdot\mathbf{v})] + \sin\theta\ \mathbf{e}\times\mathbf{v} \\ &= \cos\theta\ \mathbf{v} + (1-\cos\theta)\mathbf{e}(\mathbf{e}\cdot\mathbf{v}) + \sin\theta\ \mathbf{e}\times\mathbf{v} \\ &= \mathbf{R}\cdot\mathbf{v}, \end{aligned}$$

so $\mathbf{R}$ rotates vectors about the axis $\mathbf{e}$ by an angle θ. The only vectors that don't get rotated are parallel to $\mathbf{e}$, so the normalized eigenvector of $\mathbf{R}$ is $\mathbf{e}$ with corresponding eigenvalue 1:

$$\mathbf{R}\cdot\mathbf{e} = \cos\theta\ \mathbf{e} + (1-\cos\theta)\mathbf{e} = \mathbf{e}.$$

Since $\mathbf{R}^{-1}\cdot(\mathbf{R}\cdot\mathbf{v}) = \mathbf{v}$, $\mathbf{R}^{-1}$ rotates $\mathbf{R}\cdot\mathbf{v}$ back into $\mathbf{v}$ and is obtained by replacing $\mathbf{e}$ by $-\mathbf{e}$ in (3.25):

$$\mathbf{R}^{-1} = (\cos\theta)\mathbf{1} + (1-\cos\theta)\mathbf{ee} - \sin\theta\ \mathbf{e}\times. \tag{3.27}$$

Since the transpose of (3.25) is also given by (3.27), $\mathbf{R}$ is orthogonal.

A *symmetric* tensor $\mathbf{S}$ has three real eigenvalues with corresponding eigenvectors that are mutually orthogonal. A symmetric tensor is said to be *positive definite* if its eigenvalues are all positive, *negative definite* if its eigenvalues are all negative, and *indefinite* otherwise. The *principal axes* of $\mathbf{S}$ are the Cartesian axes in the direction of its eigenvectors. Let $(\lambda_1, \lambda_2, \lambda_3)$ denote the *eigenvalues* of $\mathbf{S}$ and $(\mathbf{e}_1, \mathbf{e}_2, \mathbf{e}_3)$ denote the corresponding normalized eigenvectors:

$$\mathbf{S}\cdot\mathbf{e}_1 = \lambda_1\mathbf{e}_1, \qquad \mathbf{S}\cdot\mathbf{e}_2 = \lambda_2\mathbf{e}_2, \qquad \mathbf{S}\cdot\mathbf{e}_3 = \lambda_3\mathbf{e}_3. \tag{3.28}$$

Then any symmetric tensor $\mathbf{S}$ can be represented by the dyadic

$$\boxed{\mathbf{S} = \lambda_1\mathbf{e}_1\mathbf{e}_1 + \lambda_2\mathbf{e}_2\mathbf{e}_2 + \lambda_3\mathbf{e}_3\mathbf{e}_3.} \tag{3.29}$$

The expansion (3.29) is called the *spectral representation* of $\mathbf{S}$.

Verification of the properties of S. Dotting (3.28) with the eigenvectors yields

$$\mathbf{e}_2\cdot\mathbf{S}\cdot\mathbf{e}_1 = \lambda_1\mathbf{e}_1\cdot\mathbf{e}_2 \qquad \text{and} \qquad \mathbf{e}_2\cdot\mathbf{S}^t\cdot\mathbf{e}_1 = \lambda_2\mathbf{e}_1\cdot\mathbf{e}_2.$$

But since $\mathbf{S}^t = \mathbf{S}$,

$$(\lambda_1 - \lambda_2)\mathbf{e}_1 \cdot \mathbf{e}_2 = 0.$$

Similarly,

$$(\lambda_1 - \lambda_3)\mathbf{e}_1 \cdot \mathbf{e}_3 = 0, \qquad (\lambda_2 - \lambda_3)\mathbf{e}_2 \cdot \mathbf{e}_3 = 0.$$

There are three cases to consider:

- If $\lambda_1 \neq \lambda_2 \neq \lambda_3$, then $\mathbf{e}_1 \cdot \mathbf{e}_2 = \mathbf{e}_1 \cdot \mathbf{e}_3 = \mathbf{e}_2 \cdot \mathbf{e}_3 = 0$, so the eigenvectors are orthogonal.
- If $\lambda_1 = \lambda_2 \neq \lambda_3$, then $\mathbf{e}_1 \cdot \mathbf{e}_3 = \mathbf{e}_2 \cdot \mathbf{e}_3 = 0$, so $\mathbf{e}_1$ and $\mathbf{e}_2$ may be chosen mutually orthogonal with any orientation in a plane normal to $\mathbf{e}_3$.
- If $\lambda_1 = \lambda_2 = \lambda_3$, then no restriction is imposed on the relative orientation of the eigenvectors and they may be chosen mutually orthogonal.

You can show that the eigenvalues are real by assuming they are not and obtaining a contradiction. The representation (3.29) is obtained by applying $\mathbf{S}$ to the identity tensor:

$$\mathbf{S} = \mathbf{S} \cdot \mathbf{1} = \mathbf{S} \cdot (\mathbf{e}_i\mathbf{e}_i) = (\mathbf{S} \cdot \mathbf{e}_i)\mathbf{e}_i = \lambda_1\mathbf{e}_1\mathbf{e}_1 + \lambda_2\mathbf{e}_2\mathbf{e}_2 + \lambda_3\mathbf{e}_3\mathbf{e}_3.$$

There is one-to-one correspondence between symmetric tensors and quadratic forms. If $\mathbf{T}$ is a tensor of order two, the expression $\mathbf{r} \cdot \mathbf{T} \cdot \mathbf{r} = T_{ij}x_ix_j$ is called a *quadratic form*. Since

$$\mathbf{r} \cdot \mathbf{T} \cdot \mathbf{r} = \mathbf{r} \cdot \mathbf{S} \cdot \mathbf{r} \qquad \text{where} \qquad \mathbf{S} = \frac{1}{2}(\mathbf{T} + \mathbf{T}^t),$$

we can always assume that the tensor $\mathbf{S}$ of the quadratic form is symmetric. When the principal axes of $\mathbf{S}$ are chosen for the coordinates (x_1, x_2, x_3), the spectral representation (3.29) produces a *diagonal* quadratic form:

$$\mathbf{r} \cdot \mathbf{S} \cdot \mathbf{r} = \lambda_1 x_1^2 + \lambda_2 x_2^2 + \lambda_3 x_3^2. \tag{3.30}$$

This shows that when $\mathbf{S}$ is positive definite the quadratic form (3.30) is *positive definite* or is positive for all nonzero $\mathbf{r}$, and when $\mathbf{S}$ is indefinite the quadratic form (3.30) is *indefinite* or vanishes for some nonzero $\mathbf{r}$. The equation

$$\mathbf{r} \cdot \mathbf{S} \cdot \mathbf{r} = x_iS_{ij}x_j = 1$$

defines a *quadric surface*, whose equation in the principal axes is

$$\lambda_1 x_1^2 + \lambda_2 x_2^2 + \lambda_3 x_3^2 = 1. \tag{3.31}$$

The signs of the eigenvalues determine the type of surface given by (3.31):

- $\lambda_1 > 0, \lambda_2 > 0, \lambda_3 > 0$, ellipsoid, (3.32)

- $\lambda_1 > 0, \lambda_2 > 0, \lambda_3 < 0,$ hyperboloid with one sheet,
- $\lambda_1 > 0, \lambda_2 < 0, \lambda_3 < 0,$ hyperboloid with two sheets,
- $\lambda_1 = 0, \lambda_2 > 0, \lambda_3 > 0,$ elliptic cylinder,
- $\lambda_1 = 0, \lambda_2 > 0, \lambda_3 < 0,$ hyperbolic cylinder,
- $\lambda_1 = \lambda_2 = 0, \lambda_3 > 0,$ two parallel planes,
- $\lambda_1 < 0, \lambda_2 < 0, \lambda_3 \leq 0,$ not a real surface.

PROBLEMS 3.3

1. Check to see that the expansion of (3.21) leads to (3.22) and (3.23).

2. Find the eigenvalues and normalized eigenvectors of $\mathbf{T} = \mathbf{1} + \mathbf{e}_1\mathbf{e}_2 + \mathbf{e}_2\mathbf{e}_1$.

3. A tensor $\mathbf{T}$ has the following components with respect to a set of Cartesian coordinates:
$$T = \begin{bmatrix} 1 & 2 & 0 \\ 0 & -1 & 1 \\ 1 & 0 & 2 \end{bmatrix}.$$
Find the values of the principal invariants J_1, J_2, J_3.

4. Show that $S_{ij}T_{ji} = \text{tr}\ (\mathbf{S} \cdot \mathbf{T})$, where $\mathbf{S}$ and $\mathbf{T}$ are arbitrary second-order tensors.

5. Show that the roots of the cubic equation $x^3 + ax^2 + bx + c = 0$ are the eigenvalues of the matrix
$$T = \begin{bmatrix} 0 & 0 & -c \\ 1 & 0 & -b \\ 0 & 1 & -a \end{bmatrix}.$$

6. Show that the axis of an antisymmetric tensor $\mathbf{A}$ is given by $\omega_i = -\frac{1}{2}\epsilon_{ijk}A_{jk}$. Show that $\mathbf{1} + \mathbf{A}$ is nonsingular.

7. Given a rotation tensor $\mathbf{R}$, find the rotation angle θ and eigenvector $\mathbf{e}$ used in the representation (3.25).

8. Show that the determinant of a rotation tensor is $+1$ and the determinant of a reflection tensor is -1.

9. Verify the properties of a reflection tensor (3.26). Show that $\mathbf{H}^2 = \mathbf{1}$.

10. If $\mathbf{T}$ is any nonsingular second-order tensor, show that $\mathbf{T} \cdot \mathbf{T}^t$ is symmetric and positive definite.

11. Define a linear form and a cubic form. Sketch a linear surface and a cubic surface.

12. Show that any rotation tensor can be represented as $\mathbf{R} = e^{\theta \mathbf{e}\times}$. This is known as the *polar* representation of $\mathbf{R}$, by analogy with the polar representation $e^{i\theta} = \cos\theta + i\sin\theta$ of a complex number having absolute value 1, where $i = \sqrt{-1}$. (Note: If $\mathbf{T}$ is a second order tensor, then the exponential of $\mathbf{T}$ is defined by the infinite series expansion $e^{\mathbf{T}} = \mathbf{1} + \mathbf{T} + \frac{\mathbf{T}^2}{2} + \cdots \frac{\mathbf{T}^n}{n!} + \cdots$.)

3.4 GEOMETRIC DISTORTIONS AND THE POLAR DECOMPOSITION THEOREM

Given a region of space we may construct vectors $\mathbf{r}$ from a fixed origin O to each point in the region. A second-order tensor $\mathbf{T}$ maps every vector $\mathbf{r}$ into the image vector $\mathbf{x} = \mathbf{T} \cdot \mathbf{r}$. If the origin of the image vectors is also chosen to be O, a second-order tensor can be distinguished by the shape of the region formed by the tips of the image vectors.

Consider a cube in space with edges of length ℓ parallel to the orthonormal base vectors $(\mathbf{e}_1, \mathbf{e}_2, \mathbf{e}_3)$. This cube will be deformed by $\mathbf{T}$ into a parallelepiped with edges represented by the vectors

$$(\ell\mathbf{T} \cdot \mathbf{e}_1,\ \ell\mathbf{T} \cdot \mathbf{e}_2,\ \ell\mathbf{T} \cdot \mathbf{e}_3).$$

The volume of this parallelepiped is

$$\ell^3 |(\mathbf{T} \cdot \mathbf{e}_1) \times (\mathbf{T} \cdot \mathbf{e}_2) \cdot (\mathbf{T} \cdot \mathbf{e}_3)| = \ell^3 |\det \mathbf{T}|.$$

Now the volume of any region in space may be obtained by adding the volumes of small cubes contained within the region. Hence $|\det \mathbf{T}|$ is the ratio of the deformed volume to the undeformed volume of any region distorted by $\mathbf{T}$.

A tensor $\mathbf{T}$ is singular if and only if $\det \mathbf{T} = 0$. Hence a singular tensor maps every region of space into another region having no volume. Specifically, a singular tensor maps every region either onto a plane ($\mathbf{P}' = \mathbf{e}_1\mathbf{e}_1 + \mathbf{e}_2\mathbf{e}_2$), onto a line ($\mathbf{P} = \mathbf{e}\mathbf{e}$), or into the origin ($\mathbf{T} = \mathbf{0}$).

Let us now study the action of *nonsingular* tensors on regions of space. The key to our study is the *polar decomposition theorem*: Any nonsingular tensor $\mathbf{T}$ of order two may be decomposed into the product of an *orthogonal* tensor $\mathbf{Q}$ and a positive definite *symmetric* tensor $\mathbf{S}$ or $\mathbf{S}'$:

$$\boxed{\mathbf{T} = \mathbf{S} \cdot \mathbf{Q} = \mathbf{Q} \cdot \mathbf{S}'.} \tag{3.33}$$

As indicated in (3.33), the orthogonal tensor $\mathbf{Q}$ is unique but there are two possible symmetric tensors $\mathbf{S}$ or $\mathbf{S}'$, depending on the order of the decomposition.

Proof of (3.33): The tensor $\mathbf{T} \cdot \mathbf{T}^t$ is symmetric and positive definite, so it has a spectral representation

$$\mathbf{T} \cdot \mathbf{T}^t = \lambda_1^2 \mathbf{e}_1\mathbf{e}_1 + \lambda_2^2 \mathbf{e}_2\mathbf{e}_2 + \lambda_3^2 \mathbf{e}_3\mathbf{e}_3.$$

Then we can define

$$\mathbf{S} = (\mathbf{T}\cdot\mathbf{T}^t)^{1/2} = \lambda_1\mathbf{e}_1\mathbf{e}_1 + \lambda_2\mathbf{e}_2\mathbf{e}_2 + \lambda_3\mathbf{e}_3\mathbf{e}_3.$$

Since

$$\mathbf{S}^{-1} = \frac{1}{\lambda_1}\mathbf{e}_1\mathbf{e}_1 + \frac{1}{\lambda_2}\mathbf{e}_2\mathbf{e}_2 + \frac{1}{\lambda_3}\mathbf{e}_3\mathbf{e}_3,$$

we can define

$$\mathbf{Q} = \mathbf{S}^{-1}\cdot\mathbf{T}.$$

$\mathbf{Q}$ is orthogonal because

$$\begin{aligned}\mathbf{Q}\cdot\mathbf{Q}^t &= (\mathbf{S}^{-1}\cdot\mathbf{T})\cdot(\mathbf{T}^t\cdot\mathbf{S}^{-1})\\ &= \mathbf{S}^{-1}\cdot(\mathbf{T}\cdot\mathbf{T}^t)\cdot\mathbf{S}^{-1}\\ &= \mathbf{S}^{-1}\cdot\mathbf{S}^2\cdot\mathbf{S}^{-1} = \mathbf{1}.\end{aligned}$$

Also define

$$\mathbf{S}' = \mathbf{Q}^t\cdot\mathbf{S}\cdot\mathbf{Q}.$$

Then

$$\mathbf{T} = \mathbf{S}\cdot\mathbf{Q} = \mathbf{Q}\cdot\mathbf{S}'.$$

In light of the polar decomposition theorem, let us see what a nonsingular tensor $\mathbf{T}$ does to the unit sphere

$$\mathbf{r}\cdot\mathbf{r} = 1. \tag{3.34}$$

Setting $\mathbf{x} = \mathbf{T}\cdot\mathbf{r}$ and substituting $\mathbf{r} = \mathbf{T}^{-1}\cdot\mathbf{x}$ into (3.34), we find that the image of the unit sphere is

$$\mathbf{x}\cdot\left[(\mathbf{T}^{-1})^t\cdot\mathbf{T}^{-1}\right]\cdot\mathbf{x} = 1.$$

But by the polar decomposition theorem (3.33),

$$\mathbf{x}\cdot\mathbf{S}^{-2}\cdot\mathbf{x} = 1.$$

Using the spectral representation (3.29) for $\mathbf{S}$ and letting $\mathbf{x} = x_i\mathbf{e}_i$, we obtain

$$\frac{x_1^2}{\lambda_1^2} + \frac{x_2^2}{\lambda_2^2} + \frac{x_3^2}{\lambda_3^2} = 1. \tag{3.35}$$

Hence the unit sphere is deformed into an *ellipsoid*, as shown in Fig. 3.7. The semi-axes of the ellipsoid coincide with principal axes of $\mathbf{S}$ and have magnitude $(\lambda_1, \lambda_2, \lambda_3)$ of eigenvalues of $\mathbf{S}$. The mapping $\mathbf{x} = \mathbf{S}\cdot\mathbf{Q}\cdot\mathbf{r}$ of the sphere may be accomplished in two steps. First the orthogonal tensor $\mathbf{Q}$ is applied, causing a rotation and/or reflection of the sphere. Next the symmetric tensor $\mathbf{S}$ is applied, causing a symmetrical deformation of the sphere into an ellipsoid. Note that no distortion of the sphere will occur unless the tensor $\mathbf{S}$ is different from $\mathbf{1}$.

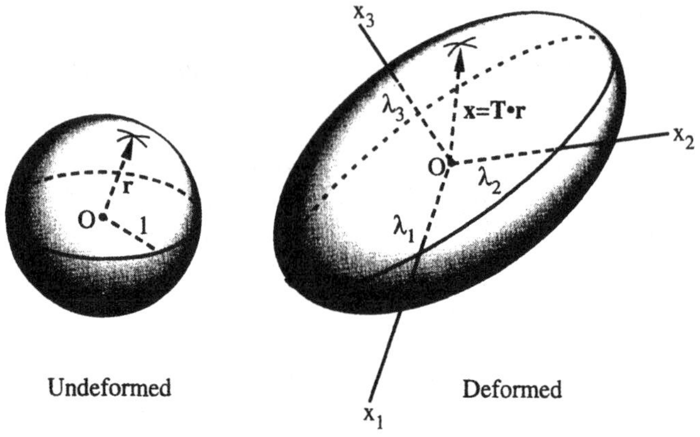

Figure 3.7. A nonsingular tensor **T** maps a sphere into an ellipsoid.

A geometrical figure may be represented in a computer by the coordinates (x_1, x_2, x_3) of many of its points. The computer can alter the shape by replacing the coordinates x_i by $y_i + T_{ij}x_j$, where **y** represents a translation and **T** a rotation and distortion. Then the computer can project the three-dimensional figure onto a plane for display on the screen by simply deleting one of the coordinates.

PROBLEMS 3.4

1. Given a region of space and a second-order tensor **T**, show that the shape of the region formed by the tips of the image vectors $\mathbf{x} = \mathbf{T} \cdot \mathbf{r}$ does not depend upon the choice of origin O for the vectors **r** and **x**.

2. What tensor will change every geometric figure into a similar one of larger dimensions?

3. Show that a second-order tensor maps three-dimensional space either onto the whole space, or onto a plane through the origin, or onto a line through the origin, or onto the origin.

4. Show that a nonsingular second-order tensor **T** maps lines into lines and planes into planes.

5. $\mathbf{T} = \mathbf{e}_1\mathbf{e}_1 + 4\mathbf{e}_2\mathbf{e}_2 + 9\mathbf{e}_3\mathbf{e}_3$. Find $\mathbf{T}^{1/2}$ and $\mathbf{T}^{-1}$.

6. Show that every cube inscribed in the unit sphere of Fig. 3.7 will be deformed by **T** into a parallelepiped inscribed in the ellipsoid of Fig. 3.7.

7. Compare the ellipsoids (3.31)–(3.32) and (3.35) associated with a positive definite symmetric tensor **S**. Can we form other ellipsoids?

8. Find the surface that gets mapped by a nonsingular tensor **T** into a unit sphere.

9. Show that every second-order tensor can be decomposed into the sum of a multiple of **1**, an antisymmetric tensor, and a traceless symmetric tensor.

10. Write a computer program that translates, rotates, and deforms three-dimensional figures, then orthogonally projects them onto a plane. You may want to remove lines that should be hidden, add shading, and so on.

3.5 PHYSICAL APPLICATIONS

In practical applications vectors and tensors are usually represented by numerical components. For instance, the vector quantity force requires three numbers for its specification. Tensor equations are usually solved in a particular coordinate system. For instance, the vector equation (1.7) is equivalent to three scalar equations each stating that the sum of the force components in one of three noncoplanar directions vanishes. Note that the Cartesian components of a physical tensor are all expressed in the same units, whereas the base vectors themselves are unitless.

Moment of inertia. The Cartesian components of the moment of inertia tensor $\mathbf{I}$ for a particle within a rigid body are obtained from (2.7) with $\mathbf{r} = x_i \mathbf{e}_i$:

$$I_{ij} = \mathbf{e}_i \cdot \mathbf{I} \cdot \mathbf{e}_j = m(x_k x_k \delta_{ij} - x_i x_j).$$

The components $I_{11} = m(x_2^2 + x_3^2)$, $I_{22} = m(x_1^2 + x_3^2)$, $I_{33} = m(x_1^2 + x_2^2)$ are called the *moments of inertia about the x_1, x_2, x_3 axes*, respectively, while the components $I_{12} = I_{21} = -m x_1 x_2$, $I_{13} = I_{31} = -m x_1 x_3$, $I_{23} = I_{32} = -m x_2 x_3$ are called the *products of inertia*. The moment of inertia components for the entire rigid body of *density* ρ (mass per unit volume) are obtained by adding up over the entire volume V the components of each small particle of mass $\rho \Delta x_1 \Delta x_2 \Delta x_3$:

$$\begin{aligned} I_{11} &= \int\int\int_V \rho(x_2^2 + x_3^2)\,dx_1 dx_2 dx_3, \\ I_{12} &= -\int\int\int_V \rho x_1 x_2\,dx_1 dx_2 dx_3, \end{aligned} \tag{3.36}$$

and so on.

Example: Figure 3.8 shows a rectangular solid of constant density with edges (ℓ_1, ℓ_2, ℓ_3) parallel to the coordinate axes. The components of the moment of inertia tensor of the entire solid about its center are

$$I_{11} = \int_{-\ell_3/2}^{\ell_3/2} \int_{-\ell_2/2}^{\ell_2/2} \int_{-\ell_1/2}^{\ell_1/2} \rho(x_2^2 + x_3^2)\,dx_1 dx_2 dx_3 = \frac{\ell_1 \ell_2 \ell_3 (\ell_2^2 + \ell_3^2)\rho}{12},$$

$$I_{12} = -\int_{-\ell_3/2}^{\ell_3/2} \int_{-\ell_2/2}^{\ell_2/2} \int_{-\ell_1/2}^{\ell_1/2} \rho x_1 x_2\,dx_1 dx_2 dx_3 = 0,$$

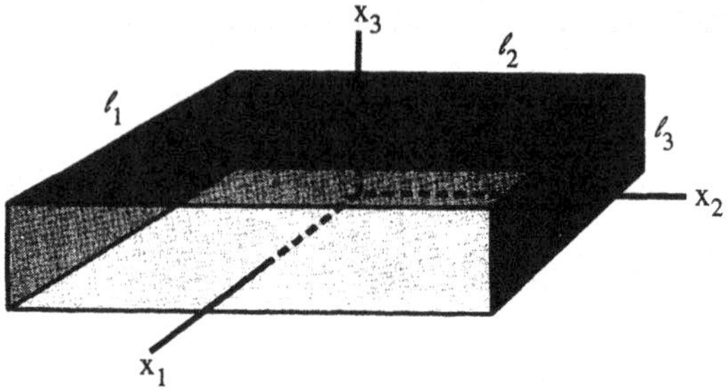

Figure 3.8. The moment of inertia tensor of this rectangular solid about its center is computed in the text.

and so on. The complete moment of inertia tensor is

$$\mathbf{I} = \frac{m}{12}\left[(\ell_2^2+\ell_3^2)\mathbf{e}_1\mathbf{e}_1 + (\ell_1^2+\ell_3^2)\mathbf{e}_2\mathbf{e}_2 + (\ell_1^2+\ell_2^2)\mathbf{e}_3\mathbf{e}_3\right], \tag{3.37}$$

where m is the total mass of the solid. Since the products of inertia are zero, these coordinate axes are the principal axes of $\mathbf{I}$.

Stress. The Cartesian components of the stress tensor $\mathbf{S}$ are

$$S_{ij} = \mathbf{e}_i \cdot \mathbf{S} \cdot \mathbf{e}_j = \mathbf{e}_i \cdot \mathbf{t}_j,$$

where $\mathbf{t}_j = \mathbf{S} \cdot \mathbf{e}_j$ denotes the traction exerted on a surface element with normal $\mathbf{e}_j$. It may be seen that the S_{ij} are components of the traction vectors acting on the faces of a small cube with edges parallel to the coordinate axes (Fig. 3.9). The

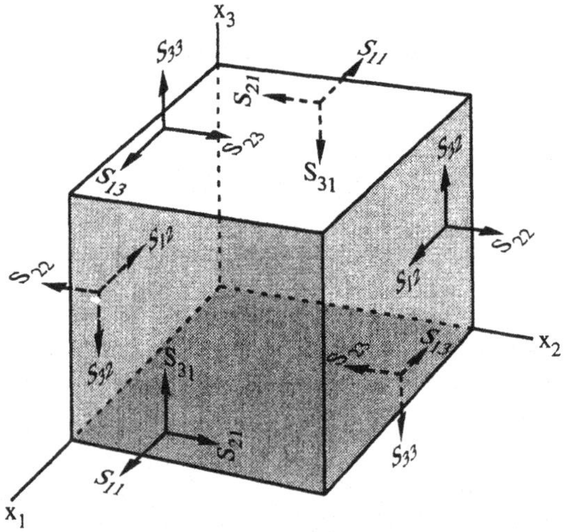

Figure 3.9. Components of the traction vectors that could act on the faces of a small cube within a continuum.

components S_{11}, S_{22}, S_{33} are called *normal* stresses, while the other components $S_{12} = S_{21}$, $S_{13} = S_{31}$, $S_{23} = S_{32}$ are called *shear* stresses. Pulling on an object creates normal stresses. Twisting an object creates shear stresses. Each of these components has the dimensions of force per unit area.

Kinematics of a rigid body. Imagine a rigid body that moves from an initial position to a final position in space. We mathematically decompose the net motion into two successive parts. For the first motion, translate the body without changing its orientation so the center (or some other point) of the body moves to its final position. For the second motion, rotate the body around its center into its final position. Let $\mathbf{r}$ denote the position vector from the center to an arbitrary place inside the object in its initial position. Then the position vector from the initial center to the same place inside the object in its final position is $\mathbf{y} + \mathbf{R} \cdot \mathbf{r}$. The vector $\mathbf{y}$ represents the displacement of the center of the body, and the rotation tensor $\mathbf{R}$ represents the rotation of the body. A rigid body has six *degrees of freedom*: three components of $\mathbf{y}$, and three arbitrary numbers that determine $\mathbf{R}$. The rotation tensor $\mathbf{R}$ is unique — it does not depend upon the point of translation or upon the sequence of rotations chosen to accomplish the net motion.

Example: A rectangular body moves from the initial to the final position shown in Fig. 3.10. The change can be accomplished by a translation of the body such that one of its points is moved into its final position, followed by a rotation with the one point fixed. The translation is not unique — two possibilities, $\mathbf{y}$ and $\mathbf{y}'$, are shown in the figure — but the rotation $\mathbf{R}$ is the same no matter which point is translated. Suppose we fix in space a Cartesian coordinate system with axes parallel to the edges of the body, as shown. Then an arbitrary vector moving with the body and denoted by $\mathbf{r} = x_i\mathbf{e}_i$ in its original position would become the vector

$$\mathbf{R} \cdot \mathbf{r} = R_{ij}x_j\mathbf{e}_i = -x_2\mathbf{e}_1 + x_3\mathbf{e}_2 - x_1\mathbf{e}_3.$$

From this we can determine the components R_{ij} of the tensor $\mathbf{R}$ characterizing the rotation:

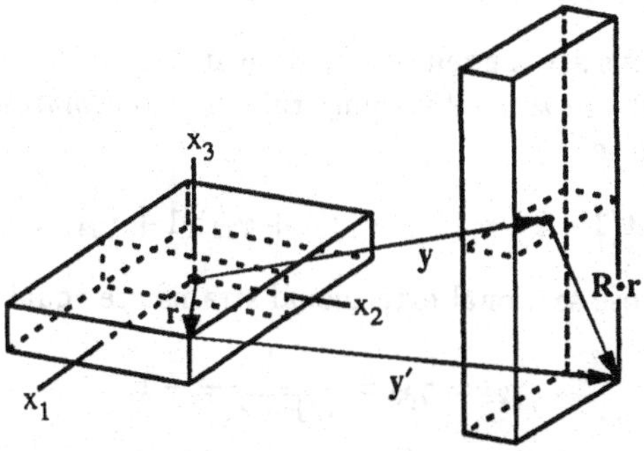

Figure 3.10. The tensor $\mathbf{R}$ characterizing this rotation is computed in the text.

$$\mathbf{R} = -\mathbf{e}_1\mathbf{e}_2 + \mathbf{e}_2\mathbf{e}_3 - \mathbf{e}_3\mathbf{e}_1. \tag{3.38}$$

Note (from Problem 7 in Section 3.3) that the rotation may be accomplished by turning the object through an angle $\theta = \frac{2\pi}{3}$ about $\mathbf{e} = \frac{1}{\sqrt{3}}(-\mathbf{e}_1 + \mathbf{e}_2 + \mathbf{e}_3)$, which is the real eigenvector of $\mathbf{R}$. Try to visualize this by rotating a rectangular ruler.

Kinematics of a continuum. Suppose each particle of a solid or fluid is displaced from its initial position $\mathbf{r}$ to a final position $\mathbf{T} \cdot \mathbf{r}$. Since the volume of a real material cannot be reduced to zero, $\det \mathbf{T} > 0$ and we can apply the polar decomposition (3.33) to $\mathbf{T}$:

$$\mathbf{T} = \mathbf{S} \cdot \mathbf{R}. \tag{3.39}$$

Here $\mathbf{R}$ is a rotation tensor that represents a rigid body rotation of the continuum. The *strain tensor*

$$\mathbf{\Gamma} = (\mathbf{T} \cdot \mathbf{T}^t)^{\frac{1}{2}} - \mathbf{1} = \mathbf{S} - \mathbf{1} \tag{3.40}$$

is a measure of the amount of distortion that occurs. The Cartesian components $\gamma_{11}, \gamma_{22}, \gamma_{33}$ are called normal or *extensional* strains, while the other components $\gamma_{12} = \gamma_{21}, \gamma_{13} = \gamma_{31}, \gamma_{23} = \gamma_{32}$ are called *shear* strains.

Example: Try stretching a rubber band. Its shape will change from a rectangular solid with edges of length (ℓ_1, ℓ_2, ℓ_3) to another rectangular solid with edges of length $(\ell_1\lambda_1, \ell_2\lambda_2, \ell_3\lambda_3)$, and a spherical region of material particles will be deformed into an ellipsoidal region, as shown in Fig. 3.11. A particle that was at position $\mathbf{r} = x_i\mathbf{e}_i$ in the undeformed band is at position

$$\mathbf{T} \cdot \mathbf{r} = x_1\lambda_1\mathbf{e}_1 + x_2\lambda_2\mathbf{e}_2 + x_3\lambda_3\mathbf{e}_3$$

in the deformed band. It follows from this that

$$\mathbf{T} = \lambda_1\mathbf{e}_1\mathbf{e}_1 + \lambda_2\mathbf{e}_2\mathbf{e}_2 + \lambda_3\mathbf{e}_3\mathbf{e}_3.$$

Since $\mathbf{T}$ is symmetric there is no rigid body rotation, and the strain tensor is

$$\mathbf{\Gamma} = (\lambda_1 - 1)\mathbf{e}_1\mathbf{e}_1 + (\lambda_2 - 1)\mathbf{e}_2\mathbf{e}_2 + (\lambda_3 - 1)\mathbf{e}_3\mathbf{e}_3.$$

The extensional strains are the relative increase in length of the edges of the band, and the shear strains are zero. Assuming rubber is *incompressible*, so its volume cannot change, leads to

$$\det(\mathbf{1} + \mathbf{\Gamma}) = (1 + \gamma_{11})(1 + \gamma_{22})(1 + \gamma_{33}) = 1.$$

By symmetry the cross-sectional extensional strains are equal so

$$\gamma_{22} = \gamma_{33} = \frac{1}{\sqrt{1 + \gamma_{11}}} - 1.$$

Thus if we double the length of a rubber band, its cross-sectional dimensions should each decrease by about 29%. Measure your rubber band to confirm this.

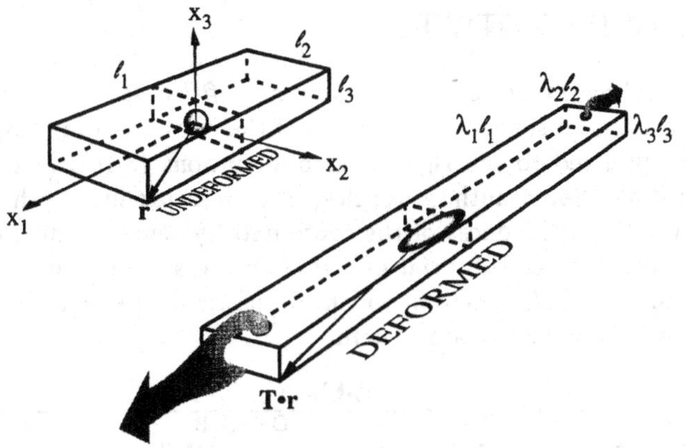

Figure 3.11. Stretching a rubber band causes its cross-sectional dimensions to decrease.

PROBLEMS 3.5

1. Find the moment of inertia tensor of a uniform spherical solid of mass m and radius a about its center.

2. Show that the rotational kinetic energy of a particle in a rotating body has the same value for all angular velocity vectors $\boldsymbol{\omega}$ that lie on the ellipsoid $\boldsymbol{\omega} \cdot \mathbf{I} \cdot \boldsymbol{\omega} = 1$.

3. Determine the Cartesian components of the stress tensor in a fluid at rest.

4. What is the locus of all the traction vectors acting on elemental surfaces passing through a given point?

5. Find the components of the momentum flux density tensor $\rho\mathbf{vv}$ along its principal axes.

6. Choosing another rotation of the rectangular body shown in Fig. 3.10, find the rotation tensor $\mathbf{R}$, its eigenvector $\mathbf{e}$, and the rotation angle θ.

7. Show that the ratio of the deformed to undeformed volume of any region is $\det(\mathbf{1} + \boldsymbol{\Gamma})$. If the components of the strain tensor $\boldsymbol{\Gamma}$ are small, show that the relative change in volume is $\operatorname{tr} \boldsymbol{\Gamma}$.

8. Sketch the effect of the deformation gradient $\mathbf{T} = \mathbf{1} + \gamma_{12}(\mathbf{e}_1\mathbf{e}_2 + \mathbf{e}_2\mathbf{e}_1)$ upon the rectangular solid of the last example in this section.

9. The skin over a finger becomes stretched when the finger is flexed from a straight position to a bent position. Develop a formula for the strain in the skin over a bent joint. Check the prediction with a cloth measuring tape placed along the top of the finger.

3.6 NOTATION OF OTHERS

If the orthogonal base vectors $(\mathbf{e}_1, \mathbf{e}_2, \mathbf{e}_3)$ form a *left-handed triad*, the components of the permutation tensor are the negative of the ones given in (3.8) and a minus sign must be prefixed to the right sides of equations (3.10), (3.11), and (3.24) (see Section 4.5). Some authors redefine the cross product with the left hand when changing the handedness of the coordinate system, in which case the cross product and other so-called *axial* or *pseudo*-vectors reverse direction. In this book all vectors are *polar* vectors, which means their direction (and magnitude) is independent of the coordinate system.

Table 3

OUR NOTATION	OUR NAME	OTHER NOTATIONS	OTHER NAMES
$\mathbf{e}_1$	Unit base vector.	$\mathbf{i}$, $\mathbf{I}$, $\mathbf{a}_1$, $\mathbf{e}_x$	
T_{ij}	Cartesian components of tensor $\mathbf{T}$.	$T_{\alpha\beta}, t_{ij}$	Cartesian tensors.
ϵ_{ijk}	Permutation symbol.	$\mathcal{E}_{ijk}, e_{ijk}$	Alternating symbol, Levi-Civita tensor.
(x_1, x_2, x_3)	Cartesian coordinates.	(x, y, z)	
$\mathbf{H} = \mathbf{1} - 2\mathbf{ee}$	Reflection tensor.	H	Householder transformation.
S_{11}	Stress component.	$S_{xx}, S_1, \sigma_{xx}, \sigma_x, \tau_{11}$	
γ_{12}	Strain component.	$E_{xy}, S_4, e_{xy}, \gamma_{xy}, \frac{1}{2}\gamma_{xy}$	Shear strain component may be twice ours.
$\mathbf{\Gamma} = (\mathbf{T} \cdot \mathbf{T}^t)^{1/2} - \mathbf{1}$	Strain tensor.	$\mathbf{\Gamma} = \frac{1}{2}(\mathbf{T} \cdot \mathbf{T}^t - \mathbf{1})$	Strain tensor may be defined in many ways but all definitions agree when the components of $\mathbf{\Gamma}$ are small.

Chapter 4

GENERAL COMPONENTS

4.1 CHANGE OF CARTESIAN BASE VECTORS AND COORDINATE SYSTEMS

Sometimes it is useful to resolve a tensor along another triad of base vectors other than a single orthonormal set $(\mathbf{e}_1, \mathbf{e}_2, \mathbf{e}_3)$. The *components* of vectors and higher-order tensors will generally change upon rotation or reflection of base vectors, but not upon a translation of base vectors. Of course, the tensors themselves remain *invariant* upon a change of basis. Let us introduce another orthonormal basis $(\bar{\mathbf{e}}_1, \bar{\mathbf{e}}_2, \bar{\mathbf{e}}_3)$, as depicted in Fig. 4.1, and determine the relationship between components of tensors in the two bases.

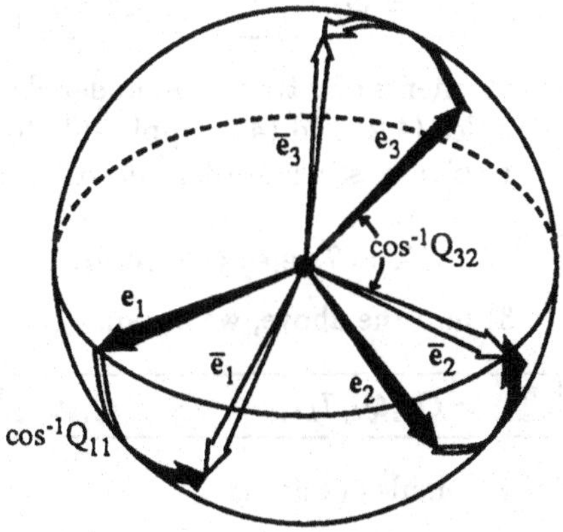

Figure 4.1. The orthonormal basis $(\mathbf{e}_1, \mathbf{e}_2, \mathbf{e}_3)$ is rotated to a new set $(\bar{\mathbf{e}}_1, \bar{\mathbf{e}}_2, \bar{\mathbf{e}}_3)$.

The "new" base vectors $(\bar{\mathbf{e}}_1, \bar{\mathbf{e}}_2, \bar{\mathbf{e}}_3)$ can be obtained by a rigid body rotation and/or reflection of the "old" base vectors $(\mathbf{e}_1, \mathbf{e}_2, \mathbf{e}_3)$. The transformation is characterized by the orthogonal tensor

$$\boxed{\mathbf{Q} = \bar{\mathbf{e}}_i \mathbf{e}_i.} \tag{4.1}$$

The new base vectors are a linear combination of the old base vectors with coefficients that are the components of $\mathbf{Q}$:

$$\bar{\mathbf{e}}_i = \mathbf{Q} \cdot \mathbf{e}_i = Q_{ji} \mathbf{e}_j. \tag{4.2}$$

The transformation from the new base vectors back to the old is characterized by $\mathbf{Q}^{-1} = \mathbf{Q}^t = \mathbf{e}_i \bar{\mathbf{e}}_i$

$$\mathbf{e}_i = \mathbf{Q}^t \cdot \bar{\mathbf{e}}_i = Q_{ij} \bar{\mathbf{e}}_j. \tag{4.3}$$

We find from this that Q_{ij} is the cosine of the angle between $\mathbf{e}_i$ and $\overline{\mathbf{e}}_j$:

$$\boxed{Q_{ij} = \mathbf{e}_i \cdot \overline{\mathbf{e}}_j .} \tag{4.4}$$

In other words, the columns of the matrix Q are the coordinates of the new base vectors relative to the old basis. The order of the indices on Q_{ij} in all of these formulas can be remembered by noting that the first index is an index of the old components whereas the second index is an index of the new components. Note that the components of $\mathbf{Q}$ are the same in either basis.

We can resolve a vector $\mathbf{v}$ along either set of base vectors:

$$\mathbf{v} = v_i \mathbf{e}_i = \overline{v}_i \overline{\mathbf{e}}_i .$$

Inserting (4.2) or (4.3) into the above, then interchanging the i and j dummy indices, we obtain the transformation laws between the components:

$$\boxed{\overline{v}_i = Q_{ji} v_j , \qquad v_i = Q_{ij} \overline{v}_j .} \tag{4.5}$$

This shows that the components of $\mathbf{v}$ transform under change of orthonormal basis according to the same rules (4.2) and (4.3) applicable to the base vectors.

Similarly, we can resolve a second-order tensor $\mathbf{T}$ along either set of base vectors:

$$\mathbf{T} = T_{ij} \mathbf{e}_i \mathbf{e}_j = \overline{T}_{ij} \overline{\mathbf{e}}_i \overline{\mathbf{e}}_j .$$

Inserting (4.2) and (4.3) into the above, we obtain

$$\boxed{\overline{T}_{ij} = Q_{ki} Q_{\ell j} T_{k\ell} , \qquad T_{ij} = Q_{ik} Q_{j\ell} \overline{T}_{k\ell} .} \tag{4.6a}$$

The matrix form of the formulas (4.6a) is

$$\overline{T} = Q^t T Q , \qquad T = Q \overline{T} Q^t . \tag{4.6b}$$

Formulas similar to (4.5) and (4.6a) hold for tensors of order three or higher.

The *coordinates* of a fixed point will generally change both upon a rotation or reflection of base vectors and upon a change of the origin of coordinates. Suppose the origin of a Cartesian coordinate system is translated a vector distance $\mathbf{y}$ and the axes are rotated by a tensor $\mathbf{R}$ (see Fig. 4.2). Now the position vector from the origin O of the old coordinate system to a point $P(x_1, x_2, x_3)$ is

$$\mathbf{r} = \vec{OP} = x_i \mathbf{e}_i .$$

The position vector from the origin $\overline{O}$ of the new coordinate system to the same point $P(\overline{x}_1, \overline{x}_2, \overline{x}_3)$ is

$$\overline{\mathbf{r}} = \vec{\overline{O}P} = \overline{x}_i \overline{\mathbf{e}}_i .$$

The two position vectors are related by

$$\overline{\mathbf{r}} = \mathbf{r} - \mathbf{y} .$$

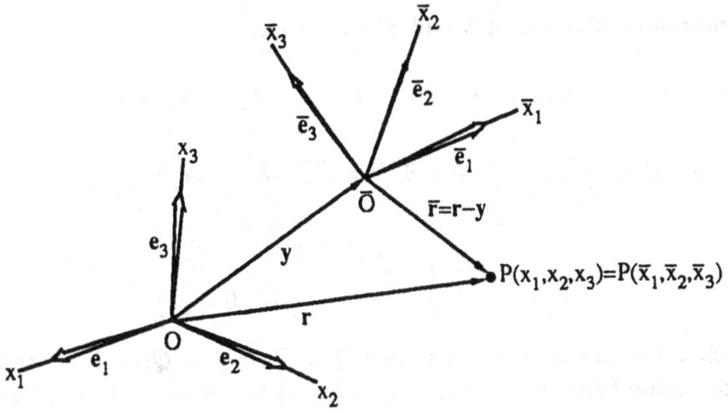

Figure 4.2. The Cartesian coordinates (x_1, x_2, x_3) are rotated and translated to a new set $(\bar{x}_1, \bar{x}_2, \bar{x}_3)$.

Using (4.2) and (4.3), we find that the old and new coordinates of P are related by

$$\bar{x}_i = R_{ji}x_j - \bar{y}_i, \qquad x_i = R_{ij}\bar{x}_j + y_i. \tag{4.7}$$

The spectral representation (3.29) of a symmetric tensor can be obtained by a rotation and/or reflection of base vectors. Thereby we can accomplish the polar decomposition (3.33) of any nonsingular second-order tensor.

Example: Consider the tensor $\mathbf{T} = T_{ij}\mathbf{e}_i\mathbf{e}_j$ whose matrix relative to a given orthonormal basis $(\mathbf{e}_1, \mathbf{e}_2, \mathbf{e}_3)$ is

$$T = \begin{bmatrix} 0 & -1 & 2 \\ 2 & -1 & 0 \\ -1 & 3 & -1 \end{bmatrix}.$$

We will find a symmetric tensor $\mathbf{S}$ and an orthogonal tensor $\mathbf{R}$ such that $\mathbf{T} = \mathbf{S}\cdot\mathbf{R}$. Now

$$S^2 = TT^t = \begin{bmatrix} 5 & 1 & -5 \\ 1 & 5 & -5 \\ -5 & -5 & 11 \end{bmatrix}.$$

The eigenvalues of S^2 with corresponding normalized eigenvectors are

$$\lambda_1^2 = 1, \qquad \bar{\mathbf{e}}_1 = \frac{\mathbf{e}_1 + \mathbf{e}_2 + \mathbf{e}_3}{\sqrt{3}},$$

$$\lambda_2^2 = 4, \qquad \bar{\mathbf{e}}_2 = \frac{\mathbf{e}_1 - \mathbf{e}_2}{\sqrt{2}},$$

$$\lambda_3^2 = 16, \qquad \bar{\mathbf{e}}_3 = \frac{\mathbf{e}_1 + \mathbf{e}_2 - 2\mathbf{e}_3}{\sqrt{6}}.$$

The spectral representations of $\mathbf{S}$ and $\mathbf{S}^{-1}$ are thus

$$\mathbf{S} = \bar{\mathbf{e}}_1\bar{\mathbf{e}}_1 + 2\bar{\mathbf{e}}_2\bar{\mathbf{e}}_2 + 4\bar{\mathbf{e}}_3\bar{\mathbf{e}}_3, \qquad \mathbf{S}^{-1} = \bar{\mathbf{e}}_1\bar{\mathbf{e}}_1 + \frac{1}{2}\bar{\mathbf{e}}_2\bar{\mathbf{e}}_2 + \frac{1}{4}\bar{\mathbf{e}}_3\bar{\mathbf{e}}_3.$$

The components of $\mathbf{S} = \bar{S}_{ij}\bar{\mathbf{e}}_i\bar{\mathbf{e}}_j$ and $\mathbf{S}^{-1} = \bar{S}_{ij}^{-1}\bar{\mathbf{e}}_i\bar{\mathbf{e}}_j$ in the basis $(\bar{\mathbf{e}}_1, \bar{\mathbf{e}}_2, \bar{\mathbf{e}}_3)$ are

$$\bar{S} = \begin{bmatrix} 1 & 0 & 0 \\ 0 & 2 & 0 \\ 0 & 0 & 4 \end{bmatrix}, \quad \bar{S}^{-1} = \begin{bmatrix} 1 & 0 & 0 \\ 0 & \frac{1}{2} & 0 \\ 0 & 0 & \frac{1}{4} \end{bmatrix}.$$

The components of the orthogonal tensor $\mathbf{Q} = Q_{ij}\mathbf{e}_i\mathbf{e}_j = Q_{ij}\bar{\mathbf{e}}_i\bar{\mathbf{e}}_j$ characterizing the transformation from $(\mathbf{e}_1, \mathbf{e}_2, \mathbf{e}_3)$ to $(\bar{\mathbf{e}}_1, \bar{\mathbf{e}}_2, \bar{\mathbf{e}}_3)$ are obtained from (4.4):

$$Q = \begin{bmatrix} \frac{1}{\sqrt{3}} & \frac{1}{\sqrt{2}} & \frac{1}{\sqrt{6}} \\ \frac{1}{\sqrt{3}} & -\frac{1}{\sqrt{2}} & \frac{1}{\sqrt{6}} \\ \frac{1}{\sqrt{3}} & 0 & -\frac{2}{\sqrt{6}} \end{bmatrix}.$$

The components of $\mathbf{S} = S_{ij}\mathbf{e}_i\mathbf{e}_j$ and $\mathbf{S}^{-1} = S_{ij}^{-1}\mathbf{e}_i\mathbf{e}_j$ in the basis $(\mathbf{e}_1, \mathbf{e}_2, \mathbf{e}_3)$ follow from (4.6b):

$$S = Q\bar{S}Q^t = \begin{bmatrix} 2 & 0 & -1 \\ 0 & 2 & -1 \\ -1 & -1 & 3 \end{bmatrix}, \qquad S^{-1} = Q\bar{S}^{-1}Q^t = \frac{1}{8}\begin{bmatrix} 5 & 1 & 2 \\ 1 & 5 & 2 \\ 2 & 2 & 4 \end{bmatrix}.$$

Finally, we obtain the components of $\mathbf{R} = R_{ij}\mathbf{e}_i\mathbf{e}_j$ in the basis $(\mathbf{e}_1, \mathbf{e}_2, \mathbf{e}_3)$:

$$R = S^{-1}T = \begin{bmatrix} 0 & 0 & 1 \\ 1 & 0 & 0 \\ 0 & 1 & 0 \end{bmatrix}.$$

PROBLEMS 4.1

1. Show that the components of the transformation tensor $\mathbf{Q} = \bar{\mathbf{e}}_i\mathbf{e}_i$ are the same in either orthonormal basis: $\mathbf{Q} = Q_{ij}\mathbf{e}_i\mathbf{e}_j = Q_{ij}\bar{\mathbf{e}}_i\bar{\mathbf{e}}_j$.

2. Show that (4.5) could also be interpreted as the relation between the components of two vectors $\mathbf{v}$ and $\bar{\mathbf{v}} = \mathbf{Q}^t \cdot \mathbf{v}$ in the same basis. Explain geometrically.

3. Show that the components of the identity tensor $\mathbf{1}$ do not change upon a rotation or reflection of axes. Show that the components of the permutation tensor $\mathbf{E}$ do not change upon a rotation but do change sign upon a reflection.

4. Show that the components of the moment of inertia $\mathbf{I}$ of a uniform cube about its center are the same in any orthogonal basis.

5. Sketch the effect of $\mathbf{T}$ defined in the example upon the unit sphere $\mathbf{r}\cdot\mathbf{r}=1$. Then find S' in the reversed polar decomposition $T = RS'$.

6. Find the spectral representation of the tensor $\mathbf{S}$ whose components in the basis $(\mathbf{e}_1, \mathbf{e}_2, \mathbf{e}_3)$ are

$$S = \begin{bmatrix} 4 & 2 & 2 \\ 2 & 4 & 2 \\ 2 & 2 & 4 \end{bmatrix}.$$

4.2 ROTATIONS IN TWO DIMENSIONS

A two-dimensional tensor is a tensor that may be resolved into components along only two orthonormal base vectors; for example,

$$\mathbf{v} = v_1\mathbf{e}_1 + v_2\mathbf{e}_2,$$
$$\mathbf{T} = T_{11}\mathbf{e}_1\mathbf{e}_1 + T_{12}\mathbf{e}_1\mathbf{e}_2 + T_{21}\mathbf{e}_2\mathbf{e}_1 + T_{22}\mathbf{e}_2\mathbf{e}_2.$$

Nearly all of the mathematical theory we have developed for three-dimensional tensors may be directly applied to two-dimensional tensors.

In particular, the equations of transformation in the preceding section are applicable to the special case of a rotation of base vectors in two dimensions. Suppose that a new set of base vectors $(\bar{\mathbf{e}}_1, \bar{\mathbf{e}}_2)$ is obtained by counterclockwise rotation of the old base vectors $(\mathbf{e}_1, \mathbf{e}_2)$ through an angle θ. This transformation is characterized by a two-dimensional rotation tensor $\mathbf{R}$ with components obtained from (4.4):

$$R_{11} = \mathbf{e}_1\cdot\bar{\mathbf{e}}_1 = \cos\theta, \quad R_{12} = \mathbf{e}_1\cdot\bar{\mathbf{e}}_2 = \cos\left(\frac{\pi}{2}+\theta\right) = -\sin\theta,$$

$$R_{21} = \mathbf{e}_2\cdot\bar{\mathbf{e}}_1 = \cos\left(\frac{\pi}{2}-\theta\right) = \sin\theta, \quad R_{22} = \mathbf{e}_2\cdot\bar{\mathbf{e}}_2 = \cos\theta.$$

$$R = \begin{bmatrix} \cos\theta & -\sin\theta \\ \sin\theta & \cos\theta \end{bmatrix}. \tag{4.8}$$

Inserting (4.8) into (4.5), we find that the old and new coordinates of a two-dimensional vector $\mathbf{v}$ are related by

$$\begin{aligned} \bar{v}_1 &= v_1\cos\theta + v_2\sin\theta, \\ \bar{v}_2 &= -v_1\sin\theta + v_2\cos\theta. \end{aligned} \tag{4.9}$$

A geometric verification of (4.9) is indicated in Fig. 4.3.

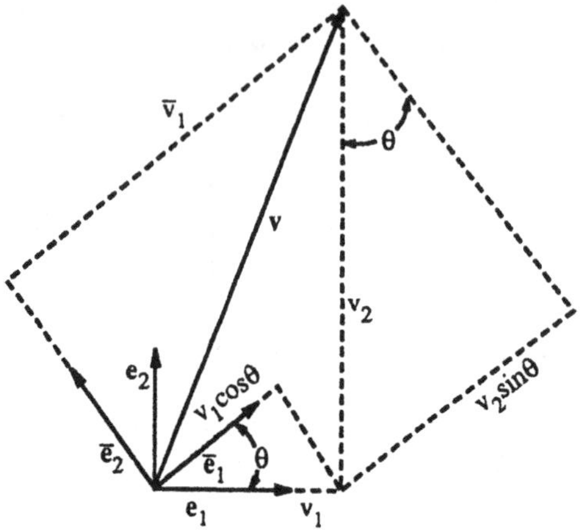

Figure 4.3. Geometrical proof of $\bar{v}_1 = v_1 \cos\theta + v_2 \sin\theta$.

Similarly, we find from (4.6) and (4.8) that the old and new components of a two-dimensional *symmetric* tensor $\mathbf{S}$ are related by

$$\left.\begin{aligned} \overline{S}_{11} &= S_{11}\cos^2\theta + S_{22}\sin^2\theta + 2S_{12}\sin\theta\cos\theta \\ \overline{S}_{12} &= (S_{22} - S_{11})\sin\theta\cos\theta + S_{12}(\cos^2\theta - \sin^2\theta) \\ \overline{S}_{22} &= S_{22}\cos^2\theta + S_{11}\sin^2\theta - 2S_{12}\sin\theta\cos\theta \end{aligned}\right\} . \tag{4.10}$$

Formulas (4.10) can be represented graphically by the *Mohr circle construction* shown in Fig. 4.4:

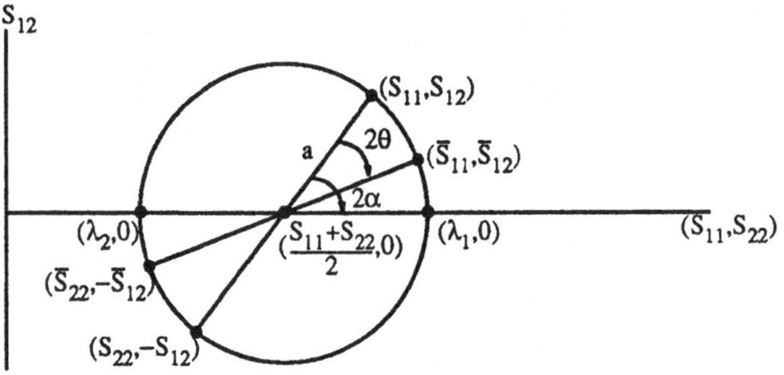

Figure 4.4. Mohr circle for relating the old (S_{11}, S_{12}, S_{22}) and new ($\bar{S}_{11}$, $\bar{S}_{12}$, $\bar{S}_{22}$) components of a symmetric two-dimensional tensor.

(a) Plot the points (S_{11}, S_{12}) and $(S_{22}, -S_{12})$ in the tensor representation plane.

(b) Connect the two points in (a) with a straight line and draw a circle of radius a equal to half the distance between the points.

(c) Draw a diameter making an angle 2θ measured *clockwise* from the line in (b). The terminal coordinates of this diameter are the points $(\overline{S}_{11}, \overline{S}_{12})$ and $(\overline{S}_{22}, -\overline{S}_{12})$.

(d) The angle from the line in (b) to the horizontal axis is two times the angle α of rotation in the $\mathbf{e}_1 - \mathbf{e}_2$ plane to the principal axes of $\mathbf{S}$. The eigenvalues (λ_1, λ_2) of $\mathbf{S}$ are the coordinates of the points of intersection of the Mohr circle with the horizontal axis.

Proof of the Mohr circle construction: By the use of trigonometric formulas, (4.10) may be given the form

$$\begin{aligned}\overline{S}_{11} &= \frac{1}{2}(S_{11}+S_{22}) + a\cos(2\alpha - 2\theta),\\ \overline{S}_{12} &= a\sin(2\alpha - 2\theta),\\ \overline{S}_{22} &= \frac{1}{2}(S_{11}+S_{22}) - a\cos(2\alpha - 2\theta),\end{aligned}$$

where

$$\begin{aligned}a &= \left[\frac{1}{4}(S_{11}-S_{22})^2 + S_{12}^2\right]^{1/2},\\ \tan\, 2\alpha &= \frac{2S_{12}}{S_{11}-S_{22}}.\end{aligned} \tag{4.11}$$

The Mohr circle construction makes the diagonalization of a symmetric tensor an easy matter.

Example: The general equation of a conic section is

$$S_{11}x_1^2 + 2S_{12}x_1x_2 + S_{22}x_2^2 + b_1x_1 + b_2x_2 + c = 0. \tag{4.12}$$

This can be written in the concise notation

$$\mathbf{r}\cdot\mathbf{S}\cdot\mathbf{r} + \mathbf{b}\cdot\mathbf{r} + c = 0.$$

We can eliminate the cross product x_1x_2 term by rotating to the principal axes of $\mathbf{S}$:

$$\overline{x}_1 = x_1\cos\alpha + x_2\sin\alpha, \qquad \overline{x}_2 = -x_1\sin\alpha + x_2\cos\alpha,$$

where the angle α of rotation is given by (4.11). Then

$$\lambda_1(\overline{x}_1)^2 + \lambda_2(\overline{x}_2)^2 + \overline{b}_1\overline{x}_1 + \overline{b}_2\overline{x}_2 + c = 0.$$

If $\lambda_1 \neq 0$ and $\lambda_2 \neq 0$, we can eliminate the linear terms by translation of axes:

$$x_1' = \overline{x}_1 + \frac{\overline{b}_1}{2\lambda_1}, \qquad x_2' = \overline{x}_2 + \frac{\overline{b}_2}{2\lambda_2}.$$

The equation of the conic section then has the canonical form

$$\lambda_1(x_1')^2 + \lambda_2(x_2')^2 + c' = 0.$$

The signs of the eigenvalues (λ_1, λ_2) determine the type of curve given by these equations.

Example: Try twisting a piece of chalk. If we cut out the small cube of material shown in Fig. 4.5a, the stress components acting on its faces would be (see Equation (7.21) and Problem 4 in Section 8.2)

$$S = \begin{bmatrix} 0 & S_{12} & 0 \\ S_{12} & 0 & 0 \\ 0 & 0 & 0 \end{bmatrix}.$$

From the Mohr circle (Fig. 4.5b), the principal axes make an angle of $\alpha = \frac{\pi}{4}$ with the original base vectors $(\mathbf{e}_1, \mathbf{e}_2)$, and the stress components in the principal axes are (Fig. 4.5c)

$$\overline{S} = \begin{bmatrix} S_{12} & 0 & 0 \\ 0 & -S_{12} & 0 \\ 0 & 0 & 0 \end{bmatrix}.$$

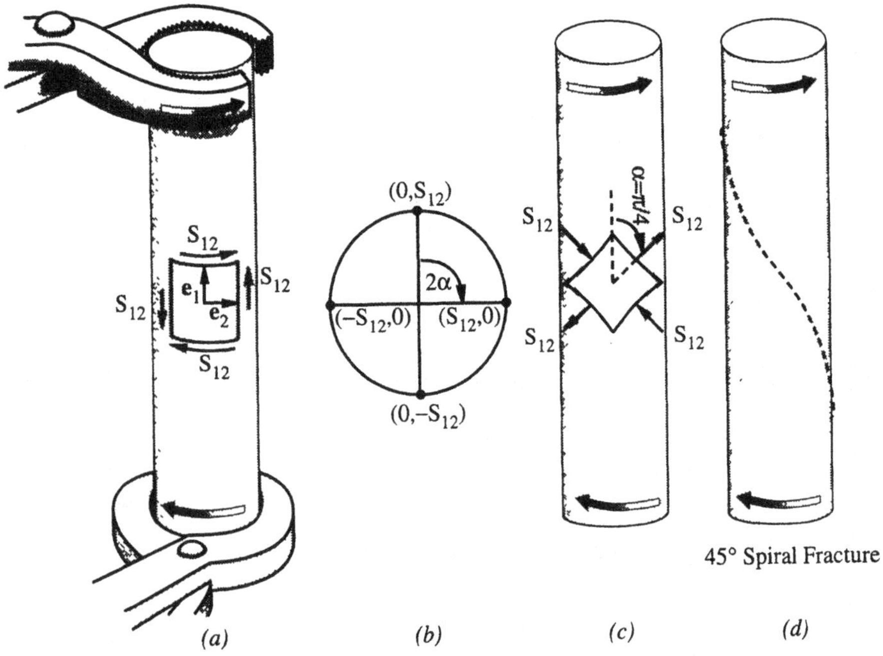

Figure 4.5. Twisting a piece of chalk creates tension on the faces of the elemental cube, and this initiates spiral fracture.

Now chalk cannot withstand a lot of tension. Hence, if you twist your chalk enough, it will crack along a 45° helix, which is the line upon which the greatest tension acts (Fig. 4.5d).

PROBLEMS 4.2

1. What happens to the Mohr circle in the case $S_{11} = S_{22}$, $S_{12} = 0$?

2. Classify the conic section (4.12) according to the sign of the determinant of **S**.

3. In the solid mechanics theory of slender beams, it is necessary to calculate various cross section properties. The *moment of inertia of a cross section* is the two-dimensional tensor whose components are (cf. 3.36)

$$I_{22} = \int_A\int x_3^2 dx_2 dx_3, \quad I_{33} = \int_A\int x_2^2 dx_2 dx_3,$$
$$I_{23} = -\int_A\int x_2 x_3 dx_2 dx_3,$$

where x_2 and x_3 are coordinates in the plane of the cross section whose area is A. Find the moment of inertia tensor of a rectangular cross section about its center, resolved along axes parallel to the sides.

4. Referring to the preceding problem, find the moment of inertia tensor of a square cross section about a *corner*, resolved along axes parallel to the sides and along principal axes.

5. Show that the maximum value of S_{12} is $\frac{1}{2}|\lambda_1 - \lambda_2|$ in axes that make an angle of $\frac{\pi}{4}$ with the principal axes of **S**, and that in these axes $S_{11} = S_{22} = \frac{1}{2}(\lambda_1 + \lambda_2)$.

6. When metals are subjected to sufficient stress, they will *yield* and then suffer a permanent plastic deformation. The Tresca yield condition predicts that yield will occur when the maximum shear stress reaches a value K. Suppose a thin rectangular plate is subjected to biaxial tension so that its only nonvanishing stress components are S_{11} and S_{22}. Calculate the critical combinations of S_{11} and S_{22} at which yield will occur.

7. The Mohr circle construction can be applied to a three-dimensional symmetric tensor **S**. Plot the values of S_{11}, S_{22}, and S_{33} on the horizontal axis, and S_{12}, S_{13}, S_{23} on the vertical axis, for the cases where at least one of the coordinates is along a principal axis.

4.3 ROTATIONS IN THREE DIMENSIONS

A three-dimensional rotation tensor **R** may be characterized by three *Euler* or *orientation angles*, which are successive angles of rotation about a set of mutually orthogonal axes. Rotations through an angle θ about the x_1, x_2, or x_3 axes may be respectively represented by the matrices, from (4.4),

$$R_1(\theta) = \begin{bmatrix} 1 & 0 & 0 \\ 0 & \cos\theta & -\sin\theta \\ 0 & \sin\theta & \cos\theta \end{bmatrix}, \qquad R_2(\theta) = \begin{bmatrix} \cos\theta & 0 & \sin\theta \\ 0 & 1 & 0 \\ -\sin\theta & 0 & \cos\theta \end{bmatrix},$$

$$R_3(\theta) = \begin{bmatrix} \cos\theta & -\sin\theta & 0 \\ \sin\theta & \cos\theta & 0 \\ 0 & 0 & 1 \end{bmatrix}. \tag{4.13}$$

The total rotation matrix R is the product of the individual rotation matrices (4.13). If the axes themselves undergo the successive rotations, the rotation matrices are multiplied in the same order, whereas if the axes about which the individual rotations occur are fixed in space, the rotation matrices are multiplied in the reverse order.

The rotation matrices (4.13) are often used in practical applications, such as converting from one coordinate system to another with (4.7).

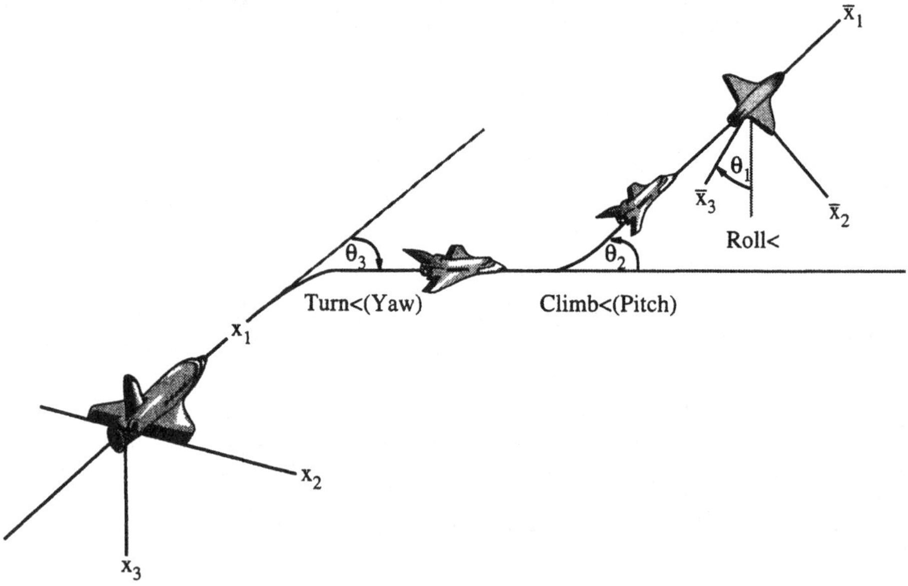

Figure 4.6. Yaw, pitch, and roll angles used to describe the rotation of an aircraft.

Turn-climb-roll angles. Pilots can describe the rotation of their aircraft by the angles $(\theta_3, \theta_2, \theta_1)$ taken successively about the *yaw-pitch-roll* axes fixed in the aircraft in the directions shown in Fig. 4.6. First the aircraft *turns* to the right through the angle θ_3, then *climbs* through the angle θ_2, and finally *rolls* about its fuselage through the angle θ_1. The total rotation matrix R, whose elements are the components of the rotation tensor $\mathbf{R}$ along either the initial or final axes, is

$$R = R_3(\theta_3)R_2(\theta_2)R_1(\theta_1) \tag{4.14}$$

$$= \begin{bmatrix} \cos\theta_2\cos\theta_3 & -\cos\theta_1\sin\theta_3 + \sin\theta_1\sin\theta_2\cos\theta_3 & \sin\theta_1\sin\theta_3 + \cos\theta_1\sin\theta_2\cos\theta_3 \\ \cos\theta_2\sin\theta_3 & \cos\theta_1\cos\theta_3 + \sin\theta_1\sin\theta_2\sin\theta_3 & -\sin\theta_1\cos\theta_3 + \cos\theta_1\sin\theta_2\sin\theta_3 \\ -\sin\theta_2 & \sin\theta_1\cos\theta_2 & \cos\theta_1\cos\theta_2 \end{bmatrix}.$$

Example: We may describe the position of a celestial object either by *geocentric* coordinates (x_1, x_2, x_3) measured from the center O of the earth or by *topocentric* coordinates $(\overline{x}_1, \overline{x}_2, \overline{x}_3)$ measured from our location $\overline{O}$ on the surface of the earth (Fig. 4.7). Let x_1 be measured in the *vernal equinox* direction ♈ along the line of intersection of the earth's equatorial plane with the earth's orbital plane about the sun, x_3 be the distance toward the earth's north pole, $\overline{x}_1$ be the object's height above us, $\overline{x}_3$ be its distance to the north of us, and approximate the shape of the earth by a sphere of radius a. Then the $(\overline{x}_1, \overline{x}_2, \overline{x}_3)$ axes can be obtained by rotating the (x_1, x_2, x_3) axes about the x_3 axis through the *right ascension* angle $\alpha = \theta_3$ followed by a rotation about the new 2-axis through the north *latitude* angle $\beta = -\theta_2$. It follows from (4.7) and (4.14) that the geocentric and

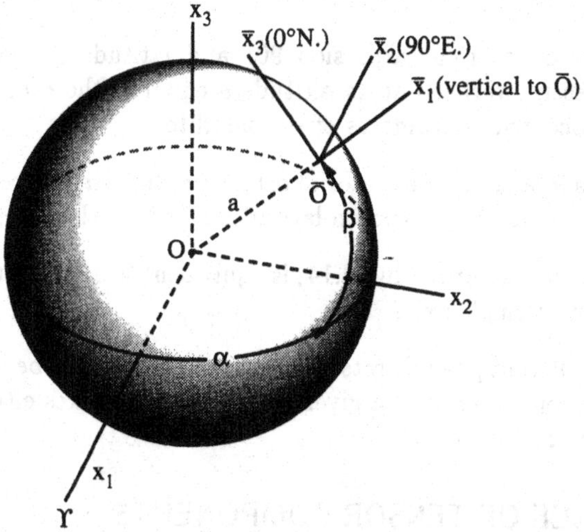

Figure 4.7. The position of a celestial object can be specified by geocentric coordinates (x_1, x_2, x_3) or topocentric coordinates $(\bar{x}_1, \bar{x}_2, \bar{x}_3)$.

topocentric coordinates are related by

$$\begin{bmatrix} \bar{x}_1 \\ \bar{x}_2 \\ \bar{x}_3 \end{bmatrix} = \begin{bmatrix} \cos\alpha\cos\beta & \sin\alpha\cos\beta & \sin\beta \\ -\sin\alpha & \cos\alpha & 0 \\ -\cos\alpha\sin\beta & -\sin\alpha\sin\beta & \cos\beta \end{bmatrix} \begin{bmatrix} x_1 \\ x_2 \\ x_3 \end{bmatrix} + \begin{bmatrix} -a \\ 0 \\ 0 \end{bmatrix}, \tag{4.15a}$$

or

$$\begin{bmatrix} x_1 \\ x_2 \\ x_3 \end{bmatrix} = \begin{bmatrix} \cos\alpha\cos\beta & -\sin\alpha & -\cos\alpha\sin\beta \\ \sin\alpha\cos\beta & \cos\alpha & -\sin\alpha\sin\beta \\ \sin\beta & 0 & \cos\beta \end{bmatrix} \begin{bmatrix} \bar{x}_1 \\ \bar{x}_2 \\ \bar{x}_3 \end{bmatrix} + \begin{bmatrix} a\cos\alpha\cos\beta \\ a\sin\alpha\cos\beta \\ a\sin\beta \end{bmatrix}. \tag{4.15b}$$

PROBLEMS 4.3

1. Decompose the rotation **R** of the rigid body shown in Fig. 3.10 into a sequence of two rotations about a set of axes fixed in the body. Then show that the rotation matrix R is the product of the appropriate rotation matrices (4.13).

2. Decompose the rotation **R** of the rigid body shown in Fig. 3.10 into a sequence of two rotations about the axes shown fixed in space. Show that the rotation matrix R is the product of the appropriate rotation matrices (4.13) multiplied in the order reversed from that in which the rotations were performed.

3. A radar station at right ascension 90° and latitude 30° north detects a satellite directly overhead at an altitude a equal to the radius of the earth. Find the geocentric coordinates of the satellite.

4. A satellite is in a circular equatorial orbit of radius $2a$, where a is the radius of the earth. Find the maximum latitudes at which the satellite is visible.

5. The rotation characterized by (4.14) is equivalent to a rotation through what angle θ about a single axis?

6. How many different possible sets of orientation angles can be used to describe a given rotation in terms of a given set of Cartesian axes either attached to the body or fixed in space?

4.4 INVARIANCE OF TENSOR COMPONENTS

We have seen that the components of tensors will in general change upon a rotation and/or reflection of base vectors. However, certain combinations of these components are the same *in every coordinate system.* Given any vector $\mathbf{v}$ and second order tensor $\mathbf{T}$, the following will be scalar invariants:

$$\left.\begin{aligned} \mathbf{v}\cdot\mathbf{v} &= v_i v_i \\ \mathbf{v}\cdot\mathbf{T}\cdot\mathbf{v} &= T_{ij} v_i v_j \\ \mathrm{tr}\,\mathbf{T} &= T_{ii} \\ \det\mathbf{T} &= \det T \end{aligned}\right\}. \tag{4.16}$$

Since the values of the principal invariants (3.23) are the same in every coordinate system, the eigenvalues that are the roots to the characteristic polynominal (3.22) are also independent of the coordinate system in which they are calculated.

Many materials possess symmetries arising from a regular microscopic structure. The components of tensors characterizing physical properties of such a material will remain the same if the base vectors undergo *particular* orthogonal transformations. We can use this invariance to reduce the number of independent components of the tensors. Let us try the procedure by finding the material properties of various idealized solids.

Constitutive relation for a linear anisotropic solid. Stress $\mathbf{S}$ in a solid can be caused by strain $\mathbf{\Gamma}$, by an electric field $\boldsymbol{\mathcal{E}}$, or by a temperature increase T. The constitutive equation relating these variables must be experimentally determined. For many materials undergoing small changes, the constitutive relation is *linear.* The physical properties of a linear material are represented by a fourth-order *elasticity* tensor $\mathbf{C}$, a third-order *piezoelectric* tensor $\mathbf{B}$, and a second-order *thermal* tensor $\mathbf{A}$. The component form of a linear constitutive relation in a given coordinate system is

$$S_{ij} = C_{ijk\ell}\gamma_{k\ell} - B_{ijk}\mathcal{E}_k - A_{ij}T. \tag{4.17}$$

Since the stress and strain tensors are symmetric, the most general *anisotropic* (unsymmetrical in material properties) solid has 36 independent elasticity constants, 18 independent piezoelectric constants, and 6 independent thermal constants satisfying

$$\left.\begin{aligned} A_{ij} &= A_{ji} \\ B_{ijk} &= B_{jik} \\ C_{ijk\ell} &= C_{jik\ell} = C_{ij\ell k} \end{aligned}\right\}. \tag{4.18}$$

In another coordinate system, the constitutive relation becomes

$$\overline{S}_{ij} = \overline{C}_{ijk\ell}\overline{\gamma}_{k\ell} - \overline{B}_{ijk}\overline{\mathcal{E}}_k - \overline{A}_{ij}T.$$

The ABC's transform according to rules like (4.6), so are in general different in any new coordinate system.

Tensor components of a cubic crystal. A crystal is a body that has a regular repeating internal arrangement in its atoms; as an example, the structure of sodium chloride is illustrated in Fig. 4.8. Crystals are divided into macroscopic

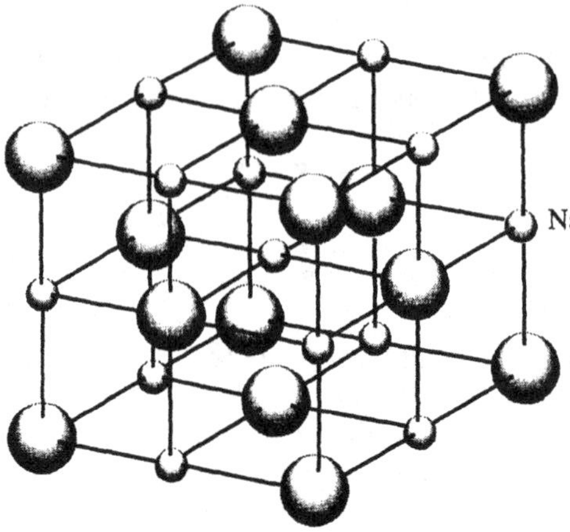

Figure 4.8. Atomic structure of sodium chloride.

symmetry classes based upon their microscopic structure. There are 32 different *crystallographic point groups* or crystal classes. Many pure metals, diamond, and sodium chloride are members of the *cubic* class. The tensor components of a cubic crystal are unchanged if any two base vectors are interchanged, or if any base vectors are reversed in direction. Remembering (4.4), we find that the orthogonal tensor characterizing a transformation from $(\mathbf{e}_1, \mathbf{e}_2, \mathbf{e}_3)$ to $(\mathbf{e}_2, \mathbf{e}_1, \mathbf{e}_3)$ has components

$$Q = \begin{bmatrix} 0 & 1 & 0 \\ 1 & 0 & 0 \\ 0 & 0 & 1 \end{bmatrix}.$$

Applying this transformation to $\mathbf{A}$ in accordance with (4.6), we find that its new components are

$$\overline{A} = \begin{bmatrix} A_{22} & A_{21} & A_{23} \\ A_{12} & A_{11} & A_{13} \\ A_{32} & A_{31} & A_{33} \end{bmatrix};$$

that is, the effect of this transformation on the components is to interchange the subscripts 1 and 2. Since any interchange of subscripts must leave the components unchanged,

$$A_{11} = A_{22} = A_{33}. \tag{4.19}$$

Similarly, the matrix characterizing a transformation from $(\mathbf{e}_1, \mathbf{e}_2, \mathbf{e}_3)$ to $(-\mathbf{e}_1, \mathbf{e}_2, \mathbf{e}_3)$ is

$$Q = \begin{bmatrix} -1 & 0 & 0 \\ 0 & 1 & 0 \\ 0 & 0 & 1 \end{bmatrix}.$$

This creates the following new components of $\mathbf{A}$:

$$\overline{A} = \begin{bmatrix} A_{11} & -A_{12} & -A_{13} \\ -A_{21} & A_{22} & A_{23} \\ -A_{31} & A_{32} & A_{33} \end{bmatrix};$$

that is, the effect of this transformation is to multiply the components by -1 for each subscript 1. Since any component with differing subscripts must be equal to its own negative,

$$A_{12} = A_{13} = A_{23} = 0. \tag{4.20}$$

It follows from (4.17), (4.19), and (4.20) that changing the temperature (while holding the strain constant) will produce hydrostatic pressure in a cubic crystal. Furthermore, since every component of a third-order tensor has an odd number of at least one subscript, no third-order tensor properties exist for cubic cystals. In particular, a cubic crystal cannot be piezoelectric:

$$B_{ijk} = 0.$$

Finally, applying the same transformations to the elasticity tensor **C**, we find that it has three independent components for a cubic crystal:

$$\left.\begin{array}{l} C_{1111} = C_{2222} = C_{3333} \\ C_{1122} = C_{2211} = C_{1133} = C_{3311} = C_{2233} = C_{3322} \\ C_{1212} = C_{1313} = C_{2323} \end{array}\right\}. \tag{4.21}$$

Only compenents with an even number of each subscript are nonzero.

Tensor components of an isotropic solid. Materials whose properties do not vary with direction are said to be *isotropic*. There is no preferred orientation in their microscopic structure. Most alloys, steel, and concrete are linear isotropic materials. Since the tensor components must be the same for *any* choice of base vectors, an isotropic material has more symmetry than a cubic crystal. Now we found for a cubic crystal that $\mathbf{A} = A_{11}\mathbf{1}$ and $\mathbf{B} = 0$, and since the components of this **A** and **B** are the same in any coordinate system, they are also the correct tensors for an isotropic material. The components of **C** must obey (4.21), but there is an additional relation for an isotropic material. This can be discovered by requiring that the tensor components remain unchanged when the base vectors undergo a rotation about the x_3 axis through an angle $\pi/4$, which is characterized by the matrix

$$Q = \begin{bmatrix} \frac{1}{\sqrt{2}} & -\frac{1}{\sqrt{2}} & 0 \\ \frac{1}{\sqrt{2}} & \frac{1}{\sqrt{2}} & 0 \\ 0 & 0 & 1 \end{bmatrix}.$$

Applying this transformation to **C** yields

$$C_{1111} = C_{1122} + 2C_{1212}.$$

Putting all the ABC's into (4.17), we can write the constitutive relation for an isotropic material in the form

$$\mathbf{S} = 2\mu\mathbf{\Gamma} + \lambda(\mathrm{tr}\,\mathbf{\Gamma})\mathbf{1} - A_{11}T\mathbf{1} \tag{4.22a}$$

or

$$\mathbf{\Gamma} = \frac{1+\nu}{Y}\mathbf{S} - \frac{\nu}{Y}(\text{tr } \mathbf{S})\mathbf{1} + \frac{(1-2\nu)A_{11}}{Y}T\mathbf{1}. \tag{4.22b}$$

Here $\mu = C_{1212} = \frac{Y}{2(1+\nu)}$ is called the *shear modulus*, $\lambda = C_{1122} = \frac{Y\nu}{(1+\nu)(1-2\nu)}$ is a *Lamé constant*, $Y = \frac{\mu(3\lambda+2\mu)}{\lambda+\mu}$ is *Young's modulus*, and $\nu = \frac{\lambda}{2(\lambda+\mu)}$ is *Poisson's ratio*. It is evident from (4.22) that the constitutive relation is the same in every coordinate system.

PROBLEMS 4.4

1. Show by directly transforming the components that (4.16) do not depend upon the choice of orthogonal base vectors.

2. What are the SI metric units of the ABC's?

3. If the constitutive relation is assumed to be linear, show that the ABC's may always be chosen to satisfy (4.18).

4. All the possible transformations that preserve the microscopic structure of a cubic crystal form a so-called algebraic *group*. How many different transformations (Q-matrices) are there in this group?

5. A cubic crystal that is completely free of stress on its boundaries is subjected to a temperature change T. Calculate the strain in the crystal.

6. For the tightrope example of Section 1.5, calculate the cross-sectional area of the wire. Assume the wire is made of steel with $E = 200$ GPa (Gigapascals).

7. Show that the relative change in volume of a linear elastic solid is

$$\frac{\text{tr } \mathbf{S}}{3B} + \frac{A_{11}T}{B},$$

where $B = \frac{Y}{3(1-2\nu)}$ is the *bulk modulus*. If the solid is incompressible, show that $\nu = \frac{1}{2}$.

8. A substance possesses a *center of symmetry* if its tensor components remain unchanged when the base vectors undergo an *inversion* $\mathbf{Q} = -\mathbf{1}$. Show that all even-order tensors are centrosymmetric, and no nonzero odd-order tensors are centrosymmetric.

9. A substance is *transversely isotropic* with respect to a line, say the x_3 axis, if its tensor properties are unchanged when the base vectors undergo any rotation about the x_3 axis. Examples are wood with all of its fibers lined up in the x_3 direction, and an originally isotropic dielectric that has been polarized in the x_3 direction. Find the nonzero components of the $\mathbf{A}$ tensor for a transversely isotropic material.

10. Sometimes an isotropic material is defined as one whose properties are unchanged under any ***rotation*** of base vectors, but whose properties may change under a ***reflection*** of base vectors. Find the constitutive relation of such a material.

11. The constitutive relation of a ***nonlinear*** isotropic elastic material is often assumed to be of the form $\mathbf{S} = f_1\mathbf{1} + f_2\mathbf{\Gamma} + f_3\mathbf{\Gamma}^2$, where (f_1, f_2, f_3) are scalar-valued functions of the principal invariants $(J_1(\mathbf{\Gamma}), J_2(\mathbf{\Gamma}), J_3(\mathbf{\Gamma}))$. Show that this constitutive relation is the same in every coordinate system, and that it has the form of (4.22a) (with $T = 0$) when $\mathbf{\Gamma}$ is small.

4.5 GENERAL BASES

Until now we have always resolved tensors along triads of base vectors that are orthonormal. However, there is no need to make this restriction. One of the remarkable facts about tensor analysis is that formulas valid for general bases are as simple algebraically as their Cartesian forms. However, it is necessary to be very careful with the bookkeeping.

We let the letter **g** ***play the role of general base vector.*** A general basis that need be neither orthogonal nor of unit length is then denoted by $(\mathbf{g}_1, \mathbf{g}_2, \mathbf{g}_3)$. In order to achieve algebraic simplicity, it is necessary to introduce a ***reciprocal*** basis $(\mathbf{g}^1, \mathbf{g}^2, \mathbf{g}^3)$ which is related to the original basis by the formulas

$$\boxed{\mathbf{g}^i \cdot \mathbf{g}_j = \delta^i_j\,;} \tag{4.23}$$

that is, the reciprocal vector $\mathbf{g}^1$ is the vector that is perpendicular to $(\mathbf{g}_2, \mathbf{g}_3)$ and that has length $\frac{1}{|\mathbf{g}_1|}$, as constructed in Fig. 4.9. Customarily $(\mathbf{g}_1, \mathbf{g}_2, \mathbf{g}_3)$ are called the ***covariant base vectors***, whereas $(\mathbf{g}^1, \mathbf{g}^2, \mathbf{g}^3)$ are called the ***contravariant base vectors.***

It will immediately be noticed that, when not using orthonormal bases, it is necessary to distinguish between subscripts and superscripts. When an index is

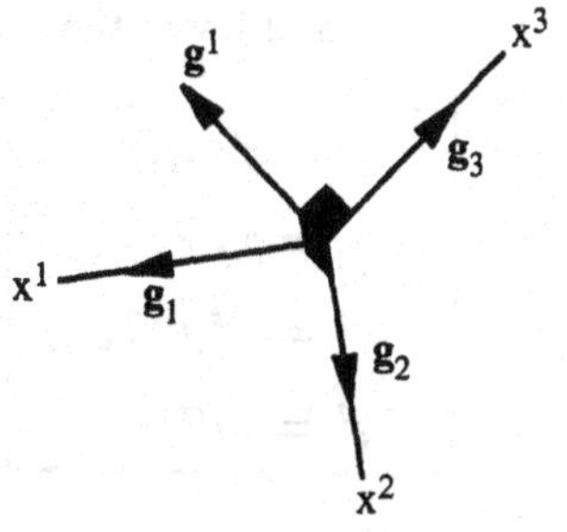

Figure 4.9. Here the base vectors $(\mathbf{g}_1, \mathbf{g}_2, \mathbf{g}_3)$ are neither orthogonal nor of unit length, but $\mathbf{g}^1$ is perpendicular to $\mathbf{g}_2$ and $\mathbf{g}_3$.

free, it must be either a subscript or superscript on both sides of an equation. Thus in (4.23) we have raised the index i on the Kronecker delta. Note that the Kronecker delta and the permutation symbol always have the meanings (3.7) and (3.8) regardless of the position of the indices. When two indices are summed, one of the indices must be a subscript and the other a superscript, as will be seen in the following equations. A vector $\mathbf{v}$ can be decomposed along either set of base vectors:

$$\boxed{\mathbf{v} = v_i \mathbf{g}^i = v^i \mathbf{g}_i .}$$

The v_i are called *covariant* components of $\mathbf{v}$, and the v^i are *contravariant* components of $\mathbf{v}$. A second-order tensor $\mathbf{T}$ has four sets of components:

$$\boxed{\mathbf{T} = T_{ij}\mathbf{g}^i\mathbf{g}^j = T^{ij}\mathbf{g}_i\mathbf{g}_j = T_i{}^j\mathbf{g}^i\mathbf{g}_j = T^i{}_j\mathbf{g}_i\mathbf{g}^j .}$$

The T_{ij} are covariant components, T^{ij} are contravariant components, and $T_i{}^j$, $T^i{}_j$ are *mixed* components of $\mathbf{T}$. Note that the mixed components $T_i{}^j$ and $T^j{}_i$ may not be equal. These representations can be generalized in an obvious way to higher-order tensors.

The components of the identity or metric tensor $\mathbf{1}$ in a general basis are

$$\boxed{\mathbf{1} = g_{ij}\mathbf{g}^i\mathbf{g}^j = g^{ij}\mathbf{g}_i\mathbf{g}_j = \mathbf{g}^i\mathbf{g}_i = \mathbf{g}_i\mathbf{g}^i .} \tag{4.24}$$

The covariant and contravariant components of $\mathbf{1}$ are given the special symbols g_{ij} and g^{ij}, respectively. They are related to the length of and angles between the basis vectors by the formulas

$$\boxed{g_{ij} = \mathbf{g}_i \cdot \mathbf{g}_j \,, \qquad g^{ij} = \mathbf{g}^i \cdot \mathbf{g}^j .} \tag{4.25}$$

The matrix with components g_{ij} is the inverse of the matrix with components g^{ij} because

$$\boxed{g_{ik}g^{kj} = \delta_i^j .} \tag{4.26}$$

The g's can be neatly used to raise and lower the indices on tensors; for example,

$$\boxed{\begin{aligned} \mathbf{g}^i &= g^{ij}\mathbf{g}_j , \\ v_i &= g_{ij}v^j , \\ T^{ij} &= g^{ik}g^{j\ell}T_{k\ell} , \\ T_i{}^j &= g_{ik}T^{kj} . \end{aligned}}$$

We use the letter e for the components of the permutation tensor $\mathbf{E}$ in a general basis:

$$\boxed{\mathbf{E} = e_{ijk}\mathbf{g}^i\mathbf{g}^j\mathbf{g}^k .} \tag{4.27}$$

These components are obtained from the base vectors by the formulas

$$e_{ijk} = \mathbf{g}_i \cdot (\mathbf{E} \cdot \mathbf{g}_k) \cdot \mathbf{g}_j = -\mathbf{g}_i \cdot \mathbf{g}_k \times \mathbf{g}_j = \mathbf{g}_i \cdot \mathbf{g}_j \times \mathbf{g}_k = \mathbf{g}_i \times \mathbf{g}_j \cdot \mathbf{g}_k = \epsilon_{ijk}\mathbf{g}_1 \times \mathbf{g}_2 \cdot \mathbf{g}_3.$$

Now from (3.12)

$$(\mathbf{g}_1 \times \mathbf{g}_2 \cdot \mathbf{g}_3)^2 = \begin{vmatrix} \mathbf{g}_1 \cdot \mathbf{g}_1 & \mathbf{g}_1 \cdot \mathbf{g}_2 & \mathbf{g}_1 \cdot \mathbf{g}_3 \\ \mathbf{g}_2 \cdot \mathbf{g}_1 & \mathbf{g}_2 \cdot \mathbf{g}_2 & \mathbf{g}_2 \cdot \mathbf{g}_3 \\ \mathbf{g}_3 \cdot \mathbf{g}_1 & \mathbf{g}_3 \cdot \mathbf{g}_2 & \mathbf{g}_3 \cdot \mathbf{g}_3 \end{vmatrix} = g, \tag{4.28}$$

where g denotes *the determinant of the matrix having g_{ij} as its components.* Note that g is just the determinant of a matrix so its value depends on the basis; that is, g is not an invariant scalar. Hence

$$e_{ijk} = \begin{cases} \sqrt{g}\,\epsilon_{ijk} & \text{if } \mathbf{g}_1 \times \mathbf{g}_2 \cdot \mathbf{g}_3 > 0 \text{ (right-handed basis)}, \\ -\sqrt{g}\,\epsilon_{ijk} & \text{if } \mathbf{g}_1 \times \mathbf{g}_2 \cdot \mathbf{g}_3 < 0 \text{ (left-handed basis)}. \end{cases}$$

A tensor formula that has been established in Cartesian components can always be directly generalized to arbitrary components. *It is necessary only to raise or lower the indices appropriately, to replace δ_{ij}* with g_{ij} (replace (3.17) with (4.24)), *and to replace ϵ_{ijk} with* e_{ijk} (replace (3.18) with (4.27)). Here is how some of our formulas from Chapter 3 might look in general components:

$$\begin{aligned} |\mathbf{v}| &= \sqrt{v_i v^i}, \\ \mathbf{u} \cdot \mathbf{v} &= u_i v^i, \\ \mathbf{u} \times \mathbf{v} &= e_{ijk} u^i v^j \mathbf{g}^k, \\ e_{ijk} e_{pqk} &= g_{ip} g_{jq} - g_{iq} g_{jp}, \\ \mathbf{S} \cdot \mathbf{T} &= S_{ik} T^{kj} \mathbf{g}^i \mathbf{g}_j, \\ \mathbf{T}^t &= T_{ji} \mathbf{g}^i \mathbf{g}^j, \\ \operatorname{tr} \mathbf{T} &= T_i^i. \end{aligned}$$

The formulas of this section can be directly applied to *oblique* coordinates, in which the coordinate axes are straight but not perpendicular. If the coordinate axes in an oblique coordinate system are chosen to coincide with the covariant base vectors $(\mathbf{g}_1, \mathbf{g}_2, \mathbf{g}_3)$, they are labeled with *superscripts* (x^1, x^2, x^3), as shown in Fig. 4.9. Then the vector from the origin O to a point is $\mathbf{r} = x^i \mathbf{g}_i$.

Our main application of the results in this section will be in subsequent work that is carried out in general *curvilinear* coordinates, in which the coordinate lines need not be straight. The following case will be of use in applying our general formulas to specific physical problems in which the coordinate lines intersect at right angles.

Orthogonal base vectors. Suppose the base vectors $(\mathbf{g}_1, \mathbf{g}_2, \mathbf{g}_3)$ are orthogonal but not necessarily of unit length. Denote the lengths of $(\mathbf{g}_1, \mathbf{g}_2, \mathbf{g}_3)$ by

(h_1, h_2, h_3), respectively. The contravariant base vectors, obtained from (4.23), are also orthogonal:

$$\mathbf{g}^1 = \frac{\mathbf{g}_1}{(h_1)^2}, \qquad \mathbf{g}^2 = \frac{\mathbf{g}_2}{(h_2)^2}, \qquad \mathbf{g}^3 = \frac{\mathbf{g}_3}{(h_3)^2}.$$

Then the covariant components of the metric tensor, their determinant, and the contravariant components are, from (4.25):

$$\begin{aligned} &g_{11} = (h_1)^2, \qquad g_{22} = (h_2)^2, \qquad g_{33} = (h_3)^2, \\ &g_{12} = g_{13} = g_{23} = 0, \qquad g = (h_1 h_2 h_3)^2, \\ &g^{11} = \frac{1}{(h_1)^2}, \qquad g^{22} = \frac{1}{(h_2)^2}, \qquad g^{33} = \frac{1}{(h_3)^2}, \\ &g^{12} = g^{13} = g^{23} = 0. \end{aligned}$$

A third orthogonal triad of base vectors $(\bar{\mathbf{e}}_1, \bar{\mathbf{e}}_2, \bar{\mathbf{e}}_3)$ that have unit length may also be introduced:

$$\bar{\mathbf{e}}_1 = \frac{\mathbf{g}_1}{h_1}, \qquad \bar{\mathbf{e}}_2 = \frac{\mathbf{g}_2}{h_2}, \qquad \bar{\mathbf{e}}_3 = \frac{\mathbf{g}_3}{h_3}. \tag{4.29}$$

The components of a tensor along the unit triad $(\bar{\mathbf{e}}_1, \bar{\mathbf{e}}_2, \bar{\mathbf{e}}_3)$ are customarily called *physical* components. A tensor may be resolved along any one of the three triads. The covariant, contravariant, and physical components of a vector $\mathbf{v}$ are related by

$$\bar{v}_1 = \frac{v_1}{h_1} = h_1 v^1, \qquad \bar{v}_2 = \frac{v_2}{h_2} = h_2 v^2, \qquad \bar{v}_3 = \frac{v_3}{h_3} = h_3 v^3.$$

Similar formulas relate the components of higher-order tensors:

$$\overline{T}_{11} = \frac{T_{11}}{(h_1)^2} = (h_1)^2 T^{11} = T_1^1, \quad \overline{T}_{12} = \frac{T_{12}}{h_1 h_2} = \frac{h_1 T_2^1}{h_2} = \frac{h_2 T_1^2}{h_1} = h_1 h_2 T^{12}. \tag{4.30}$$

The physical components of a physical tensor all have the same units.

PROBLEMS 4.5

1. Does every set of vectors $(\mathbf{g}_1, \mathbf{g}_2, \mathbf{g}_3)$ form a possible basis in three dimensions?

2. Given a basis $(\mathbf{g}_1, \mathbf{g}_2, \mathbf{g}_3)$, does a reciprocal basis $(\mathbf{g}^1, \mathbf{g}^2, \mathbf{g}^3)$ always exist and is it unique?

3. Find the contravariant components of the permutation tensor $\mathbf{E}$.

4. Show that $T_i^{\ j} = T^j_{\ i}$ if and only if $T_{ij} = T_{ji}$.

5. How many different covariant, contravariant, and mixed components are there for a general tensor of order n?

6. Given $\mathbf{g}_1 = \mathbf{e}_1$, $\mathbf{g}_2 = \mathbf{e}_2$, $\mathbf{g}_3 = \mathbf{e}_1 + \mathbf{e}_3$, $\mathbf{v} = \mathbf{e}_1 + \mathbf{e}_2 + \mathbf{e}_3$, $\mathbf{T} = \mathbf{1} + \mathbf{e}_2\mathbf{e}_1$. Find: $\mathbf{g}^i$, g_{ij}, g^{ij}, e_{ijk}, v_i, v^i, T_{ij}, T^{ij}, $T^i{}_j$, $T_i{}^j$.

7. Suppose the covariant base vectors in an oblique coordinate system (x^1, x^2, x^3) are $\mathbf{g}_1 = \mathbf{e}_1 + \mathbf{e}_2$, $\mathbf{g}_2 = \mathbf{e}_2$, $\mathbf{g}_3 = \mathbf{e}_3$. Sketch the coordinate curves x^i = constant and find the contravariant base vectors and the components of the metric tensor.

8. Suppose the old base vectors $(\mathbf{g}_1, \mathbf{g}_2, \mathbf{g}_3)$ undergo a rigid body rotation and/or reflection to a new set $(\overline{\mathbf{g}}_1, \overline{\mathbf{g}}_2, \overline{\mathbf{g}}_3)$. Find the orthogonal tensor $\mathbf{Q}$ of the transformation, the components of $\mathbf{Q}$, and the new covariant components $\overline{v}_i$ of a vector $\mathbf{v}$.

9. In a two-dimensional space, express the contravariant components of the metric tensor in terms of the covariant components.

10. Given a set of three nonplanar vectors $(\mathbf{g}_1, \mathbf{g}_2, \mathbf{g}_3)$, construct an orthonormal set $(\mathbf{e}_1, \mathbf{e}_2, \mathbf{e}_3)$ by the Gram-Schmidt process. Compare with Problem 7 in Section 1.4.

4.6 NOTATION OF OTHERS

Some authors define the transformation tensor $\mathbf{Q}$ to be the transpose of (4.1), so that their transformation matrices are the transposes of those in this book. Furthermore, whereas we have characterized a rotation tensor $\mathbf{R}$ by the parameters $\mathbf{e}$ and θ in (3.25), or by the turn-climb-roll angles $(\theta_3, \theta_2, \theta_1)$ in (4.14), several other sets of rotation parameters, and many other sequences of orientation angles, are used in the literature. No attempt is made here to catalog all the possibilities.

Table 4

OUR NOTATION	OUR NAME	OTHER NOTATIONS	OTHER NAMES
Q	Orthogonal matrix, transformation matrix.	Q^t, P	Transition matrix; same name and symbol may be transpose of ours.
B_{ijk}	Components of piezoelectric tensor.	d_{mn}	18 independent components have only two indices: m ranging from 1 to 3 and n ranging from 1 to 6.
μ, Y, ν	Elastic constants.	G, E, σ (respectively)	
v_i	Covariant components of vector $\mathbf{v}$.		Components of covariant vector.

Chapter 5

TENSOR FIELDS OF ONE VARIABLE

5.1 BASIC CALCULUS OF VECTOR FIELDS

Tensor fields are functions that associate points in three-dimensional space and/or instants of time with unique tensors. Tensor fields of a *single variable* are tensors whose domain is the points along a curve in space or the instants in an interval of time. We can label the points or instants with a single parameter t. Scalar fields $f(t)$ of a single variable t are the ordinary real-valued functions studied in elementary calculus. In this chapter we shall study the calculus of vector $\mathbf{u}(t)$ and tensor $\mathbf{T}(t)$ fields of a single variable t.

Limits, continuity, differentiation, and integration of vector fields are defined in the same way as in the case of scalar fields. The *limit* of $\mathbf{u}(t)$ as t tends to 0 is the vector approached as t approaches 0. If $\mathbf{u}(t)$ is *continuous* at t, any small increment Δt in t will produce a small change $\Delta\mathbf{u} = \mathbf{u}(t+\Delta t) - \mathbf{u}(t)$ in the vector $\mathbf{u}(t)$ (Fig. 5.1). The limit

$$\lim_{\Delta t \to 0} \frac{\mathbf{u}(t+\Delta t) - \mathbf{u}(t)}{\Delta t} = \frac{d\mathbf{u}}{dt}(t) \tag{5.1}$$

is defined as *derivative* of $\mathbf{u}(t)$ with respect to t. The second derivative of $\mathbf{u}(t)$ is just the derivative of the first derivative:

$$\frac{d^2\mathbf{u}}{dt^2} = \frac{d}{dt}\left(\frac{d\mathbf{u}}{dt}\right).$$

All tensor fields in this book are tacitly assumed to be sufficiently continuous and differentiable to satisfy the equations herein throughout their domain of definition.

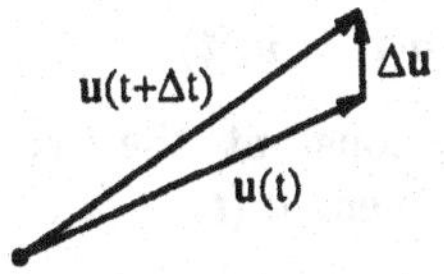

Figure 5.1. $\Delta\mathbf{u}$ is a small increment in the vector field $\mathbf{u}(t)$.

Partition an interval $[0, t]$ in the domain of $\mathbf{u}(t)$ into small segments of width $\Delta t = \frac{t}{n}$ with a set of n scalars $k\Delta t$, $k = 1, 2, \ldots, n$. Then the limit

$$\lim_{n \to \infty} \sum_{k=1}^{n} \mathbf{u}(k\Delta t)\Delta t = \int_0^t \mathbf{u}(t)dt \tag{5.2}$$

is defined as the *definite integral* of $\mathbf{u}(t)$ on the interval $[0, t]$. The fundamental theorem of calculus says that

$$\int_0^t \mathbf{u}(t)dt = \mathbf{w}(t) - \mathbf{w}(0), \tag{5.3}$$

where the *indefinite integral* $\mathbf{w}(t)$ is the vector field whose derivative is $\mathbf{u}$:

$$\frac{d\mathbf{w}}{dt} = \mathbf{u} \Longleftrightarrow \mathbf{w} = \int \mathbf{u}(t)dt + \mathbf{c}.$$

Note that $\mathbf{w}(t)$ is indefinite to the extent of an arbitrary added *constant vector* $\mathbf{c}$. A constant vector has the same value for every point in space and for all instants of time.

The rules for the differentiation and integration of combinations of tensor fields are similar to those for scalar fields. If $f(t)$ is a differentiable scalar field, $\mathbf{u}(t)$ and $\mathbf{v}(t)$ are differentiable vector fields, and c is a constant scalar, then

$$\frac{d}{dt}(f\mathbf{u} + \mathbf{v}) = \frac{df}{dt}\mathbf{u} + f\frac{d\mathbf{u}}{dt} + \frac{d\mathbf{v}}{dt},$$

$$\frac{d}{dt}(\mathbf{u} \cdot \mathbf{v}) = \frac{d\mathbf{u}}{dt} \cdot \mathbf{v} + \mathbf{u} \cdot \frac{d\mathbf{v}}{dt},$$

$$\frac{d}{dt}(\mathbf{u} \times \mathbf{v}) = \frac{d\mathbf{u}}{dt} \times \mathbf{v} + \mathbf{u} \times \frac{d\mathbf{v}}{dt},$$

$$\int (c\mathbf{u} + \mathbf{v})dt = c\int \mathbf{u}dt + \int \mathbf{v}dt,$$

$$\int \mathbf{u} \cdot \frac{d\mathbf{u}}{dt}dt = \frac{u^2}{2} + c.$$

We may, of course, decompose a vector field $\mathbf{u}(t)$ along a set of orthonormal base vectors $(\mathbf{e}_1, \mathbf{e}_2, \mathbf{e}_3)$:

$$\mathbf{u}(t) = u_i(t)\mathbf{e}_i.$$

If the base vectors $(\mathbf{e}_1, \mathbf{e}_2, \mathbf{e}_3)$ are constant, the i component of the derivative of $\mathbf{u}(t)$ is the derivative of the component $u_i(t)$:

$$\frac{d\mathbf{u}}{dt} = \frac{du_i}{dt}\mathbf{e}_i.$$

The tensor fields that occur in practical applications are usually related by differential or integral equations. The goal is to solve the equations, together with appropriate initial and boundary conditions, for the unknown fields.

PROBLEMS 5.1

1. Expand
$$\frac{d}{dt}[\mathbf{u}(t) \times \mathbf{v}(t) \cdot \mathbf{w}(t)].$$

2. Show that if $\mathbf{e}(t)$ has constant magnitude, $\frac{d\mathbf{e}}{dt}$ is either zero or perpendicular to $\mathbf{e}$.

3. Find the Cartesian components of
$$\frac{d}{dt}[\mathbf{u}(t) \times \mathbf{v}(t)]$$
in a constant basis.

4. If $\mathbf{u} = \sin t\mathbf{e}_1 + e^t\mathbf{e}_2 + \mathbf{e}_3$, find
$$\frac{d\mathbf{u}}{dt}, \qquad \frac{d^2\mathbf{u}}{dt^2}, \qquad \text{and} \qquad \int \mathbf{u}(t)dt.$$

5. Find the vector field $\mathbf{u}(t)$ that satisfies the differential equation
$$\frac{d^2\mathbf{u}}{dt^2} + \mathbf{u} = 0$$
and the initial conditions
$$\mathbf{u}(0) = \mathbf{e}_1, \qquad \frac{d\mathbf{u}}{dt}(0) = \mathbf{e}_2.$$

6. Define and develop rules for the differentiation and integration of higher-order tensor fields $\mathbf{T}(t)$ of a single variable.

5.2 MOVING BASE VECTORS

When the Cartesian base vectors may be functions of a parameter t, we emphasize this by placing a bar over them: $(\bar{\mathbf{e}}_1(t), \bar{\mathbf{e}}_2(t), \bar{\mathbf{e}}_3(t))$. If the base vectors are differentiable, they must vary smoothly with t. Then after a small interval Δt, the base vectors $(\bar{\mathbf{e}}_1(t), \bar{\mathbf{e}}_2(t), \bar{\mathbf{e}}_3(t))$ become a slightly different orthonormal set $(\bar{\mathbf{e}}_1(t+\Delta t), \bar{\mathbf{e}}_2(t+\Delta t), \bar{\mathbf{e}}_3(t+\Delta t))$. The transformation from $\bar{\mathbf{e}}_i(t)$ to $\bar{\mathbf{e}}_i(t+\Delta t)$ may be obtained by a rigid body rotation through a small angle $\Delta\theta$ about a fixed axis $\mathbf{e}$ (Fig. 5.2), so
$$\Delta\bar{\mathbf{e}}_i = \bar{\mathbf{e}}_i(t+\Delta t) - \bar{\mathbf{e}}_i(t) \approx \Delta\theta\mathbf{e} \times \bar{\mathbf{e}}_i.$$
Dividing by Δt and taking the limit (5.1), we obtain the derivative of the base vectors:
$$\boxed{\frac{d\bar{\mathbf{e}}_i}{dt} = \boldsymbol{\omega} \times \bar{\mathbf{e}}_i,} \tag{5.4}$$

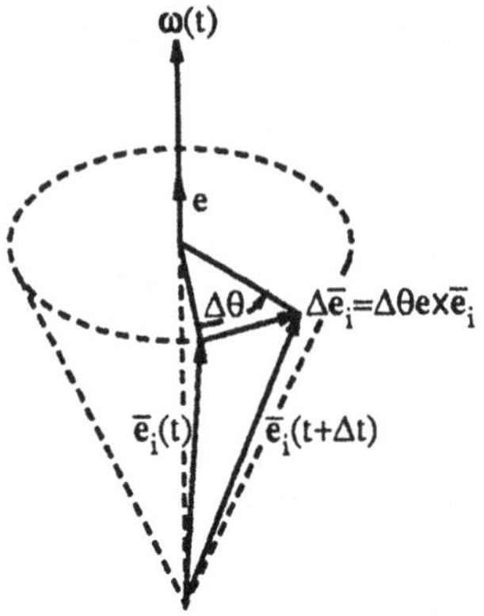

Figure 5.2. $\Delta\bar{e}_i$ is a small increment in the unit vector field $\bar{e}_i(t)$.

where

$$\boldsymbol{\omega}(t) = \frac{d\theta}{dt}\mathbf{e}.$$

If t is the time, we can interpret $\boldsymbol{\omega}(t)$ as the angular velocity of the rotating triad $(\bar{\mathbf{e}}_1, \bar{\mathbf{e}}_2, \bar{\mathbf{e}}_3)$.

The rotating basis $(\bar{\mathbf{e}}_1(t), \bar{\mathbf{e}}_2(t), \bar{\mathbf{e}}_3(t))$ may be related to a constant basis $(\mathbf{e}_1, \mathbf{e}_2, \mathbf{e}_3)$ by a finite rotation tensor $\mathbf{R}(t)$, in accordance with (4.2):

$$\bar{\mathbf{e}}_i(t) = \mathbf{R}(t) \cdot \mathbf{e}_i = R_{ji}(t)\mathbf{e}_j. \tag{5.5}$$

We obtain from (5.4) and (5.5) the relation between $\mathbf{R}(t)$ and $\boldsymbol{\omega}(t)$:

$$\boxed{\boldsymbol{\omega}\times = \frac{d\mathbf{R}}{dt} \cdot \mathbf{R}^t} \tag{5.6a}$$

or, with $\boldsymbol{\omega} = \overline{\omega}_i\bar{\mathbf{e}}_i$

$$\epsilon_{ijk}\overline{\omega}_k = \frac{dR_{ki}}{dt}R_{kj}. \tag{5.6b}$$

If we characterize the rotation tensor $\mathbf{R}(t)$ by the turn-climb-roll angles $(\theta_3(t), \theta_2(t), \theta_1(t))$ shown in Fig. 4.6, then $\boldsymbol{\omega}(t)$ can also be expressed in terms of these angles using (4.14) and (5.6):

$$\begin{bmatrix} \overline{\omega}_1 \\ \overline{\omega}_2 \\ \overline{\omega}_3 \end{bmatrix} = \begin{bmatrix} 1 & 0 & -\sin\theta_2 \\ 0 & \cos\theta_1 & \sin\theta_1\cos\theta_2 \\ 0 & -\sin\theta_1 & \cos\theta_1\cos\theta_2 \end{bmatrix} \begin{bmatrix} \dfrac{d\theta_1}{dt} \\ \dfrac{d\theta_2}{dt} \\ \dfrac{d\theta_3}{dt} \end{bmatrix}. \tag{5.7}$$

An arbitrary vector field $\mathbf{u}(t)$ may be decomposed along either the rotating or constant basis:

$$\mathbf{u}(t) = \overline{u}_i(t)\bar{\mathbf{e}}_i(t) = u_i(t)\mathbf{e}_i. \tag{5.8}$$

Differentiating $u_i = R_{ij}\overline{u}_j$, we obtain the relation between the derivatives of the components in the two systems:

$$\frac{du_i}{dt} = R_{ij}\frac{d\overline{u}_j}{dt} + \frac{dR_{ij}}{dt}\overline{u}_j. \tag{5.9}$$

Multiplying both sides of (5.9) by $\mathbf{e}_i$ and using (5.5), (5.6), and (5.8) we can write this in the concise notation

$$\boxed{\frac{d\mathbf{u}}{dt} = \frac{\overline{d}\mathbf{u}}{dt} + \boldsymbol{\omega} \times \mathbf{u}.} \tag{5.10}$$

Here we have used a bar over a symbol to denote differentiation in the rotating basis (i.e., when differentiating treat $\overline{\mathbf{e}}_i$ as constant):

$$\boxed{\frac{\overline{d}\mathbf{u}}{dt} = \frac{d\overline{u}_i}{dt}\overline{\mathbf{e}}_i.}$$

Equation (5.10) can be interpreted as saying that the change $\frac{d\mathbf{u}}{dt}$ of $\mathbf{u}$ is equal to the change $\frac{\overline{d}\mathbf{u}}{dt}$ of $\mathbf{u}$ in the moving basis plus the change $\boldsymbol{\omega} \times \mathbf{u}$ that $\mathbf{u}$ would have if it were at rest in the moving system.

PROBLEMS 5.2

1. Derive the formula (5.7).
2. Given the components $(\overline{\omega}_1(t), \overline{\omega}_2(t), \overline{\omega}_3(t))$ at the turn-climb-roll angles $(\theta_3(t), \theta_2(t), \theta_1(t))$, determine the rates of change of the angles.
3. Suppose the rotation from a constant basis $(\mathbf{e}_1, \mathbf{e}_2, \mathbf{e}_3)$ to a rotating basis $(\overline{\mathbf{e}}_1(t), \overline{\mathbf{e}}_2(t), \overline{\mathbf{e}}_3(t))$ is characterized by a *finite* angle of rotation $\theta(t)$ about a *moving* axis $\mathbf{e}(t)$. Express $\boldsymbol{\omega}(t)$ in terms of $\theta(t)$ and $\mathbf{e}(t)$.
4. Show that the equation $\frac{d'\mathbf{u}}{dt} = \frac{\overline{d}\mathbf{u}}{dt}$ is valid for all Cartesian bases $(\overline{\mathbf{e}}'_1, \overline{\mathbf{e}}'_2, \overline{\mathbf{e}}'_3)$ that are at rest with respect to the rotating basis $(\overline{\mathbf{e}}_1, \overline{\mathbf{e}}_2, \overline{\mathbf{e}}_3)$.
5. Show that $\frac{d\boldsymbol{\omega}}{dt} = \frac{\overline{d}\boldsymbol{\omega}}{dt}$.
6. The statement is often made that infinitesimal rotations add like vectors but finite rotations do not. Explain.
7. Four ants occupy the corners of a square with sides of length ℓ. Each one starts walking towards the ant on the left with a constant speed v. Find the time it takes the ants to meet at the center of the square.

5.3 NEWTONIAN MECHANICS OF A PARTICLE

The position vector $\mathbf{r}(t) = \vec{OP}$ from a fixed point O to a moving particle at point $P(t)$ will vary with the time t. Since $\Delta\mathbf{r} = \mathbf{r}\ (t + \Delta t) - \mathbf{r}(t)$ is the displacement of the particle during the time interval Δt, the limit

$$\lim_{\Delta t \to 0} \frac{\Delta \mathbf{r}}{\Delta t} = \frac{d\mathbf{r}}{dt}(t) = \mathbf{v}(t)$$

is just the linear velocity $\mathbf{v}$ of the moving particle. The vector $\mathbf{v}(t)$ is parallel to the line through $P(t)$ tangent to the trajectory C of the moving particle, and is sketched by attaching the tail of a vector $\mathbf{v}$ to the particle at $P(t)$ (Fig. 5.3). The second derivative of $\mathbf{r}(t)$ is the acceleration $\mathbf{a}$ of the moving particle:

$$\frac{d^2\mathbf{r}}{dt^2} = \mathbf{a}(t). \tag{5.11}$$

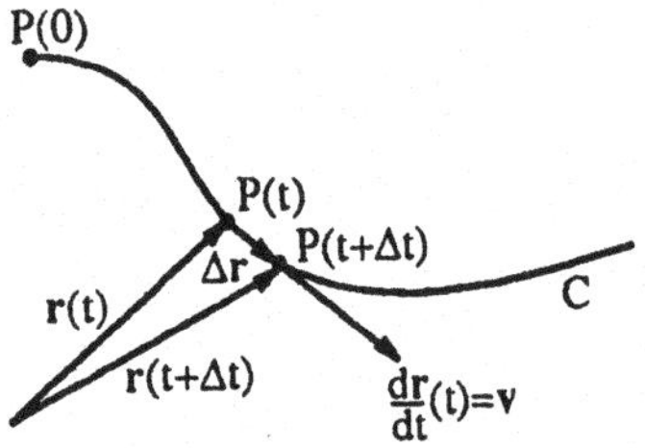

Figure 5.3. The velocity vector $\mathbf{v}(t)$ is tangent to the trajectory C of the moving particle at the point $P(t)$.

Inserting the kinematic expression (5.11) for the acceleration $\mathbf{a}$ into Newton's law (1.6) for a particle of constant mass m subjected to a force field $\mathbf{f}(t)$, we obtain

$$m\frac{d^2\mathbf{r}}{dt^2} = \mathbf{f} \tag{5.12a}$$

or, resolving along Cartesian axes (x_1, x_2, x_3) fixed in space,

$$m\frac{d^2x_i}{dt^2} = f_i. \tag{5.12b}$$

Appropriate initial conditions are to prescribe the position and velocity of the particle at an initial time $t = 0$:

$$\mathbf{r}(0) = \mathbf{r}_0, \qquad \frac{d\mathbf{r}}{dt}(0) = \mathbf{v}_0. \tag{5.13}$$

The equations (5.12) and (5.13) determine the motion of the particle in space.

To solve for the position $\mathbf{r}(t)$ of a particle, the following mathematical consequences of Newton's law (5.12) may be helpful:

- *Conservation of linear momentum theorem*: If there is no force acting on a particle, then its linear momentum is conserved:

$$\mathbf{f} = 0 \Longrightarrow m\mathbf{v} = \text{constant}.$$

- *Conservation of angular momentum theorem*: If there is no torque about a fixed point acting on a particle, then its angular momentum about the fixed point is conserved:

$$\mathbf{r} \times \mathbf{f} = 0 \Longrightarrow m\mathbf{r} \times \mathbf{v} = \text{constant}.$$

- *Energy-work theorem*: The rate of change of the kinetic energy of a particle is equal to the *power* or rate at which work is done by the force:

$$\frac{d}{dt}\left(\frac{1}{2}mv^2\right) = \mathbf{f} \cdot \mathbf{v}. \tag{5.14a}$$

Equivalently, the change in the kinetic energy of a particle between two times is equal to the total work done on the particle by the force:

$$\frac{1}{2}mv^2(t) - \frac{1}{2}mv_0^2 = \int_0^t \mathbf{f} \cdot \mathbf{v}dt. \tag{5.14b}$$

For so-called *conservative* force fields, the integrand on the right side of (5.14b) is an exact differential and can be integrated (see Sections 6.4 and 6.5):

$$\int_0^t \mathbf{f} \cdot \mathbf{v}dt = \int_{\phi(0)}^{\phi(t)} d\phi = \phi(t) - \phi(0).$$

It then follows that the *total mechanical energy*, which is the sum of the *kinetic energy* $\frac{1}{2}mv^2$ plus the *potential energy* $-\phi$, is conserved:

$$\frac{1}{2}mv^2 - \phi = \text{constant}. \tag{5.15}$$

Now suppose the origin $\overline{O}$ of a Cartesian coordinate system is translating a varying distance $\mathbf{y}(t)$ from a fixed point O and the axes are rotating with angular velocity $\boldsymbol{\omega}(t)$ relative to axes fixed in space (see Fig. 4.2). The position vector $\overline{\mathbf{r}}(t)$ from $\overline{O}$ to a moving particle and the position vector $\mathbf{r}(t)$ from O to the moving particle are related by

$$\mathbf{r}(t) = \overline{\mathbf{r}}(t) + \mathbf{y}(t). \tag{5.16}$$

Then application of (5.10) to (5.16) yields the relation between the first derivatives of the position vectors in the two systems:

$$\frac{d\mathbf{r}}{dt} = \frac{\overline{d\overline{\mathbf{r}}}}{dt} + \boldsymbol{\omega} \times \overline{\mathbf{r}} + \frac{d\mathbf{y}}{dt}. \tag{5.17}$$

Reapplication of (5.10) to (5.17) yields the relation between the second derivatives:

$$\frac{d^2\mathbf{r}}{dt^2} = \frac{\bar{d}^2\bar{\mathbf{r}}}{dt^2} + \boldsymbol{\omega} \times (\boldsymbol{\omega} \times \bar{\mathbf{r}}) + 2\boldsymbol{\omega} \times \frac{\bar{d}\bar{\mathbf{r}}}{dt} + \frac{d\boldsymbol{\omega}}{dt} \times \bar{\mathbf{r}} + \frac{d^2\mathbf{y}}{dt^2}. \tag{5.18}$$

Substituting (5.18) into (5.12a) and rearranging, we can write Newton's law for a particle in the form

$$m\frac{\bar{d}^2\bar{\mathbf{r}}}{dt^2} = \mathbf{f} - m\boldsymbol{\omega} \times (\boldsymbol{\omega} \times \bar{\mathbf{r}}) - 2m\boldsymbol{\omega} \times \frac{\bar{d}\bar{\mathbf{r}}}{dt} - m\frac{d\boldsymbol{\omega}}{dt} \times \bar{\mathbf{r}} - m\frac{d^2\mathbf{y}}{dt^2}. \tag{5.19}$$

The last four terms in (5.19) arise from the acceleration of the coordinate system but are called *fictitious forces*. In the same vein, the term $-m\dfrac{\bar{d}^2\bar{\mathbf{r}}}{dt^2}$ is sometimes called the *inertial force*.

It is clear from this that the component form of Newton's law is not invariant with respect to time-varying coordinate transformations. Those coordinate systems for which Newton's law retains the simple form (5.12b) are said to be *inertial*. Since (5.19) has the same form as (5.12) only if $\boldsymbol{\omega} = 0$ and $\dfrac{d^2\mathbf{y}}{dt^2} = 0$, only coordinate systems that are at rest or moving with a constant velocity relative to a fixed coordinate system are inertial. In Newtonian mechanics we must always postulate the location of an inertial coordinate system.

Motion of a projectile under a constant gravitational field in an inertial coordinate system. Let us develop a simple model for the motion of a projectile fired from the surface of the earth. We establish a fixed coordinate system (x_1, x_2, x_3) with origin at the projectile's initial position, x_1 measured vertically, and x_2 measured horizontally so that the x_1-x_2 plane contains the initial velocity vector (Fig. 5.4). Assuming that the projectile travels a distance that is much less than the radius of the earth, and neglecting effects such as air resistance, we can approximate the force acting upon the projectile by

$$f_1 = -mg, \qquad f_2 = 0, \qquad f_3 = 0,$$

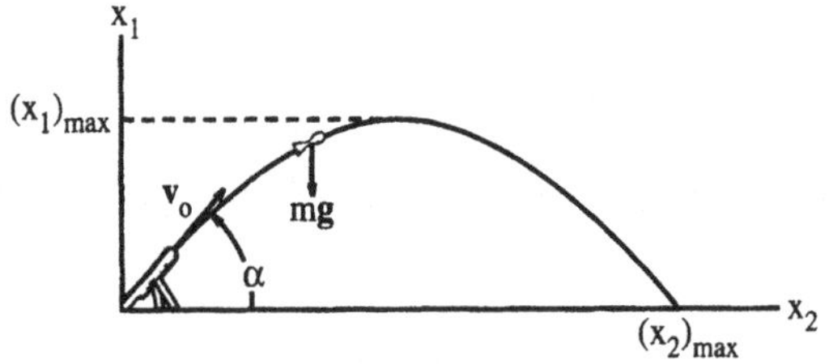

Figure 5.4. A projectile fired from the surface of the earth ideally has a parabolic trajectory.

where $g = 9.8\frac{\text{m}}{\text{s}^2}$ is the acceleration of a freely falling body near the surface of the earth. Newton's law (5.12b) then yields the equations of motion

$$\frac{d^2x_1}{dt^2} = -g, \qquad \frac{d^2x_2}{dt^2} = 0, \qquad \frac{d^2x_3}{dt^2} = 0.$$

Equations (5.13) yield the initial conditions

$$x_1(0) = x_2(0) = x_3(0) = 0,$$

$$\frac{dx_1}{dt}(0) = v_0 \sin\alpha, \qquad \frac{dx_2}{dt}(0) = v_0 \cos\alpha, \qquad \frac{dx_3}{dt}(0) = 0,$$

where α is the angle of elevation of the muzzle. Integrating and applying the initial conditions, we obtain the projectile's trajectory in the form

$$x_1 = (v_0 \sin\alpha)t - \frac{1}{2}gt^2\ , \quad x_2 = (v_0 \cos\alpha)t\ , \quad x_3 = 0. \tag{5.20}$$

These are the parametric equations of a parabola, concave downward. The maximum altitude of the projectile is $(x_1)_{\max} = \dfrac{(v_0 \sin\alpha)^2}{2g}$. The projectile will strike the ground at the range $(x_2)_{\max} = \dfrac{v_0^2 \sin 2\alpha}{g}$ and time $t_{\max} = \dfrac{2v_0 \sin\alpha}{g}$. Note that the energy-work theorem (5.14b) implies that

$$\frac{1}{2}mv^2(t) - \frac{1}{2}mv_0^2 = -mg\int_0^{x_1} dx_1 = -mgx_1,$$

so the mechanical energy $\frac{1}{2}mv^2 + mgx_1$ is conserved. As a check on the solution (5.20), we can directly verify that

$$\frac{1}{2}m\left(\frac{dx_1}{dt}\right)^2 + \frac{1}{2}m\left(\frac{dx_2}{dt}\right)^2 + mgx_1 = \frac{1}{2}mv_0^2.$$

Mechanics of a particle in a coordinate system rotating with the earth. We can account for the rotation of the earth by postulating that the coordinate system shown in Fig. 4.7 with origin at the center of the earth and axes through the vernal equinox and the north pole is inertial. Then the equation of motion in a coordinate system with origin $\overline{O}$ on the surface of the earth and rotating with the earth's angular velocity $\boldsymbol{\omega}$ is, from (5.19) with $\boldsymbol{\omega}$ assumed constant,

$$m\frac{\overline{d}^2\overline{\mathbf{r}}}{dt^2} = m\mathbf{g} - m\boldsymbol{\omega} \times (\boldsymbol{\omega} \times \overline{\mathbf{r}}) - 2m\boldsymbol{\omega} \times \frac{\overline{d}\overline{\mathbf{r}}}{dt} - m\frac{d^2\mathbf{y}}{dt^2}. \tag{5.21}$$

Here we have included the gravitational force $m\mathbf{g}$ of the earth on the particle. The quantity $\dfrac{d^2\mathbf{y}}{dt^2}$ is the acceleration of the origin $\overline{O}$ resulting from its motion about the earth's axis; this is called *centripetal* acceleration, because it is directed *towards* the axis of rotation, and is given by

$$\frac{d^2\mathbf{y}}{dt^2} = \boldsymbol{\omega} \times (\boldsymbol{\omega} \times \mathbf{y}).$$

Since $\mathbf{r} = \bar{\mathbf{r}} + \mathbf{y}$, we can write the equation of motion (5.21) as

$$m\frac{\bar{d}^2\bar{\mathbf{r}}}{dt^2} = m\mathbf{g} - m\boldsymbol{\omega} \times (\boldsymbol{\omega} \times \mathbf{r}) - 2m\boldsymbol{\omega} \times \frac{\bar{d}\bar{\mathbf{r}}}{dt}. \qquad (5.22)$$

The second and third terms on the right of (5.22) are fictitious forces arising from the rotation of the earth. The second term $-m\boldsymbol{\omega} \times (\boldsymbol{\omega} \times \mathbf{r})$ is called *centrifugal* force, because it is directed *away from* the axis of rotation (Fig. 5.5). This force is responsible for creating the earth's equatorial bulge. The first and second terms on the right of (5.22) are usually combined by defining the *effective* gravitational acceleration

$$\mathbf{g}_e = \mathbf{g} - \boldsymbol{\omega} \times (\boldsymbol{\omega} \times \mathbf{r}).$$

On the earth's surface, $\mathbf{g}_e$ has the direction of a plumb line, and the magnitude of $m\mathbf{g}_e$ is the weight of a body. The equation of motion in a rotating coordinate system thus becomes

$$m\frac{\bar{d}^2\bar{\mathbf{r}}}{dt^2} = m\mathbf{g}_e - 2m\boldsymbol{\omega} \times \frac{\bar{d}\bar{\mathbf{r}}}{dt}.$$

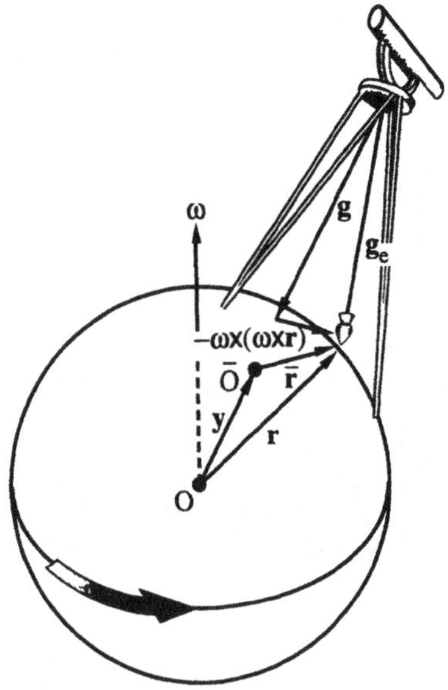

Figure 5.5. The centrifugal force $-m\boldsymbol{\omega} \times (\boldsymbol{\omega} \times \mathbf{r})$ causes a plumb bob to swing away from the earth.

The last term is the *Coriolis* force, which causes a particle moving over the earth's surface to be deflected to the right in the northern hemisphere and to the left in the southern hemisphere. The Coriolis term is important in the equations governing the large-scale motion of atmospheric winds and ocean currents. Air or water moving from high to low pressure centers will be deflected by the Coriolis force, so winds and currents circulate around in the directions sketched in Fig. 5.6 for the two hemispheres.

Figure 5.6. The Coriolis force causes large masses of air or water to circulate in the directions shown.

PROBLEMS 5.3

1. Show that there are usually two angles α at which the muzzle could be placed to reach a target. Which angle results in the shortest time of flight? What is the maximum possible range of the projectile?

2. Study the effect of air drag by adding a force $\left(-k\dfrac{dr}{dt}\right)$ to the simple model of projectile motion.

3. A projectile is fired from the north pole. Determine the motion in a coordinate system with origin at the north pole and rotating with the earth. Add only the effect of the Coriolis force to the simple model of projectile motion.

4. Water passes at a rate of 1000 m^3 each second over a waterfall 100 m high. What is the power output of a hydroelectric generator that converts all of the kinetic energy gained by the water into electrical energy? (*Note*: The density of water is roughly $10^3 \frac{kg}{m^3}$.)

5. A mass m is suspended in a frame with springs having spring constant k all around. Describe the possible motions of this three-dimensional harmonic oscillator.

6. Write Newton's law (5.19) in terms of components along the moving coordinate axes.

7. It appears to an observer rotating with the earth that the sun makes one complete circuit of the heavens in about $365\frac{1}{4}$ days. What is the angular velocity of the earth relative to the fixed stars?

8. Estimate the maximum magnitude of the centripetal acceleration and the Coriolis acceleration for a particle near the surface of the earth having velocity 100 m/sec. (*Note*: The equatorial radius of the earth is 6378 km.)

9. A weight is dropped by an observer moving with the earth. Does the trajectory follow the direction of a plumb bob?

10. The origin of a topocentric coordinate system rotating with the earth is located by its right ascension $\alpha(t)$ and latitude β from an inertial geocentric system, as shown in Fig. 4.7. Relate the first and second derivatives with respect to t of the coordinates of a celestial object in the two systems.

11. Describe the path of the stars across the night sky as seen by an observer at latitude β. Correlate this with the results of the previous problem.

12. An *inertial current* is a circular flow parallel to the earth's surface in which the acceleration is caused by a component of the Coriolis force. Determine the angular velocity of an inertial current.

13. A helicopter rotor blade is rotating about a vertical axis with constant angular speed ω. The blade is hinged at its root so that it can also rotate through an angle $\theta(t)$ about a horizontal axis. Find the fictitious forces per unit mass acting on the blade in a coordinate system rotating about the vertical axis with the blade.

14. A bucket full of water has a floating object in it which is pulled down slightly by a stretched spring. If we drop the bucket, what will happen to the object in relation to the water level?

5.4 DIFFERENTIAL GEOMETRY OF A SPACE CURVE

The differential calculus of vector fields of a single variable may be used to analyze the geometry of space curves. Consider a smooth space curve C, such as the one pictured in Fig. 5.7. To simplify the situation, let the points $P(s)$ on C be labeled by the arc length s measured along the curve from some fixed point $P(0)$.

We can consider the position vector $\mathbf{r}(s)$ from a fixed origin O to a point $P(s)$ on the curve to be a function of arc length s. Since the distance Δs between two

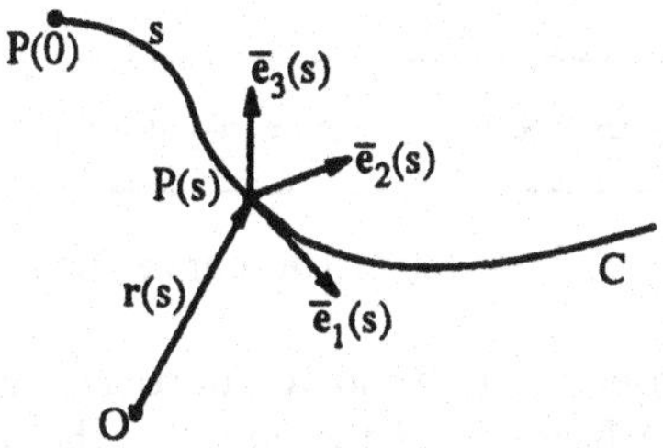

Figure 5.7. The moving trihedron $(\bar{e}_1(s), \bar{e}_2(s), \bar{e}_3(s))$ associated with a space curve C.

adjacent points is the magnitude of the vector $\Delta\mathbf{r}$ joining them, we call the ratio

$$\boxed{\bar{e}_1 = \frac{d\mathbf{r}}{ds}} \tag{5.23}$$

the ***unit tangent vector*** to the curve. The change $\Delta\bar{e}_1$ between two adjacent points is perpendicular to $\bar{e}_1$ (Fig. 5.8), so we call the unit vector

$$\bar{e}_2 = \frac{1}{\kappa}\frac{d\bar{e}_1}{ds} \tag{5.24}$$

the ***principal normal vector*** to the curve. The scalar κ (we choose $\kappa \geq 0$) is called the ***curvature*** of the space curve; and $\frac{1}{\kappa}$ is the radius of curvature of the ***osculating circle***, which is the best fitting circle tangent to the curve in the ***osculating plane*** containing $\bar{e}_1$ and $\bar{e}_2$. A third unit vector is defined by

$$\bar{e}_3 = \bar{e}_1 \times \bar{e}_2 \tag{5.25}$$

and called the ***binormal vector*** to the curve.

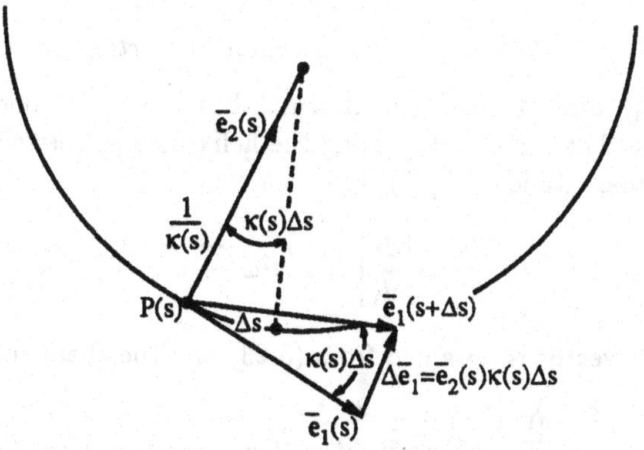

Figure 5.8. Geometrical interpretation of $\bar{e}_2 = \dfrac{1}{\kappa}\dfrac{d\bar{e}_1}{ds}$.

The derivatives of the moving trihedron $(\bar{\mathbf{e}}_1, \bar{\mathbf{e}}_2, \bar{\mathbf{e}}_3)$ are given succinctly by the *Frenet-Serret* formulas:

$$\boxed{\frac{d\bar{\mathbf{e}}_i}{ds} = \boldsymbol{k} \times \bar{\mathbf{e}}_i, \qquad \boldsymbol{k} = \tau\bar{\mathbf{e}}_1 + \kappa\bar{\mathbf{e}}_3.} \tag{5.26}$$

The *torsion* τ is a measure of how much the curve twists out of the osculating plane.

If the curve C is the trajectory of a moving particle, then we can consider the position vector $\mathbf{r}(t)$ to be a function of the time t. The linear velocity $\mathbf{v}$ of the particle is directed along $\bar{\mathbf{e}}_1$ and its magnitude is the speed $\dfrac{ds}{dt}$:

$$\mathbf{v} = \frac{d\mathbf{r}}{dt} = \frac{ds}{dt}\bar{\mathbf{e}}_1. \tag{5.27}$$

Differentiating (5.27) and using (5.24), we find that the particle's acceleration $\mathbf{a}$ has a tangential component $\dfrac{d^2s}{dt^2}$ and a normal component $\kappa\left(\dfrac{ds}{dt}\right)^2$ equal to the centripetal acceleration towards the center of the osculating circle:

$$\mathbf{a} = \frac{d^2s}{dt^2}\bar{\mathbf{e}}_1 + \kappa\left(\frac{ds}{dt}\right)^2\bar{\mathbf{e}}_2. \tag{5.28}$$

Comparing (5.4) and (5.26), we see that the angular velocity $\boldsymbol{\omega}$ of the moving trihedron is proportional to the *Darboux vector* $\boldsymbol{k}$:

$$\boldsymbol{\omega} = \frac{ds}{dt}\boldsymbol{k}.$$

Example: The equations $x_1 = a\cos\omega t$, $x_2 = a\sin\omega t$, $x_3 = ct$, where a, c, ω are constant scalars, represent a circular helix. The position vector to a point on the helix is

$$\mathbf{r}(t) = a\cos\omega t\ \mathbf{e}_1 + a\sin\omega t\ \mathbf{e}_2 + ct\mathbf{e}_3,$$

where $(\mathbf{e}_1, \mathbf{e}_2, \mathbf{e}_3)$ are constant orthonormal base vectors coinciding with the $(x_1,\ x_2,\ x_3)$ axes. The arc length s along the helix may be related to the variable t by taking the magnitude of (5.27):

$$\frac{ds}{dt} = \left|\frac{d\mathbf{r}}{dt}\right| = \sqrt{a^2\omega^2 + c^2}.$$

The unit tangent vector is obtained from (5.23) and the chain rule:

$$\bar{\mathbf{e}}_1 = \frac{\dfrac{d\mathbf{r}}{dt}}{\dfrac{ds}{dt}} = \frac{-a\omega\sin\omega t\ \mathbf{e}_1 + a\omega\cos\omega t\ \mathbf{e}_2 + c\mathbf{e}_3}{\sqrt{a^2\omega^2 + c^2}}.$$

Differentiating again yields

$$\frac{d\bar{\mathbf{e}}_1}{ds} = \frac{-a\omega^2 \cos\omega t\ \mathbf{e}_1 - a\omega^2 \sin\omega t\ \mathbf{e}_2}{a^2\omega^2 + c^2}.$$

Comparing this with (5.24), we find that the curvature is a constant

$$\kappa = \frac{a\omega^2}{a^2\omega^2 + c^2}$$

and the normal is directed towards the x_3 axis:

$$\bar{\mathbf{e}}_2 = -\cos\omega t\ \mathbf{e}_1 - \sin\omega t\ \mathbf{e}_2.$$

Here for simplicity we have assumed that $a \geq 0$. As a check on the curvature, note that it reduces to the correct values $\kappa = \frac{1}{a}$ when $c = 0$ (circle) and $\kappa = 0$ when $a = 0$ or $\omega = 0$ (straight lines). The binormal is, from (5.25),

$$\bar{\mathbf{e}}_3 = \frac{c \sin\omega t\ \mathbf{e}_1 - c\cos\omega t\ \mathbf{e}_2 + a\omega \mathbf{e}_3}{\sqrt{a^2\omega^2 + c^2}}.$$

From one of the Frenet-Serret formulas (5.26), the torsion τ is

$$\tau = \frac{d\bar{\mathbf{e}}_2}{ds} \cdot \bar{\mathbf{e}}_3 = \frac{c\omega}{a^2\omega^2 + c^2}.$$

When $c\omega$ is positive, the torsion τ is positive and the helix is *right-handed* (Fig. 5.9); when $c\omega$ is negative, the torsion τ is negative and the helix is *left-handed*.

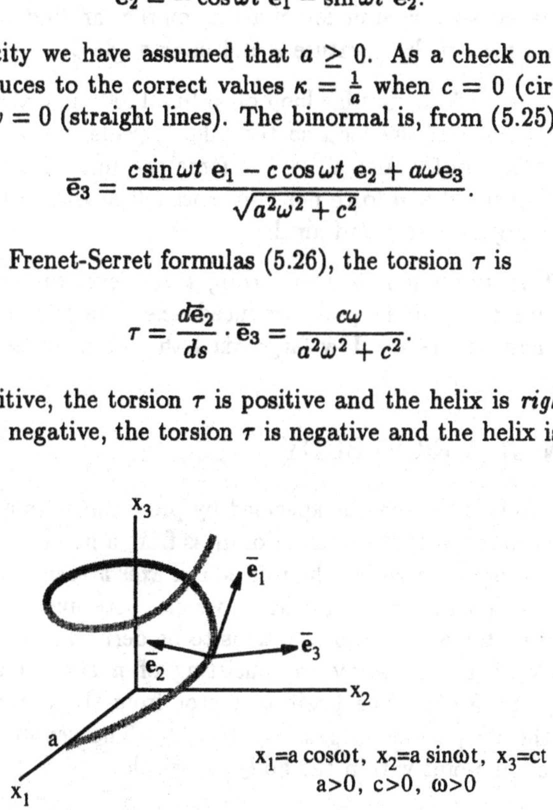

Figure 5.9. A right-handed circular helix.

PROBLEMS 5.4

1. Verify the Frenet-Serret formulas (5.26).

2. Find the angular velocity of the moving trihedron for a circular helix.

3. Find the tangent, principal normal, and binormal vectors, curvature, and torsion of the space curve $x_1 = e^t \cos t$, $x_2 = e^t \sin t$, $x_3 = e^t$.

4. Find the components of the acceleration tangential and normal to the path of the particle whose coordinates are given in the preceding problem.

5. Find the curvature and torsion of an arbitrary space curve $\mathbf{r} = \mathbf{r}(t)$.

6. A particle of mass m and charge q circulates at right angles to a uniform magnetic field $\mathcal{B}$. Find its frequency in cycles per unit time (the *cyclotron frequency*).

7. Roadways are banked on curves to prevent cars from slipping. Neglecting any frictional or aerodynamic forces acting on the car, find the correct speed to drive on a road with curvature κ and bank angle ϕ.

8. A pilot wishes to fly a circular loop of radius 1200 m. Calculate the speed needed at the top of the loop so the pilot (upside down) will feel pulled towards the floor of the aircraft with a force per unit mass of $\frac{1}{2}$ g. Find the corresponding speed and force per unit mass felt at the bottom of the loop. Neglect the engine thrust and air drag.

9. An aircraft of weight w is heading straight and level with a speed v. Find the curvature of the flight path produced when the pilot rolls the aircraft through a bank angle ϕ. How large must the lift L be to remain in level flight?

5.5 MOTION OF A RIGID BODY

The motion of a rigid body may be specified by prescribing, as a function of the time t, the displacement $\mathbf{y}(t)$ of its center of mass from a point fixed in space, and the rotation tensor $\mathbf{R}(t)$ describing the rotation of axes attached within it. Given the external forces that act on a rigid body, we can determine $\mathbf{y}(t)$ and $\mathbf{R}(t)$ by solving the set of vector differential equations to be derived in this section.

We can think of a rigid body as consisting of n small particles of mass $(m_1, \ldots, m_n)$ (Fig. 5.10). The position vector from the center of mass $O(t)$ of the body to the ith particle is denoted by $\mathbf{r}_i(t)$. The *center of mass* $O(t)$ is defined as the unique point within the body for which

$$\sum_{i=1}^{n} m_i \mathbf{r}_i(t) = 0. \tag{5.29}$$

The force acting on the ith particle is the sum of the applied external forces $\mathbf{f}_i$ and the reactive forces $\mathbf{f}_{ij}$ due to the internal constraints. If we imagine that the unvarying position of the ith particle relative to the jth particle is maintained by a rigid rod connecting them, then

$$\mathbf{f}_{ij} = -\mathbf{f}_{ji}, \qquad i \text{ or } j = 1, \ldots, n, \tag{5.30}$$

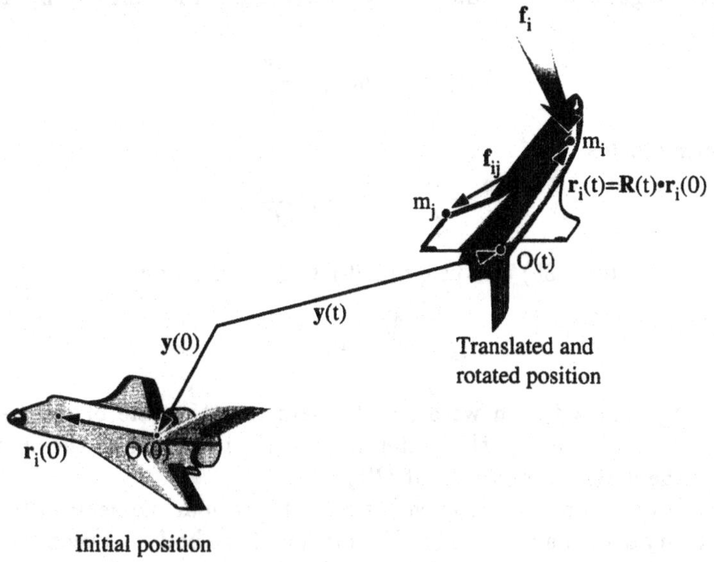

Figure 5.10. Motion of a rigid body.

$$(\mathbf{r}_i - \mathbf{r}_j) \times \mathbf{f}_{ij} = 0, \qquad i \text{ or } j = 1, \ldots, n. \tag{5.31}$$

Applying Newton's law (5.12) to the ith particle in an inertial coordinate system, we obtain

$$m_i \frac{d^2 \mathbf{r}_i}{dt^2} + m_i \frac{d^2 \mathbf{y}}{dt^2} = \mathbf{f}_i + \sum_{j=1}^{n} \mathbf{f}_{ij}, \qquad i = 1, \ldots, n. \tag{5.32}$$

(Note that our usual summation convention is not meant to be used in (5.31) and (5.32).)

Summing (5.32) over all the particles produces

$$\sum_{i=1}^{n} m_i \frac{d^2 \mathbf{r}_i}{dt^2} + m \frac{d^2 \mathbf{y}}{dt^2} = \sum_{i=1}^{n} \mathbf{f}_i + \sum_{i=1}^{n} \sum_{j=1}^{n} \mathbf{f}_{ij},$$

where $m = \sum_{i=1}^{n} m_i$ is the total mass of the body. But (5.29) and (5.30) imply that the first and last term of this equation vanish. We are left with

$$m \frac{d^2 \mathbf{y}}{dt^2} = \mathbf{f}, \tag{5.33}$$

where $\mathbf{f} = \sum_{i=1}^{n} \mathbf{f}_i$. In words, (5.33) says that the motion of the center of mass $O(t)$ is the same as that of a single particle of mass m subjected to all the external forces simultaneously.

The total angular momentum of the particles about the center of mass $O(t)$ is defined by

$$\mathbf{h}(t) = \sum_{i=1}^{n} m_i \mathbf{r}_i \times \frac{d\mathbf{r}_i}{dt}. \tag{5.34}$$

Differentiating yields

$$\frac{d\mathbf{h}}{dt} = \sum_{i=1}^{n} \mathbf{r}_i \times \left(m_i \frac{d^2\mathbf{r}_i}{dt^2} \right). \tag{5.35}$$

Inserting (5.32) into (5.35) and using (5.29)–(5.31), we obtain

$$\frac{d\mathbf{h}}{dt} = \mathbf{q}, \tag{5.36}$$

where $\mathbf{q} = \sum_{i=1}^{n} \mathbf{r}_i \times \mathbf{f}_i$. In words, (5.36) says that the rate of change of the angular momentum $\mathbf{h}$ about the center of mass $O(t)$ is equal to the sum of the moments of the external forces about $O(t)$.

Next we need to find the relation between the angular momentum $\mathbf{h}$ and the angular velocity $\boldsymbol{\omega}$ of the rigid body. The motion of the body between the times t and $t + \Delta t$ is the sum of a displacement $\mathbf{y}(t)\Delta t$ of the center of mass $O(t)$ and a rotation through an angle $\boldsymbol{\omega}(t)\Delta t$ about an axis through $O(t)$ in the direction of $\boldsymbol{\omega}(t)$. Since the vector $\mathbf{r}_i(t)$ is rotating with the body,

$$\frac{d\mathbf{r}_i}{dt} = \boldsymbol{\omega} \times \mathbf{r}_i, \qquad i = 1, \ldots, n. \tag{5.37}$$

It follows from (5.34) and (5.37) that

$$\mathbf{h} = \sum_{i=1}^{n} m_i \mathbf{r}_i \times (\boldsymbol{\omega} \times \mathbf{r}_i).$$

With the use of the vector identity (1.4), we obtain (cf. 2.6 and 2.7):

$$\mathbf{h} = \mathbf{I} \cdot \boldsymbol{\omega}, \tag{5.38}$$

where $\mathbf{I}(t)$ is the moment of inertia of the entire body about its center of mass $O(t)$.

To render the components of $\mathbf{I}$ constant, we sometimes must transform to a coordinate system attached within the body. The rates of change of angular momentum in space-fixed and body-attached systems are related by (5.10):

$$\frac{d\mathbf{h}}{dt} = \frac{\bar{d}\mathbf{h}}{dt} + \boldsymbol{\omega} \times \mathbf{h}. \tag{5.39}$$

It follows from (5.36), (5.38), and (5.39) that in a body-attached system

$$\mathbf{I} \cdot \frac{\bar{d}\boldsymbol{\omega}}{dt} + \boldsymbol{\omega} \times (\mathbf{I} \cdot \boldsymbol{\omega}) = \mathbf{q}. \tag{5.40a}$$

In Cartesian body-attached axes, the component form of (5.40a) is

$$\overline{I}_{ij}\frac{d\overline{\omega}_j}{dt} + \epsilon_{ijk}\overline{\omega}_j\overline{I}_{k\ell}\overline{\omega}_\ell = \overline{q}_i. \tag{5.40b}$$

The moment of inertia components $\overline{I}_{ij}$ for a rigid body of density ρ are given by the integrals (3.36). Equations (5.40) are often called *Euler's moment equations* of rigid body motion.

Useful consequences of these equations are the *energy-work theorems* for a rigid body:

$$\frac{d}{dt}\left(\frac{1}{2}m\frac{d\mathbf{y}}{dt}\cdot\frac{d\mathbf{y}}{dt}\right) = \mathbf{f}\cdot\frac{d\mathbf{y}}{dt} \tag{5.41}$$

$$\frac{d}{dt}\left(\frac{1}{2}\boldsymbol{\omega}\cdot\mathbf{I}\cdot\boldsymbol{\omega}\right) = \mathbf{q}\cdot\boldsymbol{\omega}. \tag{5.42}$$

Equation (5.41) says that the rate of change of *translational kinetic energy*

$$\frac{1}{2}m\frac{d\mathbf{y}}{dt}\cdot\frac{d\mathbf{y}}{dt}$$

equals the rate at which work would be done by the external forces if they acted at the center of mass. Equation (5.42) says that the rate of change of *rotational kinetic energy* $\frac{1}{2}\boldsymbol{\omega}\cdot\mathbf{I}\cdot\boldsymbol{\omega}$ equals the rate at which work is done by the moments of the external forces about the center of mass.

The motion of a rigid body is governed by the vector differential equations (5.33) for the position $\mathbf{y}(t)$ of the center of mass, (5.40) for the angular velocity $\boldsymbol{\omega}(t)$, and (5.6) for the rotation tensor $\mathbf{R}(t)$. Elegant general solutions to these equations are given in texts on rigid body mechanics. In the following we investigate three useful particular solutions.

Yo-yo. A yo-yo can be modeled as two uniform disks of radius a connected by a small axle of radius b (Fig. 5.11). A string is wound around the axle, and the loose end is held in the hand. If a yo-yo is released from rest, it will unwind itself down the string. Letting $r(t)$ denote the displacement of the center of mass, $f(t)$ denote the tension in the string, and m denote the total mass of the yo-yo (neglecting the weight of the string), we obtain from (5.33):

$$m\frac{d^2r}{dt^2} = mg - f.$$

Letting $\omega(t)$ denote the angular velocity of the yo-yo about its axle, and using (3.36) to find the moment of inertia $I = \frac{1}{2}ma^2$ of the disks about the axis of rotation (neglecting the moment of inertia of the axle), we obtain from (5.40b)

$$\frac{1}{2}ma^2\frac{d\omega}{dt} = bf.$$

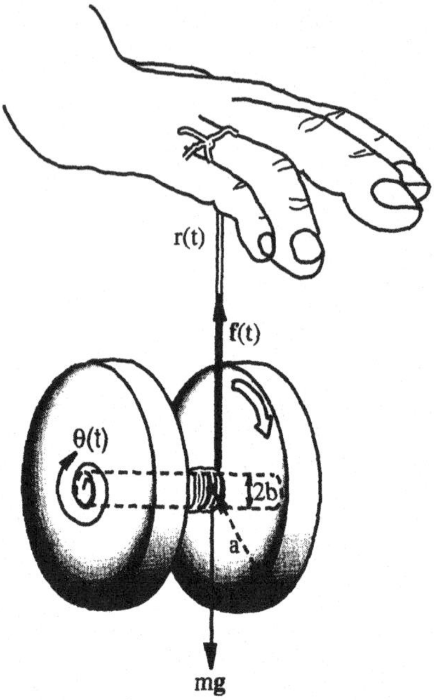

Figure 5.11. Idealized yo-yo.

Letting $\theta(t)$ denote the angle of rotation of the yo-yo from its original positi
we obtain from (5.7)

$$\omega = \frac{d\theta}{dt}.$$

Assuming the string unwinds without slipping, we obtain the kinematical relat

$$r = b\theta.$$

We now have enough equations to solve the problem. The tension in the str
has the constant value

$$f = \frac{mg}{1 + \dfrac{2b^2}{a^2}}.$$

The displacement of the yo-yo is

$$r = \frac{gt^2}{2 + \dfrac{a^2}{b^2}}.$$

Gyroscope. If a spinning gyroscope or top is subjected to a force $\mathbf{f}$ at a posit
$\mathbf{r}$ measured along its axis from its center of mass, the torque $\mathbf{r} \times \mathbf{f}$ will produc
change in the gyroscope's angular momentum $\mathbf{h}$, in accordance with (5.36):

$$\frac{d\mathbf{h}}{dt} = \mathbf{r} \times \mathbf{f}. \qquad (5.$$

If we approximate $\mathbf{h}$ by the component of the angular momentum lying along the gyroscope's axis, then from (5.38)

$$\mathbf{h} \approx I\boldsymbol{\omega}$$

where I is the moment of inertia and $\boldsymbol{\omega}$ is the angular velocity about the gyroscope's axis. Since the torque $\mathbf{r}\times\mathbf{f}$ is normal to the gyroscope's axis, the magnitude of the angular momentum $\mathbf{h}$ remains the same, so $\mathbf{h}$ just rotates with *precessional* angular velocity $\boldsymbol{\Omega}$ given by

$$\frac{d\mathbf{h}}{dt} = \boldsymbol{\Omega} \times \mathbf{h}. \tag{5.44}$$

This has been illustrated in Fig. 5.12, for a gyroscope that is subjected to a *couple* — *two* opposite forces f normal to its axis at distances r from the center of mass. In this case the above equations predict an angular speed of precession

$$\Omega = \frac{2rf}{I\omega}.$$

The angular momentum $I\omega$ can be called the *gyroscopic stiffness*. If the gyroscope is spinning rapidly and has a large moment of inertia about its axis of symmetry, the couple will cause a slow angular precession Ω in the direction shown. However, if the gyroscope were not spinning initially, the couple would cause it to rotate about a vertical axis through the center of mass with angular acceleration $\frac{d\omega'}{dt} = \frac{2rf}{I'}$, from (5.40b), with ω' being the angular speed and I' the moment of inertia about the vertical axis. A more precise solution to the equations reveals a small wobbling motion about the mean precession, a combination of the two preceding effects; in real gyroscopes this is usually damped out by frictional forces.

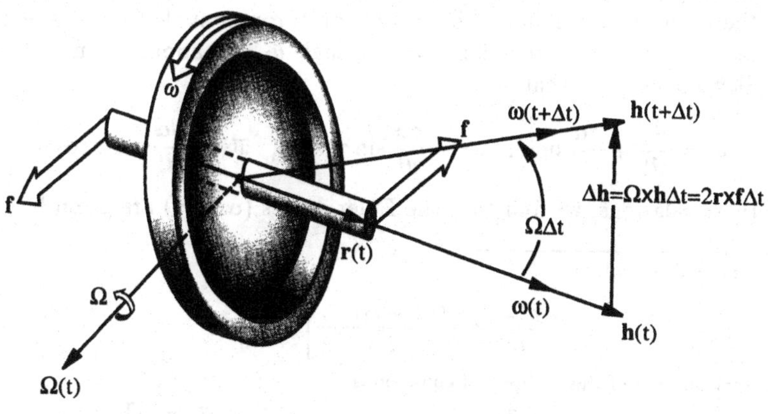

Figure 5.12. Idealized gyroscope.

Torque-free motion of an axisymmetric body. Suppose there is no net torque about the center of mass of a rigid body. Then Euler's equations (5.40) can be solved independently of the equations (5.33) for the mass center, which can still follow an arbitrary path in response to the external forces. Here we outline the solution for a body that is *inertially axisymmetrical*, meaning that two of the principal components, say $\overline{I}_{22} = \overline{I}_{33}$, of the moment of inertia tensor are equal. Then (5.40b) become in the principal axes of $\mathbf{I}$

$$\overline{I}_{11}\frac{d\overline{\omega}_1}{dt} = 0,$$

$$\overline{I}_{22}\frac{d\overline{\omega}_2}{dt} + (\overline{I}_{11} - \overline{I}_{22})\overline{\omega}_1\overline{\omega}_3 = 0,$$

$$\overline{I}_{22}\frac{d\overline{\omega}_3}{dt} + (\overline{I}_{22} - \overline{I}_{11})\overline{\omega}_1\overline{\omega}_2 = 0.$$

The first of these equations implies that $\overline{\omega}_1$ is constant, whence the last two equations can be integrated to yield[1]

$$\overline{\omega}_2 = A\sin\left[\frac{(\overline{I}_{22} - \overline{I}_{11})\overline{\omega}_1 t}{\overline{I}_{22}}\right], \qquad \overline{\omega}_3 = A\cos\left[\frac{(\overline{I}_{22} - \overline{I}_{11})\overline{\omega}_1 t}{\overline{I}_{22}}\right].$$

Here A is a constant that, along with $\overline{\omega}_1$, is determined by the initial conditions. The angular velocity vector $\boldsymbol{\omega}$ thus has constant magnitude and precesses in the body-attached system in a circle of radius A about the axis of symmetry. From (5.36) and (5.38), the angular momentum vector $\mathbf{h}$ is fixed in space and given in the body-attached system by

$$\mathbf{h} = \overline{I}_{11}\overline{\omega}_1\overline{\mathbf{e}}_1 + \overline{I}_{22}(\overline{\omega}_2\overline{\mathbf{e}}_2 + \overline{\omega}_3\overline{\mathbf{e}}_3).$$

The body-attached base vectors $(\overline{\mathbf{e}}_1, \overline{\mathbf{e}}_2, \overline{\mathbf{e}}_3)$ can be obtained from the space-fixed directions (x_1, x_2, x_3) shown in Fig. 5.13a by a rotation through angle $\theta_3 = \alpha$ about the x_3 or $\mathbf{h}$ axis, followed by a rotation through angle $\theta_2 = -\beta$ about the new 2-axis, followed by a rotation through angle $\theta_1 = \gamma$ about the new 1-axis. It then follows from (5.7) that

$$\overline{\omega}_1 = \frac{d\gamma}{dt} + \frac{d\alpha}{dt}\sin\beta, \quad \overline{\omega}_2 = \frac{d\alpha}{dt}\sin\gamma\cos\beta, \quad \overline{\omega}_3 = \frac{d\alpha}{dt}\cos\gamma\cos\beta.$$

Using these relations, we find that the Euler angles (α, β, γ) are given by

---

[1] Eliminate $\overline{\omega}_3$ to obtain

$$\frac{d^2\overline{\omega}_2}{dt^2} + \left[\frac{(\overline{I}_{22} - \overline{I}_{11})\overline{\omega}_1}{\overline{I}_{22}}\right]^2 \overline{\omega}_2 = 0.$$

The general solution of this differential equation is

$$\overline{\omega}_2 = A\sin\left[\frac{(\overline{I}_{22} - \overline{I}_{11})\overline{\omega}_1 t}{\overline{I}_{22}}\right] + B\cos\left[\frac{(\overline{I}_{22} - \overline{I}_{11})\overline{\omega}_1 t}{\overline{I}_{22}}\right].$$

Choose $B = 0$ for simplicity.

$$\alpha = \frac{\bar{I}_{11}\bar{\omega}_1 t}{\bar{I}_{22}\sin\beta}, \qquad \cot\beta = \frac{\bar{I}_{22}A}{\bar{I}_{11}\bar{\omega}_1}, \qquad \gamma = \frac{(\bar{I}_{22}-\bar{I}_{11})\bar{\omega}_1 t}{\bar{I}_{22}}. \tag{5.45}$$

Note that the angle between $\mathbf{h}$ and $\bar{\mathbf{e}}_1$ is a constant, the angle between $\mathbf{h}$ and $\boldsymbol{\omega}$ is a constant, and the three vectors $(\mathbf{h}, \boldsymbol{\omega}, \bar{\mathbf{e}}_1)$ lie in the same plane at all times. The angular velocity vector $\boldsymbol{\omega}$ traces out a ***body cone*** in the body as it precesses around the axis of symmetry and a ***space cone*** in space as it precesses around $\mathbf{h}$ (see Fig. 5.13b-c). The motion is simply described as the body cone rolling on the space cone with $\boldsymbol{\omega}$ being the line of tangency. The space cone lies outside the body cone if $\bar{I}_{11} < \bar{I}_{22}$ and inside if $\bar{I}_{11} > \bar{I}_{22}$. This ***nutational*** or ***coning*** behavior may be observed in the motion of satellites and objects tossed in the sky, since a uniform gravitational field produces no net torque about the center of mass.

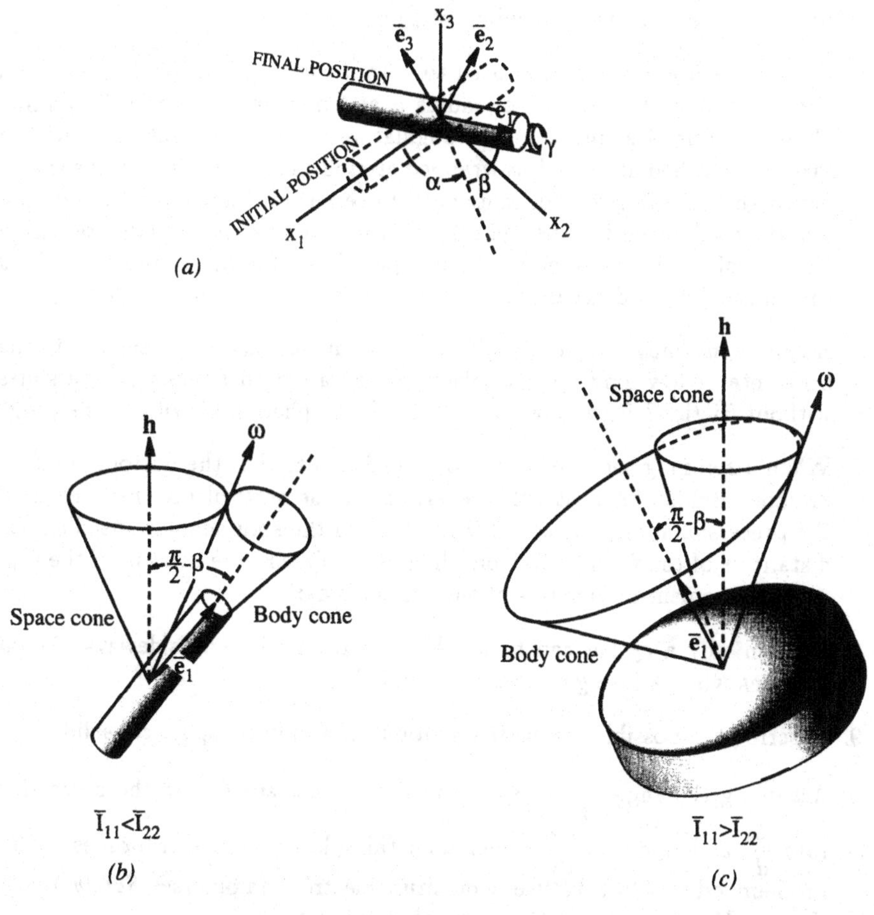

Figure 5.13. The axis of a symmetrical rigid body that is free of external torques also traces out a cone as it precesses about the angular momentum vector **h**.

PROBLEMS 5.5

1. State the theorems of conservation of linear momentum and conservation of angular momentum for a rigid body.

2. Show that if a wheel is to spin freely about a fixed axis without exerting forces or torques on its bearings, it must be statically balanced (the center of mass must lie on the axis of rotation) and dynamically balanced (the axis of rotation must be a principal axis of the inertia tensor).

3. Consider a rigid body constrained to move with one of its points fixed. Show that the equations (5.40) are also valid when the moments of inertia and torques are taken about the fixed point.

4. When a yo-yo is unwound, a sharp pull on the string will cause it to wind up again. Describe the rewinding motion.

5. A yo-yo device is sometimes used to despin rotating satellites. Consider a circular cylindrical satellite of radius a and moment of inertia I spinning about its axis of symmetry with angular velocity ω. A light cord, with a mass m attached at its end, is wrapped around the satellite's curved surface. When the mass is released, centrifugal force pulls it away from the satellite and the cord unwinds. The end of the cord is attached to the satellite with a hinge, which releases when the cord is parallel to the radial direction. Find the length ℓ of cord necessary to remove all the spin of the satellite.

6. A sphere, initially at rest, travels down a plane inclined at an angle α to the horizontal. How long does it take to travel a length ℓ if the sphere slides without friction? How long does it take if the sphere rolls without slipping?

7. When a spinning top is placed upon a pedestal so that the bottom point on its axis remains fixed, it will precess about the vertical position. Express the precessional angular speed Ω in terms of the spin angular speed ω, the distance r of the fixed point from the center of mass, the mass m of the top, and the moment of inertia I about the top's axis.

8. Explain why a cyclist can turn a bike simply by leaning sideways. What happens when a bike goes around a curve?

9. Describe the possible torque-free motions of a uniform spherical solid.

10. Assuming the angle $\frac{\pi}{2} - \beta$ shown in Fig. 5.13 is small, find the precession rate $\frac{d\alpha}{dt}$ and spin rate $\frac{d\gamma}{dt}$ of a circular thin plate (neglect all torques). This problem led to a Nobel Prize in quantum electrodynamics (see *Surely You're Joking, Mr. Feynman* in the physical references).

5.6 NOTATION OF OTHERS

Many authors use the expression "frame of reference" as a synonym for a particular coordinate system. But other authors include within a frame of reference all coordinate systems at rest with respect to each other over a period of time. To avoid confusion, the word "frame" is not introduced herein.

Table 5

OUR NOTATION	OUR NAME	OTHER NOTATIONS	OTHER NAMES
$\frac{d\mathbf{u}}{dt}$	Derivative of $\mathbf{u}(t)$.	$\dot{\mathbf{u}}, \mathbf{u}'$	
$\frac{\bar{d}\mathbf{u}}{dt}$	Derivative of $\mathbf{u}(t)$ in a rotating basis.	$\frac{d^*\mathbf{u}}{dt}, \left(\frac{d\mathbf{u}}{dt}\right)_b, \frac{\delta\mathbf{u}}{\delta t}$	
$(\bar{\mathbf{e}}_1, \bar{\mathbf{e}}_2, \bar{\mathbf{e}}_3)$	Unit tangent, normal, binormal (respectively).	$(\mathbf{T}, \mathbf{N}, \mathbf{B})$, $(\mathbf{t}, \mathbf{n}, \mathbf{b})$	
$\frac{d^2\mathbf{r}}{ds^2}$	$\kappa\bar{\mathbf{e}}_2$	$\boldsymbol{k}$	Curvature vector.

Chapter 6

TENSOR FIELDS OF MANY VARIABLES

6.1 DIFFERENTIATION OF SCALAR FIELDS

More than one variable may be required to label the spatial points and instants within the domain of a tensor field. In this chapter we label space with a set of Cartesian coordinates (x_1, x_2, x_3) and study the calculus of scalar $f(x_1, x_2, x_3)$ and vector $\mathbf{u}(x_1, x_2, x_3)$ fields of more than one spatial variable. Our developments can be extended to fields that also vary with the time t, and to higher-order tensor fields $\mathbf{T}(x_1, x_2, x_3, t)$.

The rate of change of a tensor field in one of the coordinate directions is just the *partial derivative* with respect to the coordinate. For example, to obtain the rate of change of a scalar function $f(x_1, x_2)$ of two variables in the direction x_1, apply the rule (5.1) for ordinary differentiation with x_2 treated as a constant:

$$\frac{\partial f}{\partial x_1}(x_1, x_2) = \lim_{\Delta x_1 \to 0} \frac{f(x_1 + \Delta x_1, x_2) - f(x_1, x_2)}{\Delta x_1}.$$

A geometrical interpretation of this has been sketched in Fig. 6.1. If $x_3 = f(x_1, x_2)$ represents the elevation of a mountain above points (x_1, x_2) at sea level,

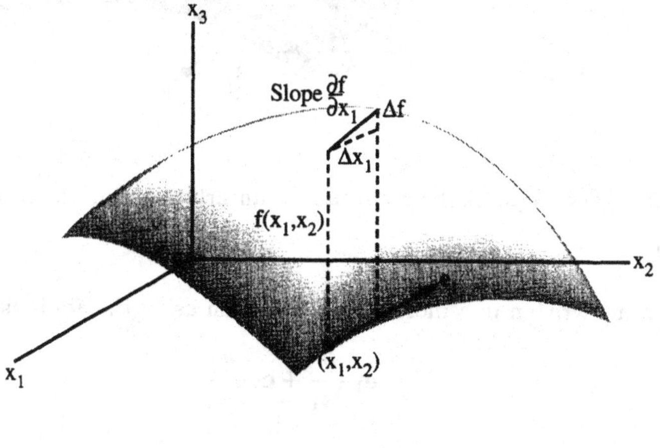

Figure 6.1. The slope of the mountain in the direction $\mathbf{e}_1$ is given by $\dfrac{\partial f}{\partial x_1}(x_1, x_2)$.

$\frac{\partial f}{\partial x_1}(x_1, x_2)$ is just the change in altitude per unit horizontal distance as a person at $(x_1, x_2, f(x_1, x_2))$ walks on the mountain in the direction of the vector $\mathbf{e}_1$.

We can generalize this and calculate the rate of change of a scalar function $f(x_1, x_2)$ in the direction of an arbitrary unit vector $\mathbf{e}$. For this purpose, draw a smooth curve C in the x_1-x_2 plane with unit tangent $\mathbf{e}$ (Fig. 6.2). Along C the position vector $\mathbf{r}(s) = x_1(s)\mathbf{e}_1 + x_2(s)\mathbf{e}_2$ is a function of arc length s and, from (5.23), $\mathbf{e}(s) = \frac{d\mathbf{r}}{ds}(s)$. The curve C is the projection of a path on the mountain having height $f(x_1(s), x_2(s))$. The change in altitude per unit horizontal distance along the path, or *directional derivative* $\frac{df}{ds}$, can be obtained from the chain rule:

$$\frac{df}{ds} = \frac{\partial f}{\partial x_1}\frac{dx_1}{ds} + \frac{\partial f}{\partial x_2}\frac{dx_2}{ds}. \tag{6.1}$$

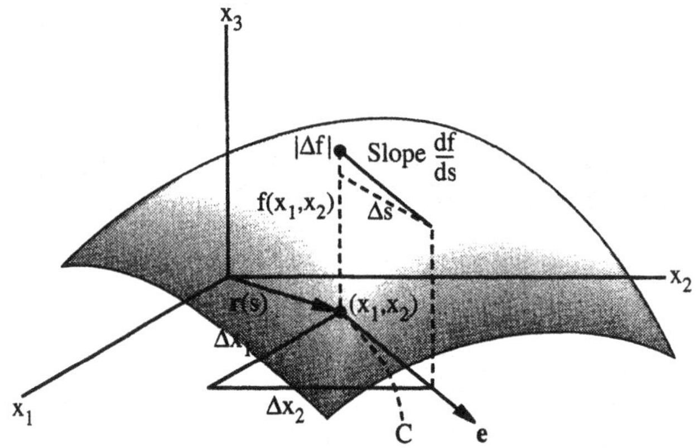

Figure 6.2. The slope of the mountain in an arbitrary direction $\mathbf{e}$ is given by $\frac{df}{ds}(x_1, x_2)$.

This can be written in a more concise notation using the abbreviation

$$\nabla = \mathbf{e}_1\frac{\partial}{\partial x_1} + \mathbf{e}_2\frac{\partial}{\partial x_2}.$$

The *del operator* ∇ represents a function that maps tensor fields into other tensor fields. The del operator ∇ maps a scalar field $f(x_1, x_2)$ into a vector field $\nabla f(x_1, x_2)$, called the *gradient* of f and abbreviated

$$\text{grad } f = \nabla f = \mathbf{e}_1\frac{\partial f}{\partial x_1} + \mathbf{e}_2\frac{\partial f}{\partial x_2}.$$

We can write the directional derivative (6.1) as

$$\boxed{\frac{df}{ds} = \mathbf{e} \cdot \nabla f.} \tag{6.2}$$

Thus, the directional derivative in a given direction is the component of ∇f in that direction.

By choosing various values of e in (6.2), we can surmise that, at a given point $(x_1, x_2, f(x_1, x_2))$ on the mountain, $\nabla f(x_1, x_2)$ points in the direction of steepest ascent, $|\nabla f(x_1, x_2)|$ equals the maximum rate of increase of f per unit distance, and $\nabla f(x_1, x_2)$ is normal to a *contour line* $f(x_1, x_2) =$ constant (Fig. 6.3).

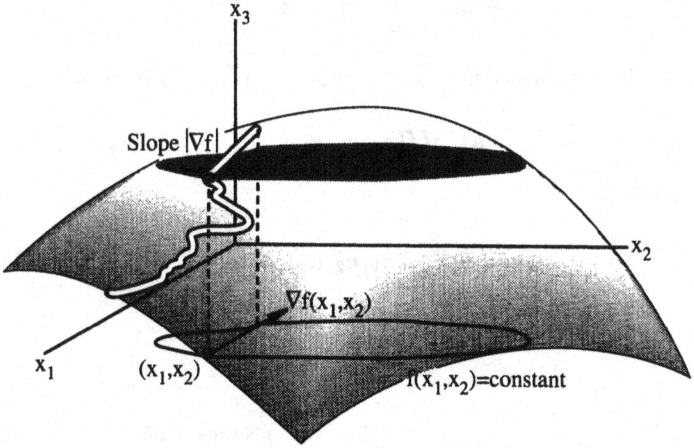

Figure 6.3. At (x_1, x_2), $\nabla f(x_1, x_2)$ is normal to a contour line $f(x_1, x_2) =$ constant and $|\nabla f(x_1, x_2)|$ is the maximum slope of the mountain.

The notions of the gradient and the directional derivative are readily generalized to scalar functions $f(x_1, x_2, x_3)$ of three variables. The del operator in three dimensions is defined by

$$\boxed{\nabla = \mathbf{e}_i \frac{\partial}{\partial x_i}.} \tag{6.3}$$

(Note we have used the summation convention to abbreviate the right side of this equation.) The del operator maps $f(x_1, x_2, x_3)$ into the gradient of f:

$$\boxed{\text{grad } f = \nabla f = \mathbf{e}_i \frac{\partial f}{\partial x_i}.} \tag{6.4}$$

The rate of change of $f(x_1, x_2, x_3)$ in an arbitrary direction specified by a unit vector e in three dimensions is still given by the formula (6.2). And, in the case of three variables, ∇f points in the direction of greatest change of f, $|\nabla f|$ equals the maximum rate of increase of f per unit distance, and ∇f is normal to the *level surfaces* $f(x_1, x_2, x_3) = \text{constant}$.

Example: We can use the gradient to derive the equations of the normal line and tangent plane to the surface $f(x_1, x_2, x_3) = \text{constant}$ at the point (a_1, a_2, a_3). (See Fig. 6.4 and review the examples in Section 3.1.) If $\mathbf{r}$ is a point on the normal line, then since $\nabla f(a_1, a_2, a_3)$ is parallel to the normal line,

$$\mathbf{r} - \mathbf{a} = t\nabla f(a_1, a_2, a_3)$$

or, in component form,

$$x_i - a_i = t\frac{\partial f}{\partial x_i}(a_1, a_2, a_3).$$

If $\mathbf{r}$ is a point on the tangent plane, then since $\nabla f(a_1, a_2, a_3)$ is perpendicular to the tangent plane,

$$(\mathbf{r} - \mathbf{a}) \cdot \nabla f(a_1, a_2, a_3) = 0$$

or, in component form,

$$(x_i - a_i)\frac{\partial f}{\partial x_i}(a_1, a_2, a_3) = 0.$$

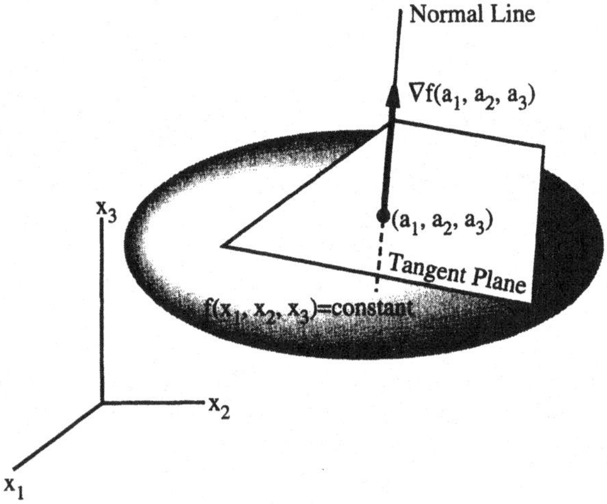

Figure 6.4. At (a_1, a_2, a_3), $\nabla f(a_1, a_2, a_3)$ is normal to the level surface $f(x_1,\ x_2,\ x_3) = f(a_1,\ a_2,\ a_3)$.

PROBLEMS 6.1

1. To graph a scalar function $f(x_1, x_2, x_3)$ of three variables in a plane, we could draw lines of constant f on a surface of constant height x_3, or we could draw contour lines on the x_1-x_2 plane for a surface of constant f. If $f = x_1^2 + x_2^2 + x_3^2$, sketch lines of constant f on the surface $x_3 = 1$, and contour lines on the x_1-x_2 plane of the surface $f = 1$.

2. What can be said about a point on a surface $x_3 = f(x_1, x_2)$ at which $\nabla f = 0$?

3. What is the rate of change of $f(x_1, x_2, x_3) = x_1 \cos x_2 + \sin x_3$ per unit distance in an arbitrary direction **e** at the origin? What is the maximum value of this change and in what direction does it occur?

4. A hiker wants to find the steepest route up a mountain of altitude $x_3 = 100 - x_1^2 - 2x_2^2$. Find the projection of the steepest path upon the x_1-x_2 plane. Assume the hike starts at the point $(\sqrt{2}, 7, 0)$.

5. If $f(x_1, x_2, x_3)$ and $h(x_1, x_2, x_3)$ are differentiable scalar fields, show that $\nabla(fh) = f\nabla h + h\nabla f$.

6. Find the equation of the normal line and tangent plane to the surface $x_3 = 1 + x_1 x_2$ at the point (1, 1, 2).

7. Find the equations of the straight lines normal and tangential to the curves $f(x_1, x_2) = \text{constant}$ at the point (a_1, a_2).

8. Find the smallest angle between the surfaces $x_1^2+x_2^2+x_3^2 = 4$ and $x_1^2+x_2^2 = 1$ at their points of intersection.

9. If **S** is a symmetric second-order tensor, show that $\mathbf{S} \cdot \mathbf{r}$ is normal to the surface $\mathbf{r} \cdot \mathbf{S} \cdot \mathbf{r} = 1$ at the point $\mathbf{r} = x_i \mathbf{e}_i$.

10. Suppose a scalar field $f(x_1, x_2)$ of two variables (x_1, x_2) obeys the partial differential equation

$$S_{11}\frac{\partial^2 f}{\partial x_1^2} + 2S_{12}\frac{\partial^2 f}{\partial x_1 \partial x_2} + S_{22}\frac{\partial^2 f}{\partial x_2^2} + b_1\frac{\partial f}{\partial x_1} + b_2\frac{\partial f}{\partial x_2} + c = 0,$$

where the coefficients S_{11}, S_{12}, S_{22}, b_1, b_2 are constants. Find a coordinate transformation that reduces this equation to the form

$$\lambda_1\frac{\partial^2 f}{\partial \bar{x}_1^2} + \lambda_2\frac{\partial^2 f}{\partial \bar{x}_2^2} + \bar{b}_1\frac{\partial f}{\partial \bar{x}_1} + \bar{b}_2\frac{\partial f}{\partial \bar{x}_2} + c = 0.$$

11. Consider the problem of finding the maximum or minimum values of a scalar function f of two or more variables subject to a constraint $h = 0$. Show that the extremum points satisfy $\nabla f = \lambda \nabla h$, where λ is a constant called the *Lagrangian multiplier*. Use this method to find the maximum values of $f(x_1, x_2) = x_1 x_2$ subject to the constraint $x_1^2 + x_2^2 = 1$.

6.2 DIFFERENTIATION OF VECTOR FIELDS

The del operator $\boldsymbol{\nabla}$ introduced in the preceding section can operate upon a vector field in the same ways as a tensor of order one. Del can be dotted into $\mathbf{u}(x_1, x_2, x_3)$ to yield a scalar field called the *divergence* of $\mathbf{u}$:

$$\boxed{\text{div } \mathbf{u} = \boldsymbol{\nabla} \cdot \mathbf{u} = \frac{\partial u_i}{\partial x_i}.} \tag{6.5}$$

The del operator $\boldsymbol{\nabla}$ can be crossed into $\mathbf{u}(x_1, x_2, x_3)$ to yield a vector field called the *curl* of $\mathbf{u}$:

$$\boxed{\text{curl } \mathbf{u} = \boldsymbol{\nabla} \times \mathbf{u} = \epsilon_{ijk} \frac{\partial u_j}{\partial x_i} \mathbf{e}_k.} \tag{6.6}$$

Del can operate upon $\mathbf{u}(x_1, x_2, x_3)$ to produce a second-order tensor field, the gradient of $\mathbf{u}$:

$$\boxed{\text{grad } \mathbf{u} = \boldsymbol{\nabla} \mathbf{u} = \frac{\partial u_j}{\partial x_i} \mathbf{e}_i \mathbf{e}_j.} \tag{6.7}$$

The operator $\boldsymbol{\nabla}$ can also be applied to a higher-order tensor field; for example, if $\mathbf{T}$ is a tensor of order two,

$$\boxed{\boldsymbol{\nabla} \cdot \mathbf{T} = \frac{\partial T_{ij}}{\partial x_i} \mathbf{e}_j.} \tag{6.8}$$

Some rules for the application of $\boldsymbol{\nabla}$ to products of differentiable fields $f(x_1, x_2, x_3)$, $\mathbf{u}(x_1, x_2, x_3)$, and $\mathbf{v}(x_1, x_2, x_3)$ are

$$\boldsymbol{\nabla} \cdot (f\mathbf{u}) = f \boldsymbol{\nabla} \cdot \mathbf{u} + \mathbf{u} \cdot \boldsymbol{\nabla} f, \tag{6.9}$$

$$\boldsymbol{\nabla} \times (f\mathbf{u}) = f \boldsymbol{\nabla} \times \mathbf{u} + \boldsymbol{\nabla} f \times \mathbf{u}, \tag{6.10}$$

$$\boldsymbol{\nabla}(\mathbf{u} \cdot \mathbf{v}) = \mathbf{u} \cdot \boldsymbol{\nabla}\mathbf{v} + \mathbf{v} \cdot \boldsymbol{\nabla}\mathbf{u} + \mathbf{u} \times (\boldsymbol{\nabla} \times \mathbf{v}) + \mathbf{v} \times (\boldsymbol{\nabla} \times \mathbf{u}), \tag{6.11}$$

$$\boldsymbol{\nabla} \cdot (\mathbf{u} \times \mathbf{v}) = \mathbf{v} \cdot \boldsymbol{\nabla} \times \mathbf{u} - \mathbf{u} \cdot \boldsymbol{\nabla} \times \mathbf{v}, \tag{6.12}$$

$$\boldsymbol{\nabla} \times (\mathbf{u} \times \mathbf{v}) = \mathbf{v} \cdot \boldsymbol{\nabla}\mathbf{u} - \mathbf{u} \cdot \boldsymbol{\nabla}\mathbf{v} + (\boldsymbol{\nabla} \cdot \mathbf{v})\mathbf{u} - (\boldsymbol{\nabla} \cdot \mathbf{u})\mathbf{v}, \tag{6.13}$$

$$\boldsymbol{\nabla} \cdot (\mathbf{u}\mathbf{v}) = (\boldsymbol{\nabla} \cdot \mathbf{u})\mathbf{v} + \mathbf{u} \cdot \boldsymbol{\nabla}\mathbf{v}. \tag{6.14}$$

The parentheses on expressions such as $\mathbf{u} \cdot \boldsymbol{\nabla}\mathbf{v}$ and $\boldsymbol{\nabla} f \times \mathbf{u}$ can be omitted because $(\mathbf{u} \cdot \boldsymbol{\nabla})\mathbf{v} = \mathbf{u} \cdot (\boldsymbol{\nabla}\mathbf{v})$ and the only possible interpretation of $\boldsymbol{\nabla} f \times \mathbf{u}$ is $(\boldsymbol{\nabla} f) \times \mathbf{u}$ Note that, in general, $\boldsymbol{\nabla}$ is not linear and products are not commutative:

$$\boldsymbol{\nabla} \cdot (f\mathbf{u}) \neq f \boldsymbol{\nabla} \cdot \mathbf{u}, \qquad \boldsymbol{\nabla} \cdot \mathbf{u} \neq \mathbf{u} \cdot \boldsymbol{\nabla}.$$

Del can also be dotted into itself to produce another differential operator $\boldsymbol{\nabla} \cdot \boldsymbol{\nabla}$ called the *Laplacian* and abbreviated

$$\boxed{\nabla^2 = \boldsymbol{\nabla} \cdot \boldsymbol{\nabla} = \frac{\partial^2}{\partial x_i \partial x_i}.} \tag{6.15}$$

The Laplacian operator ∇^2 maps a scalar field $f(x_1, x_2, x_3)$ into another scalar field $\nabla^2 f$ and a vector field $\mathbf{u}(x_1, x_2, x_3)$ into another vector field $\nabla^2\mathbf{u}$. *Laplace's equation*

$$\nabla^2 f = 0$$

and *Poisson's equation*

$$\nabla^2 f = g$$

are two of the most important partial differential equations of engineering and physics.

Some identities involving second derivatives are

$$\nabla \times \nabla f = 0, \tag{6.16}$$

$$\nabla \cdot \nabla \times \mathbf{u} = 0, \tag{6.17}$$

$$\nabla \times (\nabla \times \mathbf{u}) = \nabla\nabla \cdot \mathbf{u} - \nabla^2\mathbf{u}. \tag{6.18}$$

The operator

$$\nabla\nabla = \mathbf{e}_i\mathbf{e}_j \frac{\partial^2}{\partial x_i \partial x_j} \tag{6.19}$$

is known as the *Hessian*.

You can easily establish these identities by going to their Cartesian component form.

Proof of (6.16): If the second derivatives of $f(x_1, x_2, x_3)$ exist and are continuous, the order of differentiation of the mixed partials may be reversed:

$$\frac{\partial^2 f}{\partial x_i \partial x_j} = \frac{\partial^2 f}{\partial x_j \partial x_i}.$$

Hence,

$$\nabla \times \nabla f = \epsilon_{ijk} \frac{\partial^2 f}{\partial x_i \partial x_j} \mathbf{e}_k = 0.$$

A vector field $\mathbf{u}(x_1, x_2, x_3)$ is usually graphed by attaching the tail of an arrow $\mathbf{u}$ to its corresponding point (x_1, x_2, x_3) in space. It is also illustrative to sketch *field lines* — curves to which the arrows are tangent. If $\mathbf{r} = x_i\mathbf{e}_i$ is the position vector to a field line, then the requirement that the tangent $d\mathbf{r}$ and the vector $\mathbf{u}$ be parallel can be written

$$d\mathbf{r} \times \mathbf{u} = 0. \tag{6.20a}$$

This implies that the field lines are solutions to the equations

$$\frac{dx_1}{u_1} = \frac{dx_2}{u_2} = \frac{dx_3}{u_3}. \tag{6.20b}$$

Example: Consider the two-dimensional vector field

$$\mathbf{u} = -\omega x_2\mathbf{e}_1 + \omega x_1\mathbf{e}_2, \qquad \omega \text{ a positive constant.}$$

Even if the domain of $\mathbf{u}(x_1, x_2)$ is considered to be all of three-dimensional space, the field is the same in any plane perpendicular to the x_3-axis, so we have in Fig. 6.5 sketched it only in the x_1-x_2 plane. From (6.20b), the field lines of $\mathbf{u}$ satisfy

$$x_1 dx_1 + x_2 dx_2 = \frac{1}{2} d(x_1^2 + x_2^2) = 0.$$

Integrating yields

$$x_1^2 + x_2^2 = \text{constant}.$$

Thus the field lines are circles surrounding the origin. From (6.5) and (6.6), the divergence and curl of $\mathbf{u}$ are

$$\nabla \cdot \mathbf{u} = 0, \qquad \text{and} \qquad \nabla \times \mathbf{u} = 2\omega \mathbf{e}_3,$$

respectively. This vector field could represent the velocity of the particles in a rigid body that is rotating about the x_3-axis with angular velocity $\boldsymbol{\omega} = \omega \mathbf{e}_3$.

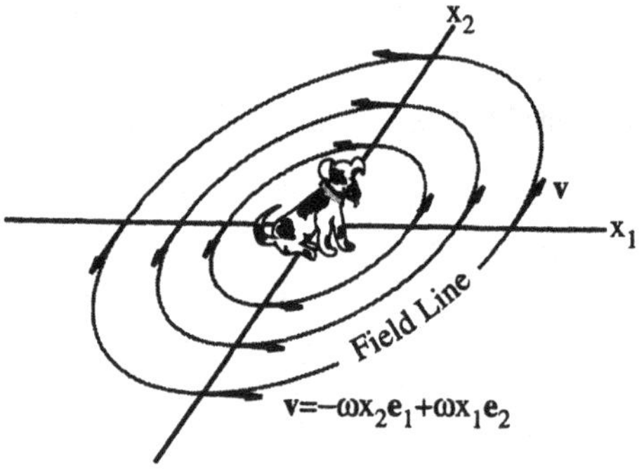

Figure 6.5. Velocity field of a rotating phonograph record.

Example: Consider the vector field

$$\mathbf{u} = K r^n \mathbf{r}, \qquad K \text{ and } n \text{ constants}, \qquad \mathbf{r} = x_i \mathbf{e}_i.$$

The domain of $\mathbf{u}(x_1, x_2, x_3)$ is all of three-dimensional space, excluding the origin in the case $n < 1$. All of the $\mathbf{u}$ vectors are pointing either away from the origin ($K > 0$ — see Fig. 6.6) or towards the origin ($K < 0$). From (6.20b), the field lines of $\mathbf{u}$ satisfy

$$\frac{dx_1}{x_1} = \frac{dx_2}{x_2} = \frac{dx_3}{x_3}.$$

Integrating yields

$$c_1 x_1 + c_2 x_2 = 0, \qquad c_3 x_1 + c_4 x_3 = 0,$$

where c_1 through c_4 are constants. Thus the field lines are straight lines radiating from the origin. From (6.5) and (6.6), the divergence and curl of $\mathbf{u}$ are

$$\nabla \cdot \mathbf{u} = (n+3)Kr^n, \qquad \nabla \times \mathbf{u} = 0.$$

Note that $\nabla \cdot \mathbf{u} =$ constant when $n = 0$, and $\nabla \cdot \mathbf{u} = 0$ when $n = -3$, which is the inverse-square law (Fig. 6.6).

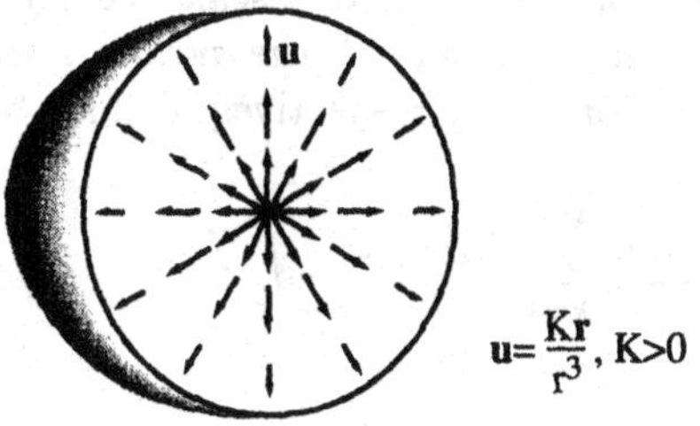

Figure 6.6. Inverse square law vector field.

PROBLEMS 6.2

1. If $\mathbf{u}(x_1, x_2, x_3) = x_1x_3\mathbf{e}_1 + x_1^2\mathbf{e}_2 + 3x_1x_2\mathbf{e}_3$, find $\nabla \cdot \mathbf{u}$, $\nabla \times \mathbf{u}$, $\nabla\mathbf{u}$, and $\nabla^2\mathbf{u}$. Verify that (6.17) and (6.18) are satisfied.

2. Show that tr $(\nabla\mathbf{u}) = \nabla \cdot \mathbf{u}$.

3. Show that $\mathbf{u} \cdot \nabla\mathbf{u} = \nabla\left(\dfrac{u^2}{2}\right) + (\nabla \times \mathbf{u}) \times \mathbf{u}$.

4. Show that $\nabla \cdot (\nabla f \times \nabla g) = 0$ and $\nabla \times (f\nabla g + g\nabla f) = 0$.

5. Show that $\nabla^2(fg) = f\nabla^2 g + g\nabla^2 f + 2\nabla f \cdot \nabla g$.

6. Find the rate of change of a vector field $\mathbf{u}(x_1, x_2, x_3)$ in the direction of a unit vector $\mathbf{e}$.

7. Compute $\nabla \cdot \mathbf{r}$, $\nabla \times \mathbf{r}$, ∇r^n, and $\nabla^2 r^n$, where $\mathbf{r} = x_i\mathbf{e}_i$.

8. Find and sketch the field lines of $\mathbf{u} = 2Kx_1\mathbf{e}_1 - Kx_2\mathbf{e}_2 - Kx_3\mathbf{e}_3$.

9. Find and sketch the field lines of $\mathbf{u} = \mathbf{e}_1 + x_2\mathbf{e}_2 + x_3\mathbf{e}_3$.

10. Show that the Taylor series expansion of the function $f(x_1, x_2, x_3)$ about the point (a_1, a_2, a_3) can be written as

$$\begin{aligned} f(x_1, x_2, x_3) &= f(a_1, a_2, a_3) + (\mathbf{r} - \mathbf{a}) \cdot \nabla f(a_1, a_2, a_3) \\ &\quad + \frac{1}{2}(\mathbf{r} - \mathbf{a}) \cdot \nabla\nabla f(a_1, a_2, a_3) \cdot (\mathbf{r} - \mathbf{a}) + \cdots . \end{aligned}$$

6.3 LINE, SURFACE, AND VOLUME INTEGRALS

The integral of a tensor field of many variables along a space curve C, over a surface S, or within a volume V of three-dimensional space may be defined in the same way as the ordinary integral (5.2). Divide C, S, or V into n small segments, and let Δs_k, ΔS_k, or ΔV_k denote the respective length, area, or volume of the kth piece (Fig. 6.7). Let f_k denote the value of a scalar field $f(x_1, x_2, x_3)$ on the kth piece. Then the *line integral of f along C*, the *surface integral of f over S*, and the *volume integral of f within V* are respectively defined by

$$\lim_{n\to\infty}\sum_{k=1}^{n} f_k \Delta s_k = \int_C f\,ds,$$

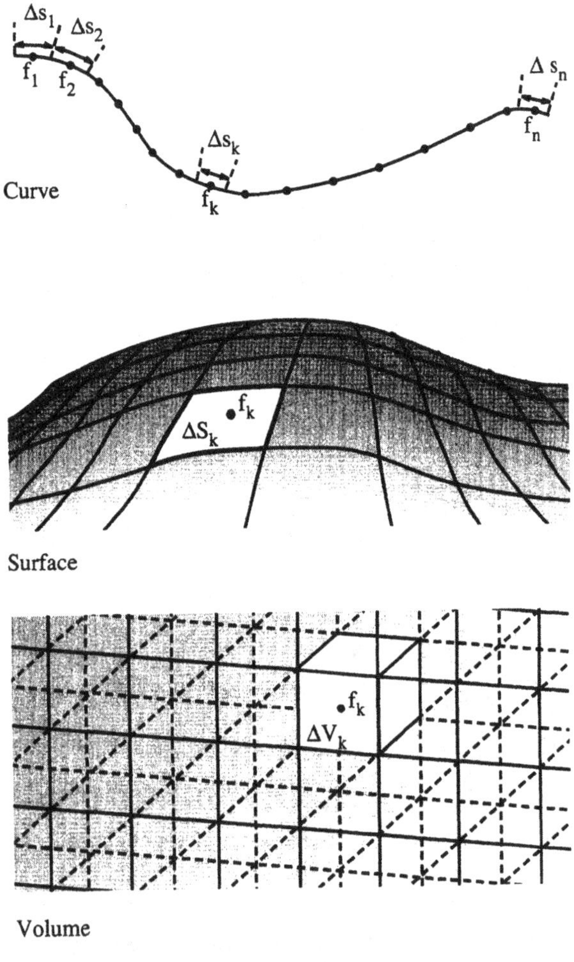

Figure 6.7. Division of curve, surface, and volume into elemental pieces.

$$\lim_{n\to\infty}\sum_{k=1}^{n} f_k \Delta S_k = \int\int_S f dS,$$

$$\lim_{n\to\infty}\sum_{k=1}^{n} f_k \Delta V_k = \int\int\int_V f dV.$$

ll of the regions C, S, and V in this book are tacitly assumed to be topologically mple enough to ensure that the integrals exist and the equations herein are valid.

In many physical applications of line, surface, and volume integrals, the in- grand arises from a vector field $\mathbf{u}(x_1, x_2, x_3)$. The integrand in a line integral often the component of $\mathbf{u}$ in the direction of the unit tangent vector $\frac{d\mathbf{r}}{ds}$ to C. ıch an integral may be written in any of the following equivalent forms:

$$\int_C \mathbf{u}\cdot d\mathbf{r} = \int_C \mathbf{u}\cdot\frac{d\mathbf{r}}{ds}ds = \int_0^t \mathbf{u}\cdot\frac{d\mathbf{r}}{dt}dt \tag{6.21a}$$

:

$$\int_C u_i dx_i = \int_C u_i \frac{dx_i}{ds}ds = \int_0^t u_i \frac{dx_i}{dt}dt. \tag{6.21b}$$

he first line integral in (6.21) can be used when the position vector $\mathbf{r} = x_i\mathbf{e}_i$ to C a known function of one of the coordinates (x_1, x_2, x_3); the second line integral (6.21) can be used when $\mathbf{r}(s) = x_i(s)\mathbf{e}_i$ is a known function of the arc length along C; and the ordinary integral in (6.21) can be used when $\mathbf{r}(t) = x_i(t)\mathbf{e}_i$ a known function of a parameter t. Note that if we reverse the direction of ıtegration along the path C, this changes the sign of $d\mathbf{r}$ and hence of the line ıtegral. When the path of integration is closed, the line integral is written

$$\oint_C \mathbf{u}\cdot d\mathbf{r}$$

nd called the *circulation of* $\mathbf{u}$ *around* C.

The integrand in a surface integral is often the component of $\mathbf{u}$ in the direction f the unit normal vector $\mathbf{n}$ to S. Such a surface integral is written as

$$\int\int_S \mathbf{u}\cdot d\mathbf{S} = \int\int_S \mathbf{u}\cdot\mathbf{n}dS$$

nd called the *flux of* $\mathbf{u}$ *through* S. To evaluate a surface integral, we can project he surface S upon one of the coordinate planes (Fig. 6.8). The projected area ΔA of an elemental patch ΔS is smaller by a factor of $|\cos\alpha|$, where α is the ngle between the normal $\mathbf{e}_i$ to the coordinate plane and the surface normal $\mathbf{n}$:

$$\Delta A = |\cos\alpha|\Delta S = |\mathbf{e}_i\cdot\mathbf{n}|\Delta S.$$

Thus

$$\boxed{\int\int_S \mathbf{u}\cdot d\mathbf{S} = \int\int_A \frac{\mathbf{u}\cdot\mathbf{n}}{|\mathbf{e}_i\cdot\mathbf{n}|}dA,} \tag{6.22}$$

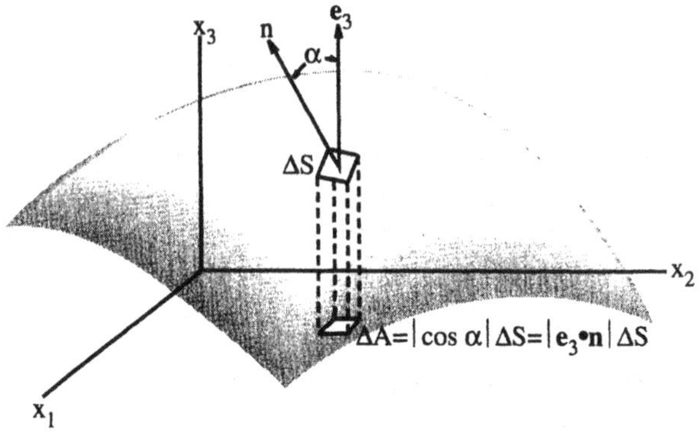

Figure 6.8. Area ΔA is the area of the projection of the surface patch ΔS upon the x_1-x_2 plane.

where the latter integrand is expressed in terms of coordinates on the projected area A of the surface S. Note that if we reverse the direction of the unit normal $\mathbf{n}$ to the surface S, this changes the sign of the surface integral. When the surface of integration is closed, the surface integral is written

$$\oiint_S \mathbf{u} \cdot d\mathbf{S}.$$

and the direction of the unit normal $\mathbf{n}$ is chosen outward from the enclosed volume.

Example: Let us calculate the line integral of the tangential component of

$$\mathbf{u} = \frac{-x_2\mathbf{e}_1 + x_1\mathbf{e}_2}{x_1^2 + x_2^2}$$

along the straight path C_1, the right-handed circular helix C_2, and the left-handed circular helix C_3 sketched in Fig. 6.9. We describe each space curve by the following equations:

$$C_1 : x_2 = x_3 = 1 - x_1, \qquad 1 \geq x_1 \geq 0,$$
$$C_2 : x_1 = \cos t, \qquad x_2 = \sin t, \qquad x_3 = \frac{2t}{\pi}, \qquad 0 \leq t \leq \frac{\pi}{2},$$
$$C_3 : x_1 = \cos t, \qquad x_2 = -\sin t, \qquad x_3 = \frac{2t}{3\pi}, \qquad 0 \leq t \leq \frac{3\pi}{2}.$$

The integrals are each evaluated using the appropriate form of (6.21):

$$\int_{C_1} u_i dx_i = \int_{C_1} \frac{-x_2 dx_1 + x_1 dx_2}{x_1^2 + x_2^2} = \int_1^0 \frac{-dx_1}{2x_1^2 - 2x_1 + 1}$$
$$= \left[\tan^{-1}(1 - 2x_1)\right]_1^0 = \frac{\pi}{2}, \tag{6.23}$$

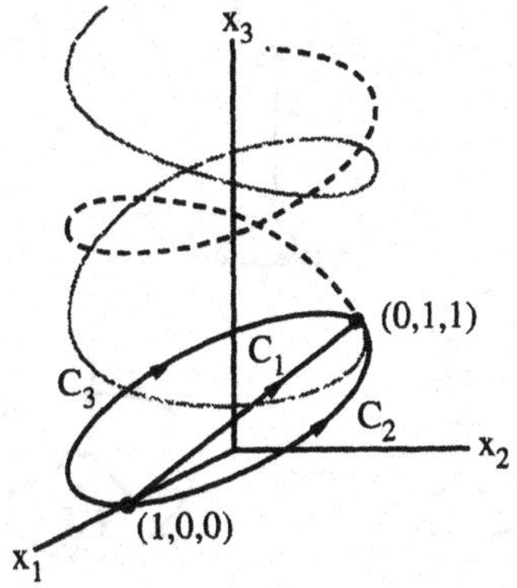

Figure 6.9. Line C_1 is a segment of a straight line, C_2 is a segment of a right-handed circular helix, and C_3 is a segment of a left-handed circular helix.

$$\int_0^{\pi/2} u_i \frac{dx_i}{dt} dt = \int_0^{\pi/2} \frac{\sin^2 t + \cos^2 t}{\sin^2 t + \cos^2 t} dt = \frac{\pi}{2}, \tag{6.24}$$

$$\int_0^{3\pi/2} u_i \frac{dx_i}{dt} dt = -\int_0^{3\pi/2} dt = -\frac{3\pi}{2}. \tag{6.25}$$

Example: Let us calculate the flux of

$$\mathbf{u} = \frac{\mathbf{r}}{r^3}, \qquad \mathbf{r} = x_i \mathbf{e}_i$$

through the circular paraboloid S_1, the upper hemisphere S_2, and the lower hemisphere S_3 sketched in Fig. 6.10. We describe each surface by the following equations:

$$\begin{aligned} S_1 &: x_3 = 1 - x_1^2 - x_2^2, && x_3 \geq 0, \\ S_2 &: x_1^2 + x_2^2 + x_3^2 = 1, && x_3 \geq 0, \\ S_3 &: x_1^2 + x_2^2 + x_3^2 = 1, && x_3 \leq 0. \end{aligned}$$

A unit normal vector to each surface is obtained using the gradient:

$$\mathbf{n}_1 = \frac{\nabla(x_1^2 + x_2^2 + x_3)}{|\nabla(x_1^2 + x_2^2 + x_3)|} = \frac{2x_1\mathbf{e}_1 + 2x_2\mathbf{e}_2 + \mathbf{e}_3}{\sqrt{1 + 4x_1^2 + 4x_2^2}},$$

$$\mathbf{n}_2 = \frac{\nabla(x_1^2 + x_2^2 + x_3^2)}{|(\nabla(x_1^2 + x_2^2 + x_3^2)|} = \frac{\mathbf{r}}{r},$$

$$\mathbf{n}_3 = -\frac{\nabla(x_1^2 + x_2^2 + x_3^2)}{|\nabla(x_1^2 + x_2^2 + x_3^2)|} = -\frac{\mathbf{r}}{r}.$$

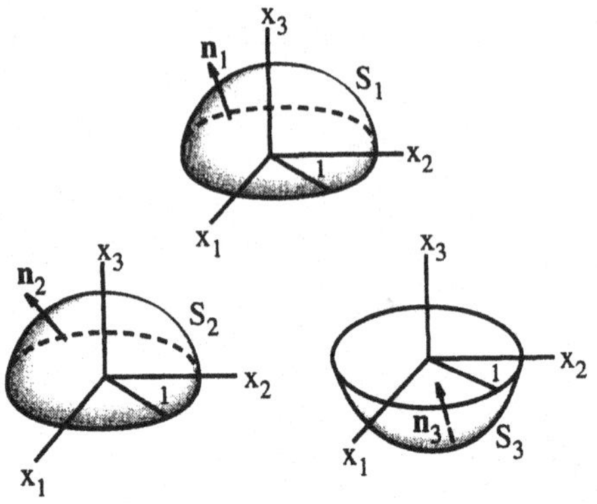

Figure 6.10. Surface S_1 is a section of a circular paraboloid; S_2 and S_3 are unit hemispheres.

Here the sign of each normal has been chosen so as to produce a nonnegative vertical component on the surface. The flux of **u** through the paraboloidal surface S_1 is evaluated by projecting S_1 onto the circular area A in the x_1-x_2 plane, in accordance with (6.22), and transforming to polar coordinates via $x_1 = \rho\cos\theta$, $x_2 = \rho\sin\theta$:

$$\begin{aligned}\int\!\!\int_{S_1} \mathbf{u}\cdot d\mathbf{S} &= \int\!\!\int_A \frac{x_1^2+x_2^2+1}{[(x_1^2+x_2^2)^2-(x_1^2+x_2^2)+1]^{3/2}}dx_1dx_2\\ &= \int_0^{2\pi}\int_0^1 \frac{\rho^2+1}{(\rho^4-\rho^2+1)^{3/2}}\rho d\rho d\theta\\ &= 2\pi\left[\frac{\rho^2-1}{(\rho^4-\rho^2+1)^{1/2}}\right]_0^1 = 2\pi. \end{aligned} \tag{6.26}$$

The flux of **u** through the hemispherical surfaces S_2 or S_3 is equal to plus or minus the surface area of a unit hemisphere, which is 2π:

$$\int\!\!\int_{S_2} \mathbf{u}\cdot d\mathbf{S} = \int\!\!\int_{S_2} dS = 2\pi, \tag{6.27}$$

$$\int\!\!\int_{S_3} \mathbf{u}\cdot d\mathbf{S} = -\int\!\!\int_{S_3} dS = -2\pi. \tag{6.28}$$

PROBLEMS 6.3

1. Calculate the line integral of the tangential component of $\mathbf{u} = -x_2\mathbf{e}_1 + x_1\mathbf{e}_2 + x_1\mathbf{e}_3$ along the paths C_1 through C_3 in the text.

2. Calculate the line integrals of the tangential component of $\mathbf{u} = -x_2\mathbf{e}_1 + x_1\mathbf{e}_2 + x_1\mathbf{e}_3$ along the straight path $\{x_1 = x_2 = x_3, 0 \le x_1 \le 1\}$, the twisted path $\{x_1 = t, x_2 = t^2, x_3 = t^3, 0 \le t \le 1\}$, and the circular path $\{x_1 = \cos s, x_2 = \sin s, x_3 = 1, 0 \le s \le 2\pi\}$.

3. Calculate the circulation of $\mathbf{u} = -x_2\mathbf{e}_1 + x_1\mathbf{e}_2 + x_1\mathbf{e}_3$ and $\mathbf{u} = x_i\mathbf{e}_i$ around the ellipse $x_1 = \cos t$, $x_2 = 3\sin t$, $x_3 = 1$, $0 \le t \le 2\pi$.

4. Calculate the flux of $\mathbf{r} = x_i\mathbf{e}_i$ through the surfaces S_1-S_3 in the text.

5. Calculate the flux of $\mathbf{r} = x_i\mathbf{e}_i$ through the triangular surface $\{2x_1+2x_2+x_3 = 2, x_1 \ge 0, x_2 \ge 0, x_3 \ge 0\}$, the cylindrical surface $\{x_1^2 + x_2^2 = 1, x_1 \ge 0, x_2 \ge 0, 0 \le x_3 \le 2\}$, and the cubical surface $\{x_1 = 0 \text{ or } 1, 0 \le x_2 \le 1, 0 \le x_3 \le 1\} + \{x_2 = 0 \text{ or } 1, 0 \le x_1 \le 1, 0 \le x_3 \le 1\} + \{x_3 = 0 \text{ or } 1, 0 \le x_1 \le 1, 0 \le x_2 \le 1\}$.

6. Calculate the flux of $\mathbf{u} = -x_2\mathbf{e}_1 + x_1\mathbf{e}_2 + x_1\mathbf{e}_3$ and $\mathbf{r} = x_i\mathbf{e}_i$ outward through the spherical surface $x_1^2 + x_2^2 + x_3^2 = a^2$.

7. Find the volume of the region $0 \le x_3 \le 1 - x_1^2 - x_2^2$ between the circular paraboloid S_1 in the text and the x_1-x_2 plane.

6.4 GREEN'S THEOREM AND POTENTIAL FIELDS OF TWO VARIABLES

In this section we focus our attention on scalar fields and two-dimensional vector fields of two variables (x_1, x_2).

The extension of the fundamental theorem of calculus (5.3) to fields of two variables is known as *Green's theorem.* This states that a line integral of two scalar functions $u_1(x_1, x_2)$ and $u_2(x_1, x_2)$ around a closed curve C in the x_1-x_2 plane is related to a double integral of their derivatives over the area A enclosed by C:

$$\boxed{\oint_C (u_1 dx_1 + u_2 dx_2) = \int\int_A \left(\frac{\partial u_2}{\partial x_1} - \frac{\partial u_1}{\partial x_2}\right) dx_1 dx_2.} \qquad (6.29)$$

Here, and throughout this section, a closed curve C is traversed in the direction which keeps the enclosed area A on the left-hand side.

Proof of (6.29): Consider first a rectangular area with edges of length ℓ_1 and ℓ_2 parallel to the x_1 and x_2 axes, respectively (Fig. 6.11a). We can show that (6.29) is valid for such a region by direct integration of the right side:

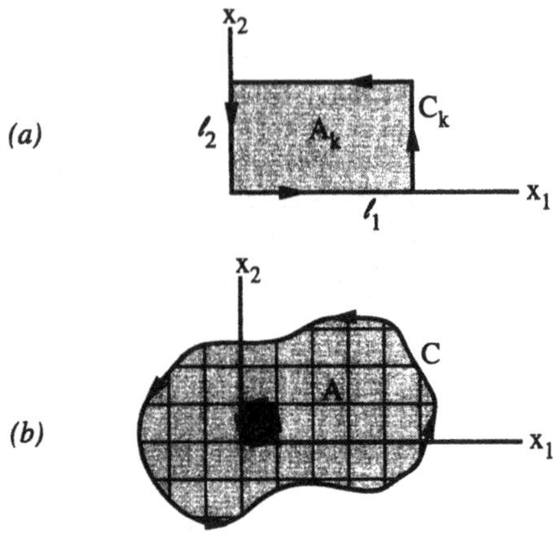

Figure 6.11. A line integral around the perimeter C of the large area is the sum of the line integrals around the perimeters C_k of the rectangular patches.

$$\int_0^{\ell_2}\int_0^{\ell_1}\left(\frac{\partial u_2}{\partial x_1}-\frac{\partial u_1}{\partial x_2}\right)dx_1dx_2 = \int_0^{\ell_2}[u_2(\ell_1,x_2)-u_2(0,x_2)]dx_2$$

$$-\int_0^{\ell_1}[u_1(x_1,\ell_2)-u_1(x_1,0)]dx_1 = \int_0^{\ell_1} u_1(x_1,0)dx_1$$

$$+\int_0^{\ell_2} u_2(\ell_1,x_2)dx_2 + \int_{\ell_1}^{0} u_1(x_1,\ell_2)dx_1$$

$$+\int_{\ell_2}^{0} u_2(0,x_2)dx_2 = \oint_{C_k}(u_1dx_1+u_2dx_2).$$

Now we divide the general area into small patches each of which is rectangular (Fig. 6.11b). The double integral over the entire area is equal to the sum of the double integrals over each rectangular patch and thus to the sum of the line integrals around each rectangular patch. But the contributions to the line integral along each of the internal edges will be zero, since two line integrals in opposite directions are added along each of these edges. Hence the grand summation equals the line integral around the perimeter of the entire area.

For our proof of Green's theorem to be valid, the area A within the curve C must be ***simply connected*** or ***singly connected***. A region of space is simply connected if any closed curve within the region can be continuously shrunk to a point within the region; that is, a simply connected area has no holes. However, Green's theorem can be extended to multiply connected regions. The area sketched in Fig. 6.12 can be made into a simply connected region with a ***single cut*** C_3 joining its outer perimeter C_1 to its inner perimeter C_2, so the area is said to be

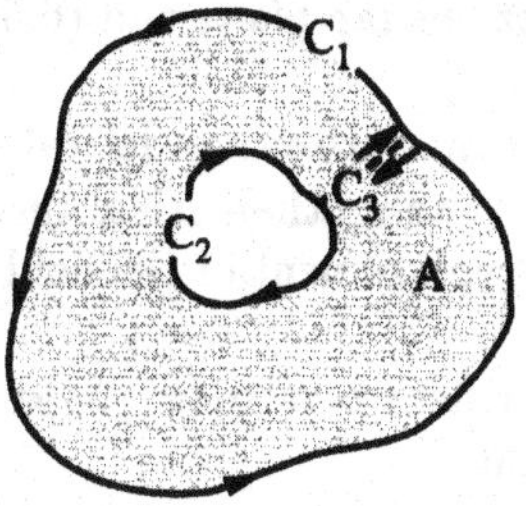

Figure 6.12. The doubly connected area A can be rendered singly connected by making the cut C_3.

doubly connected. Then Green's theorem can be applied to the simply connected region whose perimeter is $C_1 + C_2 + C_3$. Since the line integrals along the cut C_3 cancel, the curve C in Green's theorem is the sum of the outer perimeter C_1 and the inner perimeter C_2, traversed so as to keep the area on the left-hand side.

If we introduce the two-dimensional vector field $\mathbf{u}(x_1, x_2) = u_1(x_1, x_2)\mathbf{e}_1 + u_2(x_1, x_2)\mathbf{e}_2$, Green's theorem (6.29) can be stated in either of the following equivalent forms:

$$\oint_C \mathbf{u} \cdot d\mathbf{r} = \int\int_A \nabla \times \mathbf{u} \cdot \mathbf{e}_3 dx_1 dx_2 \tag{6.30a}$$

$$\oint_C \mathbf{u} \cdot \mathbf{n} ds = \int\int_A \nabla \cdot \mathbf{u} dx_1 dx_2. \tag{6.30b}$$

Here

$$\mathbf{n} = \frac{dx_2}{ds}\mathbf{e}_1 - \frac{dx_1}{ds}\mathbf{e}_2$$

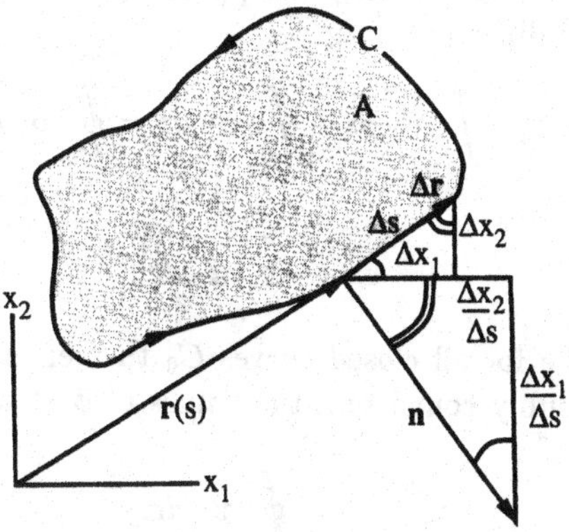

Figure 6.13. Geometrical proof that $\mathbf{n} = \dfrac{dx_2}{ds}\mathbf{e}_1 - \dfrac{dx_1}{ds}\mathbf{e}_2$ is normal to C.

is the unit outward normal vector to the curve C (Fig. 6.13). Equation (6.30a) is referred to as *Stokes' theorem in the plane* and (6.30b) as the *divergence theorem in the plane.*

Very often vector fields $\mathbf{u}$ that arise in practical applications have the property that $\nabla \times \mathbf{u} = 0$ or $\nabla \cdot \mathbf{u} = 0$. The mathematical theory governing these fields may be greatly simplified by the replacement of the field $\mathbf{u}$ by derivatives of a *potential function* ϕ or ψ.

First, suppose the vector field $\mathbf{u} = u_1(x_1, x_2)\mathbf{e}_1 + u_2(x_1, x_2)\mathbf{e}_2$ is *irrotational* or *conservative*, meaning that

$$\nabla \times \mathbf{u} = 0 \Longleftrightarrow \frac{\partial u_2}{\partial x_1} = \frac{\partial u_1}{\partial x_2}.$$

We can then define the *scalar potential* ϕ at an arbitrary point (x_1, x_2) within the domain of $\mathbf{u}$ by

$$\phi(x_1, x_2) - \phi(a_1, a_2) = \int_C \mathbf{u} \cdot d\mathbf{r}, \tag{6.31}$$

where C is a path joining some fixed point (a_1, a_2) to (x_1, x_2). For any two curves C_1 and C_2 that join (a_1, a_2) to (x_1, x_2) and which form the perimeter of a simply connected area, Green's theorem (6.29) implies that

$$\int_{C_1} \mathbf{u} \cdot d\mathbf{r} = \int_{C_2} \mathbf{u} \cdot d\mathbf{r},$$

that is, the line integral is independent of the path. Thus, if the domain of $\mathbf{u}$ is simply connected, the definition (6.31) associates a single value of ϕ with every point (x_1, x_2). In a multiply connected domain, if we choose different paths that circle around a hole, the resulting values of ϕ that we get from (6.31) can differ, so that ϕ may be a multiple-valued function. For instance, by making cuts we can show that the line integrals along the paths C_3 and C_1 in the doubly connected region of Fig. 6.14 differ by

$$\int_{C_3} \mathbf{u} \cdot d\mathbf{r} - \int_{C_1} \mathbf{u} \cdot d\mathbf{r} = \oint_{C_0} \mathbf{u} \cdot d\mathbf{r},$$

where

$$\oint_{C_0} \mathbf{u} \cdot d\mathbf{r}$$

has the same value for all closed curves C_0 that encircle the hole. Thus, if the domain of $\mathbf{u}$ is doubly connected, the values of ϕ at a given point can differ by integer multiples of

$$\oint_{C_0} \mathbf{u} \cdot d\mathbf{r}.$$

In any domain, the change $\Delta\phi$ in the potential function along a space curve with tangent $\Delta\mathbf{r}$ is

$$\Delta\phi = \nabla\phi \cdot \Delta\mathbf{r} = \mathbf{u} \cdot \Delta\mathbf{r}. \tag{6.32}$$

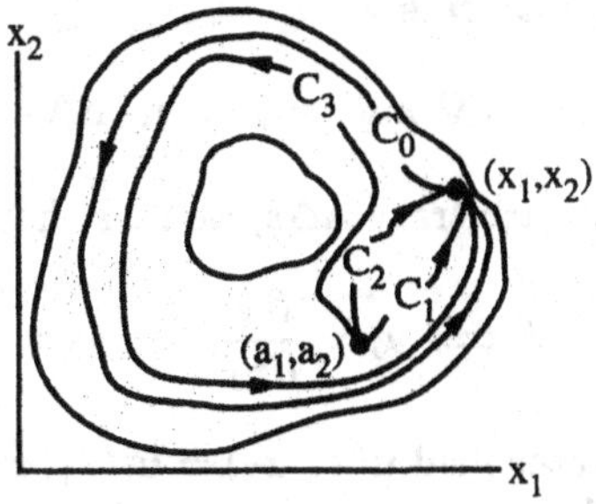

Figure 6.14. If $\nabla \times \mathbf{u} = 0$ in the doubly connected region,

$$\int_{C_1} \mathbf{u}\cdot d\mathbf{r} = \int_{C_2} \mathbf{u}\cdot d\mathbf{r} = \int_{C_3} \mathbf{u}\cdot d\mathbf{r} - \oint_{C_0} \mathbf{u}\cdot d\mathbf{r}.$$

Since this relation holds for arbitrary $\Delta\mathbf{r}$, we must have[1]

$$\mathbf{u} = \nabla\phi \Longleftrightarrow u_1 = \frac{\partial\phi}{\partial x_1}, \qquad u_2 = \frac{\partial\phi}{\partial x_2}. \tag{6.33}$$

Next, suppose a vector field $\mathbf{u} = u_1(x_1,x_2)\mathbf{e}_1 + u_2(x_1,x_2)\mathbf{e}_2$ is *solenoidal* or *rotational*, meaning that

$$\nabla\cdot\mathbf{u} = 0 \Longleftrightarrow \frac{\partial u_1}{\partial x_1} = -\frac{\partial u_2}{\partial x_2}.$$

We can then define the *stream function* ψ at an arbitrary point (x_1, x_2) within the domain of $\mathbf{u}$ by

$$\psi(x_1,x_2) - \psi(a_1,a_2) = \int_C \mathbf{u}\cdot\mathbf{n}ds, \tag{6.34}$$

where C is a path joining some fixed point (a_1,a_2) to (x_1,x_2). The *vector potential* field is

$$\boldsymbol{\psi}(x_1,x_2) = \psi(x_1,x_2)\mathbf{e}_3.$$

For any two curves C_1 and C_2 that join (a_1,a_2) to (x_1,x_2) and which form the perimeter of a simply connected area, Green's theorem (6.30b) implies that

$$\int_{C_1} \mathbf{u}\cdot\mathbf{n}ds = \int_{C_2} \mathbf{u}\cdot\mathbf{n}ds,$$

that is, the line integral is independent of the path. Thus, if the domain of $\mathbf{u}$ is simply connected, the definition (6.34) associates a single value of ψ with every

[1] Choose $\Delta\mathbf{r} = \mathbf{e}_1\Delta x_1$ with $\Delta x_1 \neq 0$. Then (6.32) implies $(\mathbf{u}-\nabla\phi)\cdot\Delta\mathbf{r} = \left(u_1 - \frac{\partial\phi}{\partial x_1}\right)\Delta x_1 = 0$. Hence, $u_1 = \frac{\partial\phi}{\partial x_1}$. A similar argument proves $u_2 = \frac{\partial\phi}{\partial x_2}$.

point (x_1, x_2). The change $\Delta\psi$ in the stream function moving a distance Δs alon a space curve with unit normal $\mathbf{n}$ is

$$\Delta\psi = \nabla \times \boldsymbol{\psi} \cdot \mathbf{n}\Delta s = \mathbf{u} \cdot \mathbf{n}\Delta s.$$

Since this relation holds for arbitrary $\mathbf{n}\Delta s$, we must have

$$\mathbf{u} = \nabla \times \boldsymbol{\psi} \Longleftrightarrow u_1 = \frac{\partial \psi}{\partial x_2}, \qquad u_2 = -\frac{\partial \psi}{\partial x_1}. \tag{6.35}$$

Finally, suppose the vector field $\mathbf{u}(x_1, x_2) = u_1(x_1, x_2)\mathbf{e}_1 + u_2(x_1, x_2)\mathbf{e}_2$ is *bot* irrotational and solenoidal:

$$\nabla \times \mathbf{u} = \nabla \cdot \mathbf{u} = 0.$$

Then we can introduce either a scalar potential ϕ or a vector potential $\boldsymbol{\psi} = \psi\mathbf{e}$ such that

$$\mathbf{u} = \nabla\phi \quad \text{or} \quad \mathbf{u} = \nabla \times \boldsymbol{\psi},$$

and both the scalar potential and the stream function satisfy Laplace's equation

$$\nabla^2\phi = 0, \qquad \nabla^2\psi = 0.$$

The *equipotential lines* $\phi(x_1, x_2) =$ constant for an irrotational vector fiel $\mathbf{u}(x_1, x_2)$ are orthogonal to the field lines of $\mathbf{u}$ because the normal to the equipo tential lines is $\nabla\phi = \mathbf{u}$. The *streamlines* $\psi(x_1, x_2) =$ constant for a solenoida vector field $\mathbf{u}(x_1, x_2)$ coincide with the field lines of $\mathbf{u}$ because $\nabla\psi \cdot \mathbf{u} = 0$. Th equipotential lines and streamlines for an irrotational and solenoidal vector fiel are thus mutually orthogonal, except possibly at points where $\mathbf{u} = 0$.

Example: Consider the vector field

$$\mathbf{u} = Kx_1\mathbf{e}_1 - Kx_2\mathbf{e}_2, \qquad K \text{ a constant.}$$

Since $\nabla \times \mathbf{u} = \nabla \cdot \mathbf{u} = 0$, the field $\mathbf{u}(x_1, x_2)$ is both irrotational and solenoidal. Le us find the scalar potential ϕ by directly solving the differential equations (6.33)

$$\frac{\partial \phi}{\partial x_1} = Kx_1, \qquad \frac{\partial \phi}{\partial x_2} = -Kx_2. \tag{6.36}$$

The first equation in (6.36) can be integrated to yield

$$\phi = \frac{Kx_1^2}{2} + f(x_2), \tag{6.37}$$

where f is an arbitrary function of x_2. Substitution of (6.37) into the secon equation in (6.36) produces

$$\frac{df}{dx_2} = -Kx_2.$$

This equation can be integrated to yield $f = -\frac{Kx_2^2}{2}$. The scalar potential is thus

$$\phi = \frac{K}{2}(x_1^2 - x_2^2),$$

where we have for simplicity dropped a possible additive constant. Similarly, we can integrate the differential equations (6.35),

$$\frac{\partial \psi}{\partial x_1} = Kx_2, \qquad \frac{\partial \psi}{\partial x_2} = Kx_1,$$

to obtain the stream function

$$\psi = Kx_1x_2.$$

The equipotential lines are the hyperbolas $x_1^2 - x_2^2 = \text{constant}$, and the streamlines are the hyperbolas $x_1x_2 = \text{constant}$ (Fig. 6.15). Note that the equipotential lines and streamlines form an orthogonal net, except at the origin where $\mathbf{u} = 0$. The line integrals of the tangential and normal components of $\mathbf{u}$ along any curve C joining the point (a_1, a_2) to (x_1, x_2) are, from (6.31) and (6.34),

$$\int_C \mathbf{u} \cdot d\mathbf{r} = \frac{K}{2}(x_1^2 + x_2^2) - \frac{K}{2}(a_1^2 + a_2^2),$$

$$\int_C \mathbf{u} \cdot \mathbf{n} ds = Kx_1x_2 - Ka_1a_2.$$

Thus the line integrals around any closed curve are zero:

$$\oint \mathbf{u} \cdot d\mathbf{r} = \oint \mathbf{u} \cdot \mathbf{n} ds = 0.$$

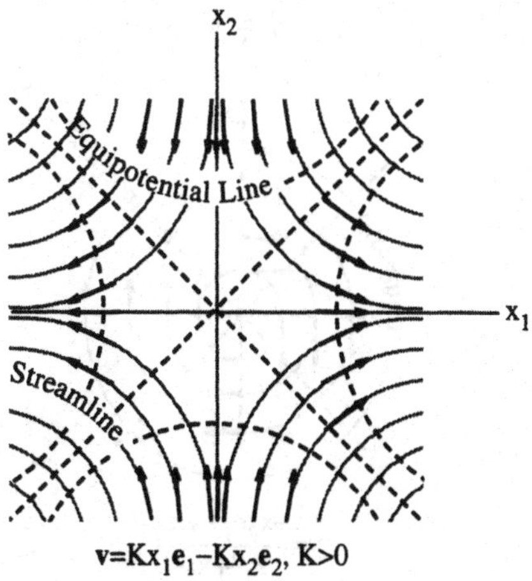

Figure 6.15. Streamlines and equipotential lines of a two-dimensional stagnation point flow.

In fluid mechanics, this particular $\mathbf{u}(x_1, x_2)$ is the velocity field of a two-dimensional *stagnation point flow*, and the origin where $\mathbf{u} = 0$ is called a *stagnation point*. The field could simulate the impingement of two opposing streams of fluid.

Example: Consider the vector field

$$\mathbf{u} = \frac{-x_2\mathbf{e}_1 + x_1\mathbf{e}_2}{x_1^2 + x_2^2}.$$

Since $\mathbf{u}$ is not defined along the x_3-axis, the domain of $\mathbf{u}$ is doubly connected. Off the x_3-axis, $\nabla \times \mathbf{u} = \nabla \cdot \mathbf{u} = 0$, so the field $\mathbf{u}(x_1, x_2)$ is both irrotational and solenoidal. The scalar potential ϕ and the stream function ψ are determined by the differential equations:

$$\frac{\partial\phi}{\partial x_1} = \frac{-x_2}{x_1^2 + x_2^2}, \qquad \frac{\partial\phi}{\partial x_2} = \frac{x_1}{x_1^2 + x_2^2},$$

$$\frac{\partial\psi}{\partial x_1} = -\frac{x_1}{x_1^2 + x_2^2}, \qquad \frac{\partial\psi}{\partial x_2} = -\frac{x_2}{x_1^2 + x_2^2}.$$

Solving, we find that

$$\phi = \tan^{-1}\frac{x_2}{x_1}, \qquad \psi = -\frac{1}{2}\ln(x_1^2 + x_2^2).$$

The equipotential lines are the straight lines $c_1x_1 + c_2x_2 = 0$, where c_1 and c_2 are constants; and the streamlines are the circles $x_1^2 + x_2^2 =$ constant (Fig. 6.16). The scalar potential $\phi(x_1, x_2)$ is a multiple-valued function, as can easily be seen by introducing polar coordinates via $x_1 = r\cos\theta$, $x_2 = r\sin\theta$:

$$\phi = \theta.$$

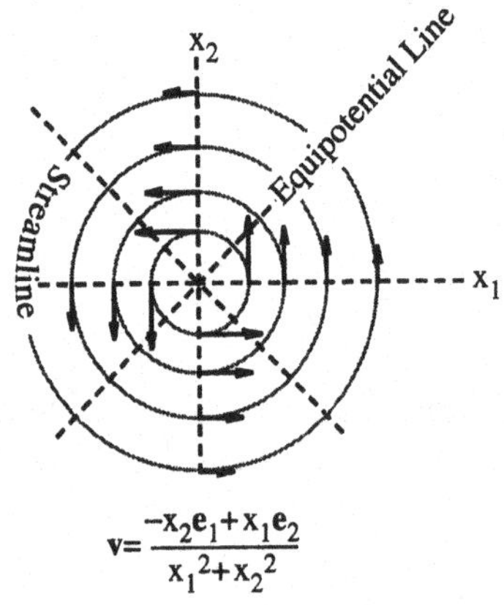

Figure 6.16. Streamlines and equipotential lines of a line vortex.

The circulation of $\mathbf{u}$ around any closed path C is, from (6.31):

$$\oint_C \mathbf{u}\cdot d\mathbf{r} = \begin{cases} 0 & \text{if } C \text{ does not encircle the } x_3\text{-axis,} \\ 2\pi & \text{if } C \text{ encircles the } x_3\text{-axis.} \end{cases}$$

We can now generalize our previous results (6.23) through (6.25) for the specific paths C_1 through C_3 (Fig. 6.9), and calculate the line integral of the tangential component of $\mathbf{u}$ along *any* curve C joining the point (1, 0, 0) to the point (0, 1, 1):

$$\int_C \mathbf{u}\cdot d\mathbf{r} = \begin{cases} \dfrac{\pi}{2} & \text{if } C \text{ can be continuously deformed within the domain of } \mathbf{u} \text{ into the right-handed helix } C_2, \\ -\dfrac{3\pi}{2} & \text{if } C \text{ can be continuously deformed within the domain of } \mathbf{u} \text{ into the left-handed helix } C_3. \end{cases}$$

In fluid mechanics, this particular $\mathbf{u}(x_1, x_2)$ is the velocity field of a *line vortex*. If there are many vortices, the circulation

$$\oint_C \mathbf{u}\cdot d\mathbf{r}$$

is the net algebraic strength of all the vortex filaments encircled by C.

PROBLEMS 6.4

1. Verify Green's theorem for $\mathbf{u} = -x_2\mathbf{e}_1 + x_1\mathbf{e}_2$ and C the unit circle $x_1^2 + x_2^2 = 1$ by directly integrating the left and right sides of (6.30a).
2. Verify Green's theorem for $\mathbf{u} = x_1\mathbf{e}_1 + x_2\mathbf{e}_2$ and C the unit circle $x_1^2 + x_2^2 = 1$ by directly integrating the left and right sides of (6.30b).
3. Compute
$$\oint_C \mathbf{u}\cdot\mathbf{n}\,ds$$
for the two examples in the text and C an arbitrary closed curve.
4. Compute
$$\int_C \mathbf{u}\cdot d\mathbf{r} \quad \text{and} \quad \int_C \mathbf{u}\cdot\mathbf{n}\,ds$$
for $\mathbf{u} = x_2\mathbf{e}_1 + x_1\mathbf{e}_2$ and C the hypocycloid arc $\{x_1 = \cos^3 t,\ x_2 = \sin^3 t,\ 0 \le t \le \frac{\pi}{4}\}$.
5. Find the stream function for the field
$$\mathbf{u}(x_1, x_2) = \begin{cases} -x_2\mathbf{e}_1 + x_1\mathbf{e}_2, & x_1^2 + x_2^2 \le 1, \\ \dfrac{-x_2\mathbf{e}_1 + x_1\mathbf{e}_2}{x_1^2 + x_2^2}, & x_1^2 + x_2^2 \ge 1. \end{cases}$$

6. The various possible values of the stream function $\psi(x_1, x_2)$ at a point (x_1, x_2) within a doubly connected region may all differ by integral multiples of what constant?

7. Consider the field
$$\mathbf{u} = \frac{x_1\mathbf{e}_1 + x_2\mathbf{e}_2}{x_1^2 + x_2^2},$$
the velocity field of a *line source* in fluid mechanics. Find the potential functions, sketch the equipotential lines and streamlines, and find
$$\oint_C \mathbf{u} \cdot d\mathbf{r} \qquad \text{and} \qquad \oint_C \mathbf{u} \cdot \mathbf{n} ds$$
for every closed curve C.

8. Show that the area A of the region within a closed curve C in the x_1-x_2 plane is
$$A = \frac{1}{2} \oint_C (x_1 dx_2 - x_2 dx_1).$$
Find the area enclosed by the ellipse
$$\frac{x_1^2}{a^2} + \frac{x_2^2}{b^2} = 1.$$

9. It can be shown that the real and imaginary parts of a function $f(z) = \phi(x_1, x_2) + i\psi(x_1, x_2)$ of a complex variable $z = x_1 + ix_2$ are the scalar potential and stream function for an irrotational solenoidal vector field $\mathbf{u}(x_1, x_2)$. Find the vector fields corresponding to the functions $f(z) = z$, $f(z) = \frac{1}{2}z^2$, and $f(z) = -i \ln z$.

10. The potential function idea can be extended to higher-order tensors. Show that the equation $\nabla \cdot \mathbf{T} = 0$, where $\mathbf{T}(x_1, x_2)$ is a symmetric two-dimensional tensor field of order two, can be satisfied by introducing an *Airy potential function* $\Phi(x_1, x_2)$ such that
$$T_{11} = \frac{\partial^2 \Phi}{\partial x_2^2}, \quad T_{12} = -\frac{\partial^2 \Phi}{\partial x_1 \partial x_2}, \quad T_{22} = \frac{\partial^2 \Phi}{\partial x_1^2}.$$

6.5 STOKES' AND DIVERGENCE THEOREMS AND POTENTIAL FIELDS OF THREE VARIABLES

The generalization of the results in the previous section to scalar and vector fields of three variables (x_1, x_2, x_3) is straightforward.

Stokes' theorem states that the circulation of a vector field $\mathbf{u}$ (x_1, x_2, x_3) around a closed space curve C is equal to the flux of $\nabla \times \mathbf{u}$ through *any* open surface S whose perimeter is C:

$$\oint_C \mathbf{u} \cdot d\mathbf{r} = \int\!\!\int_S \nabla \times \mathbf{u} \cdot d\mathbf{S},$$

or (6.38)

$$\oint_C u_i dx_i = \int\!\!\int_S \epsilon_{ijk} \frac{\partial u_j}{\partial x_i} n_k dS$$

We require that the surface S be simply connected, so that its perimeter C can be continuously contracted across S to encircle any point P on S. The direction of the normal $\mathbf{n}$ at P is the direction of the extended thumb when the fingers of the right hand are wrapped along the contracted perimeter in the direction of integration.

Proof of (6.38): Divide the surface S into small patches each of which is approximately rectangular. The flux of $\nabla \times \mathbf{u}$ through S is equal to the sum of the fluxes through each rectangular patch and, by Green's theorem (6.30a), to the sum of the circulations of $\mathbf{u}$ around the perimeter of each rectangular patch. But the contributions to the line integral along each of the internal edges will be zero, so the grand summation equals the circulation of $\mathbf{u}$ around the perimeter C of the entire surface.

The *divergence theorem* or *Gauss's theorem* states that the flux of a vector field $\mathbf{u}(x_1, x_2, x_3)$ out of a closed surface S is equal to the integral of $\nabla \cdot \mathbf{u}$ within the volume V enclosed by S:

$$\oint\!\!\oint_S \mathbf{u} \cdot d\mathbf{S} = \int\!\!\int\!\!\int_V \nabla \cdot \mathbf{u} dV,$$

or (6.39)

$$\oint\!\!\oint_S u_i n_i dS = \int\!\!\int\!\!\int_V \frac{\partial u_i}{\partial x_i} dV.$$

We require that the volume V have no holes, so that its surface S can be continuously contracted through V so as to surround any point in V. The direction of the normal $\mathbf{n}$ to S is outward from the enclosed volume.

Proof of (6.39): You can show that (6.39) is valid for a rectangular parallelepiped with edges parallel to the coordinate axes by direct integration of the right side. Divide the general volume V into small rectangular blocks. The integral of $\nabla \cdot \mathbf{u}$ within V is equal to the sum of the integrals within each rectangular block and thus to the sum of the fluxes of $\mathbf{u}$ through the surfaces of each rectangular block. But the contributions to the surface integral along each of the internal faces will be zero, since two fluxes in opposite directions are added along each of those faces. Hence the grand summation equals the flux of $\mathbf{u}$ through the outer surface S.

Stokes' theorem can be extended to multiply connected surfaces S by including all the perimeters of S in C; the correct direction of integration around the

(a)

(b)

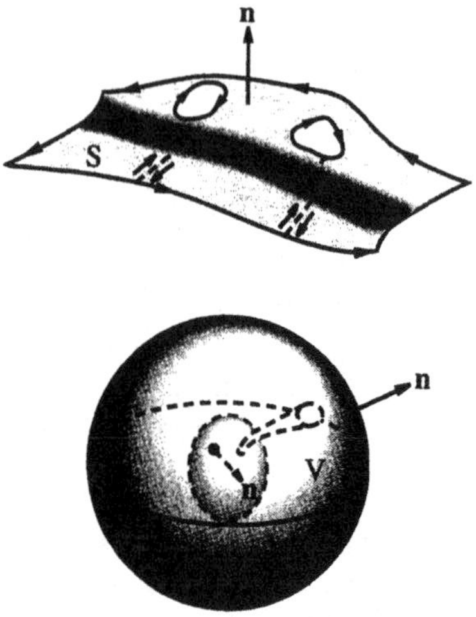

Figure 6.17. (a) The multiply connected surface S can be rendered simply connected by making the two cuts shown. (b) The volume V can be made holeless by cutting out the narrow tube.

perimeters may be determined by making the surface into a simply connected region with cuts joining the perimeters (Fig. 6.17a). The divergence theorem can be extended to volumes V with holes by including all the boundary surfaces of V in S; this can be shown by cutting the volume V so narrow tubes join the boundary surfaces, applying the divergence theorem to the resulting holeless volume, then noting that the net flux of $\mathbf{u}$ out of the tubular surfaces is zero (Fig. 6.17b).

An important consequence of the divergence and Stokes' theorems is obtained by applying them to small regions of space. By applying (6.38) to a very small area A with perimeter C, we find that the normal component of the curl of $\mathbf{u}$ is equal to the circulation of $\mathbf{u}$ per unit area:

$$\nabla \times \mathbf{u} \cdot \mathbf{n} = \lim_{A \to 0} \frac{\oint_C \mathbf{u} \cdot d\mathbf{r}}{A}.$$

By applying (6.39) to a very small volume V enclosed by a surface S, we find that the divergence of $\mathbf{u}$ is equal to the outward flux of $\mathbf{u}$ per unit volume:

$$\nabla \cdot \mathbf{u} = \lim_{V \to 0} \frac{\oint\!\!\oint_S \mathbf{u} \cdot d\mathbf{S}}{V}.$$

Now suppose the vector field $\mathbf{u}(x_1, x_2, x_3)$ is *irrotational* or *conservative*, meaning that $\nabla \times \mathbf{u} = 0$. We can then define the *scalar potential* ϕ at an arbitrary point (x_1, x_2, x_3) within the domain of $\mathbf{u}$ by

$$\phi(x_1, x_2, x_3) - \phi(a_1, a_2, a_3) = \int_C \mathbf{u} \cdot d\mathbf{r}, \tag{6.40}$$

where C is a path within the domain of $\mathbf{u}$ joining some fixed point (a_1, a_2, a_3) to (x_1, x_2, x_3). Stokes' theorem implies that the line integral is the same for any two curves that form the perimeter of a simply connected surface. Thus, if the domain of $\mathbf{u}$ is simply connected, the definition (6.40) associates a single value of ϕ with every point (x_1, x_2, x_3). In the same manner as in the previous section, we can show that

$$\mathbf{u} = \nabla\phi \Longleftrightarrow u_i = \frac{\partial\phi}{\partial x_i}. \tag{6.41}$$

Next suppose the vector field $\mathbf{u}(x_1, x_2, x_3)$ is *solenoidal* or *rotational*, meaning that

$$\nabla \cdot \mathbf{u} = 0 \Longleftrightarrow \frac{\partial u_i}{\partial x_i} = 0. \tag{6.42}$$

Then we can define a *vector potential* by

$$\begin{aligned} \boldsymbol{\psi}(x_1, x_2, x_3) = {} & \mathbf{e}_2 \int_0^{x_1} u_3(x_1', x_2, x_3)\,dx_1' \\ & -\mathbf{e}_3 \left[\int_0^{x_2} u_1(0, x_2', x_3)\,dx_2' + \int_0^{x_1} u_2(x_1', x_2, x_3)\,dx_1' \right] + \nabla f. \end{aligned} \tag{6.43}$$

By differentiating (6.43) and using (6.42), you can directly verify that

$$\mathbf{u} = \nabla \times \boldsymbol{\psi} \Longleftrightarrow u_i = \epsilon_{ijk}\frac{\partial\psi_k}{\partial x_j}. \tag{6.44}$$

Note $\boldsymbol{\psi}$ is not unique — we can add to $\boldsymbol{\psi}$ the gradient of an arbitrary scalar function $f(x_1, x_2, x_3)$ — and $\boldsymbol{\psi}$ may not be defined at all points within the domain of $\mathbf{u}$.

Example: Consider the vector field

$$\mathbf{u} = 2Kx_1\mathbf{e}_1 - Kx_2\mathbf{e}_2 - Kx_3\mathbf{e}_3, \qquad K \text{ a constant.}$$

Since $\nabla \times \mathbf{u} = \nabla \cdot \mathbf{u} = 0$, the field $\mathbf{u}(x_1, x_2, x_3)$ is both irrotational and solenoidal. We can introduce either a scalar potential ϕ or a vector potential $\boldsymbol{\psi}$. The scalar potential ϕ is determined by the differential equations (6.41):

$$\frac{\partial\phi}{\partial x_1} = 2Kx_1, \qquad \frac{\partial\phi}{\partial x_2} = -Kx_2, \qquad \frac{\partial\phi}{\partial x_3} = -Kx_3.$$

Solving, we find that

$$\phi = Kx_1^2 - \frac{K}{2}x_2^2 - \frac{K}{3}x_3^2.$$

Let us find a vector potential $\boldsymbol{\psi}$ directly by solving the differential equations (6.44):

$$\begin{aligned} &\frac{\partial\psi_3}{\partial x_2} - \frac{\partial\psi_2}{\partial x_3} = 2Kx_1, \qquad \frac{\partial\psi_1}{\partial x_3} - \frac{\partial\psi_3}{\partial x_1} = -Kx_2, \\ &\frac{\partial\psi_2}{\partial x_1} - \frac{\partial\psi_1}{\partial x_2} = -Kx_3. \end{aligned} \tag{6.45}$$

To make the solution as simple as possible, choose $\psi_1 = 0$. The second and third equations in (6.45) can then be integrated to yield

$$\psi_2 = -Kx_1x_3 + g(x_2, x_3), \qquad \psi_3 = Kx_1x_2 + h(x_2, x_3) \tag{6.46}$$

where g and h are arbitrary functions of x_2 and x_3. Substitution of (6.46) into the first equation in (6.45) produces

$$\frac{\partial h}{\partial x_2} - \frac{\partial g}{\partial x_3} = 0.$$

The simplest possible solution to this equation is $g = h = 0$. A vector potential is thus $\boldsymbol{\psi} = -Kx_1x_3\mathbf{e}_2 + Kx_1x_2\mathbf{e}_3$. This is the same result that we would have obtained from the formula (6.43). In fluid mechanics, this particular $\mathbf{u}(x_1, x_2, x_3)$ is the velocity field of a three-dimensional *stagnation point flow.*

Example: Consider the inverse square law

$$\mathbf{u} = \frac{\mathbf{r}}{r^3}, \qquad \mathbf{r} = x_i\mathbf{e}_i.$$

Note that the origin is excluded from the domain of $\mathbf{u}$. Except at the origin $\nabla \times \mathbf{u} = \nabla \cdot \mathbf{u} = 0$, so the field $\mathbf{u}(x_1, x_2, x_3)$ is both irrotational and solenoidal. The scalar potential ϕ is determined by the differential equations (6.41):

$$\frac{\partial \phi}{\partial x_i} = \frac{x_i}{(x_1^2 + x_2^2 + x_3^2)^{3/2}}.$$

Solving, we find that

$$\phi(x_1, x_2, x_3) = -\frac{1}{(x_1^2 + x_2^2 + x_3^2)^{1/2}}.$$

The *equipotential surfaces* $\phi(x_1, x_2, x_3)$ = constant are the spherical surfaces $x_1^2 + x_2^2 + x_3^2$ = constant, which have as normal lines the field lines of $\mathbf{u}$ (Fig. 6.6). The divergence theorem predicts that the flux of $\mathbf{u}$ out of any closed surface not enclosing the origin is zero, and by cutting tubes we can show that the flux of $\mathbf{u}$ out of any closed surface enclosing the origin is the same as the flux of $\mathbf{u}$ out of a spherical surface centered at the origin:

$$\oint\!\!\oint_S \mathbf{u} \cdot d\mathbf{S} = \begin{cases} 0 & \text{if origin is not within } S, \\ 4\pi & \text{if origin is within } S. \end{cases}$$

We can now generalize our previous results (6.26)–(6.28) for the specific surfaces S_1 through S_3 (Fig. 6.10), and calculate the flux of $\mathbf{u}$ through *any* surface S whose perimeter is the unit circle $x_1^2 + x_2^2 = 1$, $x_3 = 0$:

$$\int\!\!\int_S \mathbf{u} \cdot d\mathbf{S} = \begin{cases} 2\pi & \text{if } S \text{ can be continuously deformed within the domain of } \mathbf{u} \text{ into the upper hemisphere } S_2, \\ -2\pi & \text{if } S \text{ can be continuously deformed within the domain of } \mathbf{u} \text{ into the lower hemisphere } S_3. \end{cases} \tag{6.47}$$

In fluid mechanics, this particular $\mathbf{u}(x_1, x_2, x_3)$ is the velocity field of a *point source*.

Example: A more general vector field is obtained by superimposing a continuous distribution of sources occupying a volume V of space:

$$\mathbf{u}(x_1, x_2, x_3) = \iiint_V \frac{K(x_1', x_2', x_3')(\mathbf{r} - \mathbf{r}')}{|\mathbf{r} - \mathbf{r}'|^3} dV'. \tag{6.48}$$

Here $K(x_1', x_2', x_3')$ is the *source density* at the point $\mathbf{r}' = x_i' \mathbf{e}_i$. The flux of $\mathbf{u}$ through any surface S is the net algebraic strength of all the sources enclosed by S:

$$\oiint_S \mathbf{u} \cdot d\mathbf{S} = 4\pi \iiint_{V_1} K(x_1, x_2, x_3) dV. \tag{6.49}$$

Here the integration proceeds over the entire volume V_1 occupied by the sources within S. Application of (6.49) to an arbitrary small region, and use of the divergence theorem, leads to

$$\nabla \cdot \mathbf{u}(x_1, x_2, x_3) = 4\pi K(x_1, x_2, x_3). \tag{6.50}$$

The scalar potential corresponding to (6.48) is obtained by superimposing the potential of each point source:

$$\phi(x_1, x_2, x_3) = -\iiint_V \frac{K(x_1', x_2', x_3')}{|\mathbf{r} - \mathbf{r}'|} dV'. \tag{6.51}$$

It follows from (6.50) that this scalar potential obeys Poisson's equation:

$$\nabla^2 \phi(x_1, x_2, x_3) = 4\pi K(x_1, x_2, x_3). \tag{6.52}$$

The *Helmholtz theorem* or the *fundamental theorem of vector analysis* states that *any* continuous vector field may be decomposed into the sum of the gradient of a scalar potential function and the curl of a vector potential field:

$$\boxed{\mathbf{u}(x_1, x_2, x_3) = \nabla \phi(x_1, x_2, x_3) + \nabla \times \boldsymbol{\psi}(x_1, x_2, x_3).} \tag{6.53}$$

Proof of (6.53): Let

$$\mathbf{w}(x_1, x_2, x_3) = -\frac{1}{4\pi} \iiint_V \frac{\mathbf{u}(x_1', x_2', x_3')}{|\mathbf{r} - \mathbf{r}'|} dV'. \tag{6.54}$$

Here V denotes the entire domain of $\mathbf{u}$, or if the domain of $\mathbf{u}$ is infinite and the integral is infinite, V can be chosen arbitrarily large but finite. Since each component has the same form as (6.51), we conclude from (6.52) that $\mathbf{w}$ obeys the *vector* Poisson equation

$$\nabla^2 \mathbf{w} = \mathbf{u}. \tag{6.55}$$

But the vector identity (6.18) implies that

$$\nabla^2 \mathbf{w} = \nabla(\nabla \cdot \mathbf{w}) - \nabla \times (\nabla \times \mathbf{w}).$$

Hence set

$$\phi = \nabla \cdot \mathbf{w}, \qquad \boldsymbol{\psi} = -\nabla \times \mathbf{w}.$$

PROBLEMS 6.5

1. Verify Stokes' theorem for $\mathbf{u} = -x_2\mathbf{e}_1 + x_1\mathbf{e}_2 + x_1\mathbf{e}_3$ and the paraboloidal surface $x_3 = x_1^2 + x_2^2$ bounded by the circular path in Problem 2 of Section 6.3 by directly integrating the right side of (6.38).

2. Verify the divergence theorem for $\mathbf{r} = x_i\mathbf{e}_i$ and the cubical surface in Problem 5 of Section 6.3 by directly integrating the right side of (6.39).

3. Obtain the planar versions (6.30a and b) of Stokes' and the divergence theorems directly from the three-dimensional versions (6.38) and (6.39).

4. Show that $\dfrac{\mathbf{r} \cdot \Delta\mathbf{S}}{r^3}$, where $\mathbf{r} = x_i\mathbf{e}_i$, is plus or minus the solid angle subtended by ΔS.

5. Find scalar and vector potentials for a constant vector field $\mathbf{u} = \mathbf{c} = c_i\mathbf{e}_i$.

6. Compute the line integral of

$$\mathbf{u} = 2x_1 e^{x_2} \sin\frac{\pi x_3}{2}\mathbf{e}_1 + x_1^2 e^{x_2} \sin\frac{\pi x_3}{2}\mathbf{e}_2 + \frac{\pi}{2}x_1^2 e^{x_2} \cos\frac{\pi x_3}{2}\mathbf{e}_3$$

along any path joining the point (0, 0, 0) to (1, 1, 1).

7. Show that there is no vector potential $\boldsymbol{\psi}$ satisfying $\nabla \times \boldsymbol{\psi} = \dfrac{\mathbf{r}}{r^3}$ at all points in space except the origin.

8. Show that

$$\oiint_S f\mathbf{n}\,dS = \iiint_V \nabla f\,dV, \qquad \oiint_S \mathbf{u} \times d\mathbf{S} = -\iiint_V \nabla \times \mathbf{u}\,dV,$$

and

$$\oiint_S (f\nabla g - g\nabla f) \cdot d\mathbf{S} = \iiint_V (f\nabla^2 g - g\nabla^2 f)dV,$$

$$\oiint_S \mathbf{T} \cdot d\mathbf{S} = \iiint_V \nabla \cdot \mathbf{T}\,dV,$$

where S includes all the boundary surfaces of V.

9. If $\mathbf{u} = \nabla\phi + \nabla \times \boldsymbol{\psi}$, find differential equations for each of the potentials ϕ and $\boldsymbol{\psi}$.

10. Show that any continuous two-dimensional vector field

$$\mathbf{u} = u_1(x_1, x_2)\mathbf{e}_1 + u_2(x_1, x_2)\mathbf{e}_2$$

can be written as

$$\mathbf{u}(x_1, x_2) = \nabla\phi(x_1, x_2) + \nabla \times \boldsymbol{\psi}(x_1, x_2),$$

where

$$\phi = \nabla \cdot \mathbf{w}, \qquad \boldsymbol{\psi} = -\nabla \times \mathbf{w},$$

and

$$\mathbf{w}(x_1, x_2) = \frac{1}{2\pi} \int\int_A \mathbf{u}(x_1', x_2') \ln |\mathbf{r} - \mathbf{r}'| dx_1' dx_2'.$$

6.6 NOTATION OF OTHERS

Mathematicians have developed a compact notation for antisymmetrical tensors known as *differential forms.* The mathematical formalism for manipulating differential forms, called *exterior calculus*, unifies and generalizes tensor calculus. Rather than adopting two different notations, we use in this book only the classical notation familar to most engineers and physicists. Tensor calculus may not be so elegant, but it is adequate for modeling physical phenomena.

Table 6

OUR NOTATION	OUR NAME	OTHER NOTATIONS	OTHER NAMES
$\mathbf{u}(x_1, x_2, x_3)$	Vector field.	$\mathbf{u}(\mathbf{x})$	
$\dfrac{\partial \mathbf{u}}{\partial x_1}$	Partial derivative of **u**.	$\partial_x \mathbf{u},\ \mathbf{u}_{,1}$	
$\nabla \times \mathbf{u}$	Curl **u**.		Rot **u**.
$\nabla \mathbf{u}$	Gradient of **u**.	$\dfrac{\partial \mathbf{u}}{\partial \mathbf{r}}$	
∇^2	Laplacian operator.	Δ	Harmonic operator.
$\mathbf{u} = \nabla\phi$	ϕ is scalar potential.	$\mathbf{u} = -\nabla\phi$	Potential may be negative of ours.
$\phi = -\dfrac{1}{\lvert\mathbf{r} - \mathbf{r}'\rvert^{1/2}}$	Scalar potential of point source.	$G(\mathbf{r}, \mathbf{r}')$	Green's function satisfying $\nabla^2\phi = 4\pi\delta(\mathbf{r} - \mathbf{r}')$, where δ is the Dirac delta function.
$\int_C \mathbf{u} \cdot d\mathbf{r}$	Line integral.	$\int_\sigma \mathbf{u} \cdot d\mathbf{s}$	
C	Perimeter of a region S.	∂S	
$\int\int_S \mathbf{u} \cdot d\mathbf{S}$	Surface integral.	$\int_S u_n dS$	
$\int\int\int_V f dV$	Volume integral.	$\int_V f d^3x$	

Chapter 7

APPLICATIONS

7.1 HEAT CONDUCTION

There are so many fruitful applications of tensor calculus that we devote an entire chapter to them. Even this is only a brief introduction to some of the most basic equations of engineering and physics. To model reality accurately, it may be necessary to account for more effects by adding more terms to the equations herein, making the equations more complex and coupled together. Moreover, there is a vast literature developing analytical and numerical techniques for solving these equations. All we do in this book is briefly glance at the tip of the iceberg of all the useful equations and solutions.

One of the most familiar physical examples of a scalar field is the temperature $T(x_1, x_2, x_3, t)$ associated with each point (x_1, x_2, x_3) in space and instant of time t. The weather map sketched in Fig. 7.1 shows the air temperature at a fixed time and various longitudes and latitudes on the earth's surface. The contour lines of constant temperature are called *isothermal lines*.

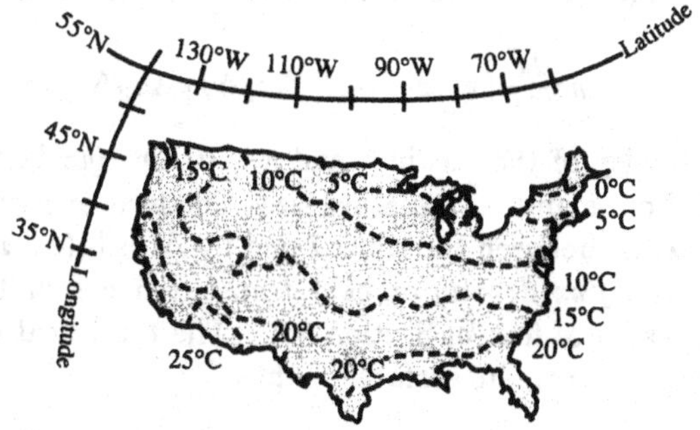

Figure 7.1. Typical air temperature distribution over the earth's surface.

A gradient in the temperature field of a solid or fluid at rest will result in the transfer of heat energy by *conduction*. The greater motion of molecules or atomic particles at higher temperature will impart energy to adjacent particles at lower temperatures. We can derive an equation for the heat transfer by enforcing the conservation of heat energy for a *control volume*, which is a fixed region within the conducting medium. The net heat flowing into a control volume across its surface must equal the increase in the thermal energy stored by the material within the control volume. Depending on whether the control volume is a small differential volume or a big finite volume, we obtain the conservation equation in either differential or integral form.

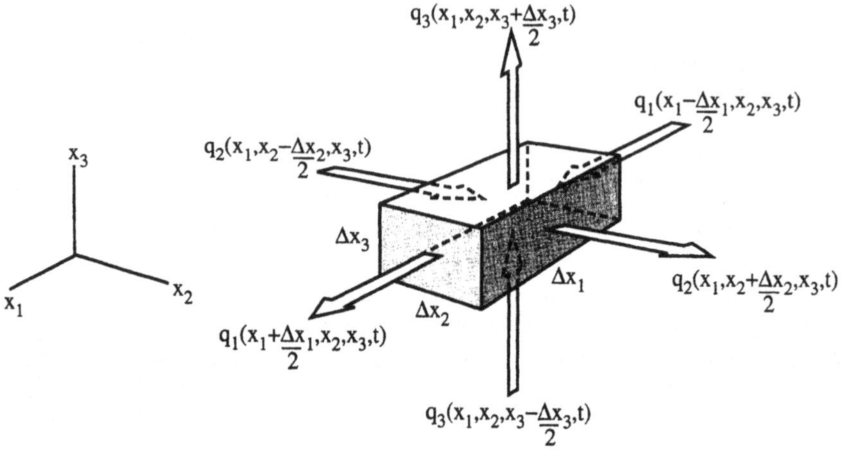

Figure 7.2. Control volume showing the heat per unit time and area passing through its surfaces.

Consider first a control volume that is a small rectangular parallelepiped with center at the point (x_1, x_2, x_3) and edges of length $(\Delta x_1, \Delta x_2, \Delta x_3)$ parallel to the coordinate axes (Fig. 7.2). The increase in energy stored within the parallelepiped is

$$\rho c \frac{\partial T}{\partial t}(x_1, x_2, x_3, t)\, \Delta x_1 \Delta x_2 \Delta x_3 \Delta t, \tag{7.1}$$

where ρ is the density of the medium and c is its *specific heat*, that is, the heat stored per unit of mass and temperature rise. The heat energy passing through a surface element $\mathbf{n}\,\Delta A$ between the times t and $t+\Delta t$ is $[\mathbf{q}(x_1, x_2, x_3, t)\cdot\mathbf{n}]\Delta A \Delta T$, where $\mathbf{q}$ is the *heat flux density* or heat transferred per unit of time and area (review the discussion of flux in Section 2.4). The total heat energy passing out of the parallelepiped across its surfaces is thus

$$\begin{aligned}
&\left[q_1\left(x_1 + \frac{\Delta x_1}{2}, x_2, x_3, t\right) - q_1\left(x_1 - \frac{\Delta x_1}{2}, x_2, x_3, t\right)\right]\Delta x_2 \Delta x_3 \Delta t \\
&+ \left[q_2\left(x_1, x_2 + \frac{\Delta x_2}{2}, x_3, t\right) - q_2\left(x_1, x_2 - \frac{\Delta x_2}{2}, x_3, t\right)\right]\Delta x_1 \Delta x_3 \Delta t \\
&+ \left[q_3\left(x_1, x_2, x_3 + \frac{\Delta x_3}{2}, t\right) - q_3\left(x_1, x_2, x_3 - \frac{\Delta x_3}{2}, t\right)\right]\Delta x_1 \Delta x_2 \Delta t.
\end{aligned} \tag{7.2}$$

Conservation of heat energy for the control volume requires that (7.1) plus (7.2) equal zero, if there are no sources of heat within the control volume. Dividing by $\Delta x_1 \Delta x_2 \Delta x_3 \Delta t$ and proceeding to the limit by shrinking the parallelepiped down to the point (x_1, x_2, x_3), we obtain

$$\rho c \frac{\partial T}{\partial t}(x_1, x_2, x_3, t) + \nabla \cdot \mathbf{q}(x_1, x_2, x_3, t) = 0. \tag{7.3a}$$

Alternatively, suppose the control volume is an arbitrary volume V bounded by a surface S within the conducting medium. The rate of increase of thermal energy stored by the material within V plus the rate of heat flowing out of V across S must equal zero:

$$\frac{d}{dt} \int\int\int_V \rho c T dV + \oiint_S \mathbf{q} \cdot d\mathbf{S} = 0. \tag{7.3b}$$

Application of the divergence theorem yields

$$\int\int\int_V \left[\rho c \frac{\partial T}{\partial t} + \nabla \cdot \mathbf{q}\right] dV = 0.$$

Since this integral is zero for any region V, the integrand must vanish everywhere within the conductor. Thus the differential form (7.3a) and the integral form (7.3b) of the heat energy conservation law are equivalent.

For a thermally linear isotropic medium, heat flows from hot to cold in a direction perpendicular to the isothermal lines or surfaces, in accordance with the constitutive relation known as *Fourier's law of heat conduction*:

$$\mathbf{q} = -k\nabla T. \tag{7.4}$$

Here k is an experimentally determined scalar called the *thermal conductivity*, which we assume to be constant. Then the heat flux $\mathbf{q}$ is a conservative vector field and its scalar potential is proportional to the temperature T. Substitution of (7.4) into (7.3a) results in the *heat equation*

$$\frac{\partial T}{\partial t} = \alpha \nabla^2 T, \tag{7.5}$$

where $\alpha = \dfrac{k}{\rho c}$ is the *thermal diffusivity*.

The relation (7.5) is the basic differential equation of heat conduction analysis. It may also be applied to diffusion processes such as the transport of smoke particles through air or of salt through water. In this context, T is the *concentration* of the introduced species and α is the *diffusivity*.

The solution of (7.5) for $T(x_1, x_2, x_3, t)$ depends upon conditions existing in the conductor at some initial time and on the boundaries. A common initial condition is to specify the value of T:

$$T = T_0 \quad \text{at} \quad t = 0. \tag{7.6}$$

A general boundary condition is to prescribe the heat transferred from the surface of a conductor to an adjacent medium such as a fluid. If T is the temperature at the surface of the conductor, and T_∞ is the temperature of the adjacent medium a relatively large distance away from the surface, the heat transferred per unit of time and surface area from the conductor to the adjacent medium is given by *Newton's law of cooling*:

$$q = h(T - T_\infty) \quad \text{on} \quad S. \tag{7.7}$$

The ***heat transfer coefficient h*** depends upon the surface geometry, properties of the adjacent medium, flow of adjacent fluid, and so on. Since the heat transferred to the surface by conduction equals the heat leaving the surface, we must have

$$-k\frac{\partial T}{\partial n} = h(T - T_\infty) \quad \text{on} \quad S. \tag{7.8}$$

where $\dfrac{\partial T}{\partial n} = \mathbf{n} \cdot \nabla T$ is the ***normal derivative*** or directional derivative in the direction of the unit normal $\mathbf{n}$ to S. The boundary condition (7.8) encompasses the common special cases of ***isothermal boundaries*** ($h \to \infty$), over which the temperature remains constant, and ***insulated boundaries*** ($h \to 0$), across which no heat conduction occurs.

We can determine the way the solution depends upon the parameters by ***nondimensionalizing*** the variables. A dimensionless temperature field, spatial coordinates, and time may be defined by

$$T^* = \frac{T - T_\infty}{T_0 - T_\infty}, \qquad x_i^* = \frac{x_i}{\ell}, \qquad t^* = \frac{\alpha t}{\ell^2}.$$

Here ℓ denotes a characteristic length of the conducting medium. The differential equation (7.5), initial condition (7.6), and boundary condition (7.8) then become

$$\frac{\partial T^*}{\partial t^*} = \nabla^{*2} T^*, \tag{7.9}$$

$$T^* = 1 \quad \text{at} \quad t^* = 0, \tag{7.10}$$

$$-\frac{\partial T^*}{\partial n^*} = \text{Bi}\, T^* \quad \text{on} \quad S^*, \tag{7.11}$$

where $\nabla^* = \ell\nabla$, S, and S^* are geometrically similar (same shape and ratio of corresponding body dimensions is ℓ), and Bi is the dimensionless ***Biot number***

$$\text{Bi} = \frac{h\ell}{k}.$$

It may be seen from (7.11) that the Biot number measures the relative magnitude of the temperature gradients within the conductor near its surface compared to the temperature difference between the surface and its surroundings. The solution $T^*(x_1^*, x_2^*, x_3^*, t^*, \text{Bi})$ to the equations (7.9) through (7.11) depends only upon the dimensionless variables and the Biot number. If we can obtain an analytical or numerical solution to a specific example, the solution in the nondimensionalized variables will be identical for all conductors with the same shape (but a possibly different size) and same Biot number.

In the following we obtain simple solutions to a couple of practical problems.

Steady-state conduction. If we restrict our study to steady-state conditions, meaning the temperature is independent of the time t, the heat equation (7.5) reduces to Laplace's equation

$$\nabla^2 T = 0.$$

In the example shown in Fig. 7.3, a plane wall separates two regions of differing temperatures T_1 and T_2. We assume that the heat flow in the wall has reached a steady state, so that the temperature field $T(x_1)$ depends only upon the spatial coordinate x_1, and the heat equation simplifies to

$$\frac{d^2T}{dx_1^2} = 0.$$

We also assume the Biot number is large, so the temperature at the surfaces of the wall are the same as the outside environment:

$$T(0) = T_1, \qquad T(\ell) = T_2.$$

Integrating and applying the boundary conditions, we find that the temperature field varies linearly with x_1:

$$T = T_1 - \left(\frac{T_1 - T_2}{\ell}\right) x_1. \tag{7.12}$$

The temperature drop is related to the heat q transferred across the wall by, from (7.4) and (7.12):

$$T_1 - T_2 = qR, \tag{7.13}$$

where $R = \dfrac{\ell}{k}$ is the *thermal resistance* of a wall of width ℓ. The insulation used in the walls of houses may have an R-value as large as 30 $\frac{\text{ft}^2 \cdot {}^\circ\text{F} \cdot \text{hr}}{\text{Btu}}$. From (7.13),

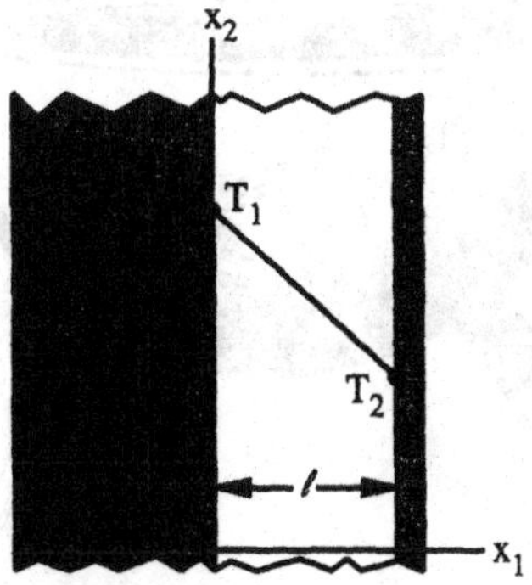

Figure 7.3. The temperature field in a plane wall varies linearly through the thickness.

R30 means that a square foot of wall would loose heat by conduction at a rate of $\frac{1}{30}\,\frac{\text{Btu}}{\text{hr}}$ for every 1°F difference in temperature between the two sides of the wall. (1 Btu = 252 cal = 1055 J is the heat necessary to raise one pound of water 1°F.)

Newtonian transient heating or cooling. In Fig. 7.4 a small hot metal casting at temperature T_0 is quenched by immersing it at time $t = 0$ in a liquid of temperature $T_\infty < T_0$. We assume the Biot number is small, so the temperature gradients within the metal are negligible and the temperature $T(t)$ of the metal depends only on the time t. Then Fourier's law (7.4) and the heat equation (7.5) are not applicable, but we can derive an equation for the temperature field by applying the energy balance (7.3b) to the entire solid of volume V and surface area S. Using (7.7) and assuming h, ρ, c are constants, we thereby obtain

$$\rho c V \frac{dT}{dt} = -hS(T - T_\infty).$$

Integrating and applying the initial condition (7.6), we find that the temperature of the solid decays exponentially with t:

$$T = T_\infty + (T_0 - T_\infty)e^{-hSt/\rho c V}. \tag{7.14}$$

Thus in a *decay time* $\tau = \dfrac{\rho c V}{hS}$ the temperature of the solid will decay to $\frac{1}{e}$ of its original value.

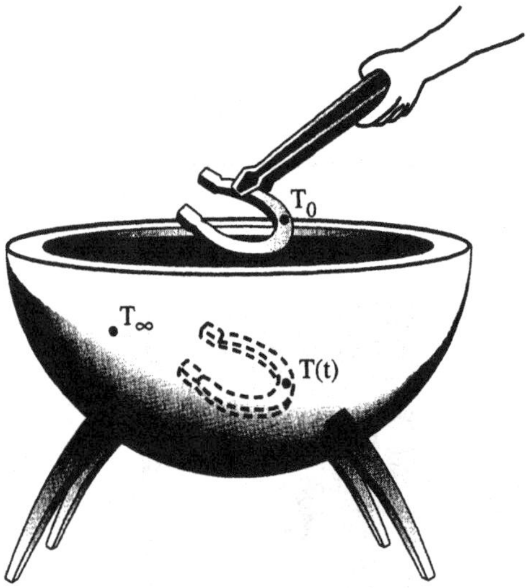

Figure 7.4. The temperature field in a quenched forging decays exponentially with time.

PROBLEMS 7.1

1. If heat ΔQ is added reversibly to a body at temperature T, the increase in entropy of the body is $\Delta S = \dfrac{\Delta Q}{T}$. Calculate the change in entropy when a substance of mass m having a constant specific heat c is heated from T_0 to T_1.

2. Find the path followed by a heat-seeking particle that is released at the point (1, 1, 1) into a temperature field $T = 100 - x_1^2 - 2x_2^2 - 3x_3^2$.

3. Show that the Biot number is the ratio of the thermal resistance of the conductor to the surface resistance.

4. Express the solutions (7.12) and (7.14) in terms of dimensionless variables.

5. Determine how the cooking time per unit weight of turkeys varies with their weight. Assume the turkeys have similar shapes, constant density, and the Biot number is large.

6. Steel balls 1 cm in diameter are annealed by heating to a high temperature and then slowly cooled in an air environment. Determine the decay time for the cooling process. Take

$$h = 20\,\frac{\mathrm{W}}{\mathrm{m}^2\cdot\mathrm{K}}, \quad k = 40\,\frac{\mathrm{W}}{\mathrm{m}\cdot\mathrm{K}}, \quad \rho = 8000\,\frac{\mathrm{kg}}{\mathrm{m}^3}, \quad c = 600\,\frac{\mathrm{J}}{\mathrm{kg}\cdot\mathrm{K}}.$$

7. Show that the thermal resistance of a plane wall made up of several layers is the sum of the resistances of each layer.

8. A plane wall is held at constant temperature T_1 on $x_1 = 0$ and exposed to a fluid on $x_1 = \ell$. Show that the temperature drop in the wall relative to the temperature difference between the surface $x_1 = \ell$ and the fluid is the Biot number.

9. Find an expression for the heat lost per unit of time and area through a glass window of thickness ℓ if the inside air temperature is $T_{1\infty}$ and the outside air temperature is $T_{2\infty}$.

10. Determine the temperature distribution $T(x_1, t)$ in a plane wall for the initial condition $T(x_1, 0) = T_0 \sin \dfrac{\pi x}{\ell}$ and boundary conditions $T(0, t) = T(\ell, t) = 0$.

11. Determine the temperature distribution $T(x_1, t)$ in a plane wall for the initial condition $T(x_1, 0) = T_0 \cos \dfrac{3\pi x}{\ell}$ and insulated boundaries.

12. If a conductor contains a *heat source* generating heat at the rate per unit volume $K(x_1, x_2, x_3, t)$, show that the temperature field obeys the equation

$$\frac{\partial T}{\partial t} = \alpha \nabla^2 T + \frac{K}{\rho c}.$$

Determine the temperature distribution around a point source radiating heat at the constant rate Q.

7.2 SOLID MECHANICS

The primary field quantity in solid mechanics is the *displacement* $\mathbf{u}(x_1, x_2, x_3, t)$ of a solid particle from its position (x_1, x_2, x_3) in the undeformed body. We assume that the solid is restrained at its boundaries and subjected to relatively small loads, so that the displacement field $\mathbf{u}(x_1, x_2, x_3, t)$ is very small.

A solid particle that was at the point $\mathbf{r}(0) = x_i \mathbf{e}_i$ in the undeformed body at the initial time $t = 0$ will be at the position $\mathbf{r}(t) = \mathbf{r}(0) + \mathbf{u}(x_1, x_2, x_3, t)$ at time t (Fig. 7.5). A neighboring particle that was at the point $\mathbf{r}(0) + \delta\mathbf{r} = (x_i + \delta x_i)\mathbf{e}_i$ will be at the position $\mathbf{r}(0) + \delta\mathbf{r} + \mathbf{u}(x_1 + \delta x_1, x_2 + \delta x_2, x_3 + \delta x_3, t)$ at time t. The vector $\delta\mathbf{r}$ joining the two particles will become the vector

$$\delta\mathbf{r} + \mathbf{u}(x_1 + \delta x_1, x_2 + \delta x_2, x_3 + \delta x_3, t) - \mathbf{u}(x_1, x_2, x_3, t)$$

$$\approx \delta\mathbf{r} + \frac{\partial \mathbf{u}}{\partial x_i}(x_1, x_2, x_3, t)\delta x_i = \mathbf{T} \cdot \delta\mathbf{r},$$

where

$$\mathbf{T} = \mathbf{1} + (\nabla\mathbf{u})^t.$$

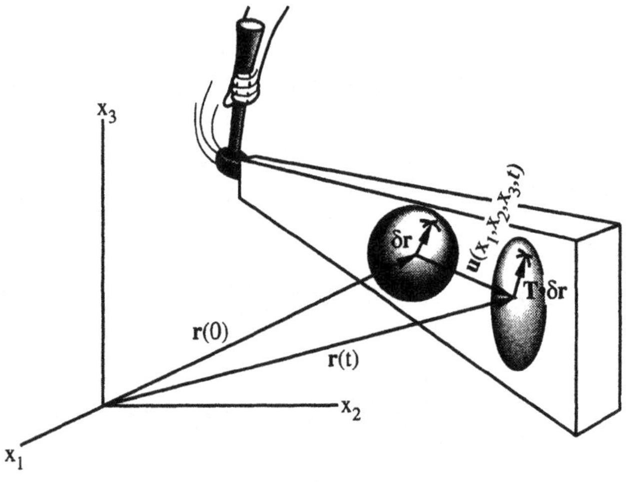

Figure 7.5. Plane longitudinal wave traveling down a rod.

Here we have retained only the lowest-order terms in the Taylor series expansion of $\mathbf{u}$ about the point (x_1, x_2, x_3).

Remembering the analysis in Chapter 3, we recognize $\mathbf{T}(x_1, x_2, x_3, t)$ as the tensor characterizing the kinematical properties of the deformation. The tensor $\nabla\mathbf{u}$ is often called the *deformation gradient tensor*. Since, neglecting products of the derivatives $\dfrac{\partial u_i}{\partial x_j}$,

$$\det \mathbf{T} = 1 + \nabla \cdot \mathbf{u},$$

we can interpret $\nabla \cdot \mathbf{u}$ as the relative change of volume or *dilatation* of the continuum.

Separating $\mathbf{T}$ into the sum of a symmetrical and an antisymmetrical tensor

$$\mathbf{T} = \mathbf{1} + \frac{1}{2}\left[(\nabla\mathbf{u})^t + \nabla\mathbf{u}\right] + \frac{1}{2}\left[(\nabla\mathbf{u})^t - \nabla\mathbf{u}\right]$$

we easily see that the polar decomposition (3.39) is obtained by letting

$$\mathbf{R} = \mathbf{1} + \frac{1}{2}\left[(\nabla\mathbf{u})^t - \nabla\mathbf{u}\right],$$

$$\mathbf{S} = \mathbf{1} + \frac{1}{2}\left[(\nabla\mathbf{u})^t + \nabla\mathbf{u}\right]. \tag{7.15}$$

Since

$$\mathbf{R} = \mathbf{1} + \frac{1}{2}(\nabla \times \mathbf{u}) \times$$

we can interpret $\frac{1}{2}\nabla \times \mathbf{u}$ as the vector representing the rigid body rotation part of the deformation (cf. 3.25). It follows from (3.40) and (7.15) that the strain within the continuum is given by

$$\mathbf{\Gamma} = \frac{1}{2}\left[\nabla\mathbf{u} + (\nabla\mathbf{u})^t\right] \tag{7.16a}$$

or

$$\gamma_{ij} = \frac{1}{2}\left(\frac{\partial u_i}{\partial x_j} + \frac{\partial u_j}{\partial x_i}\right). \tag{7.16b}$$

To recapitulate the kinematics, visualize a small sphere of solid particles in the undeformed solid (as in Fig. 7.5). The divergence $\nabla \cdot \mathbf{u}$ is the change in the volume occupied by the particles in the deformed solid divided by the volume of the original sphere. The curl $\nabla \times \mathbf{u}$ is twice the vector representing the amount of rotation of the sphere. The strain $\mathbf{\Gamma}$ is a measure of the amount of distortion of the rotated sphere into an ellipsoid.

Next let us apply Newton's law to the solid material within a small rectangular parallelepiped with center at the point (x_1, x_2, x_3) and edges of length $(\Delta x_1, \Delta x_2, \Delta x_3)$ parallel to inertial coordinate axes. In Fig. 7.6 we have shown only the x_1 components S_{1i} of the traction vectors acting on the center of each face of the element. The sum of all the forces acting on the element in the

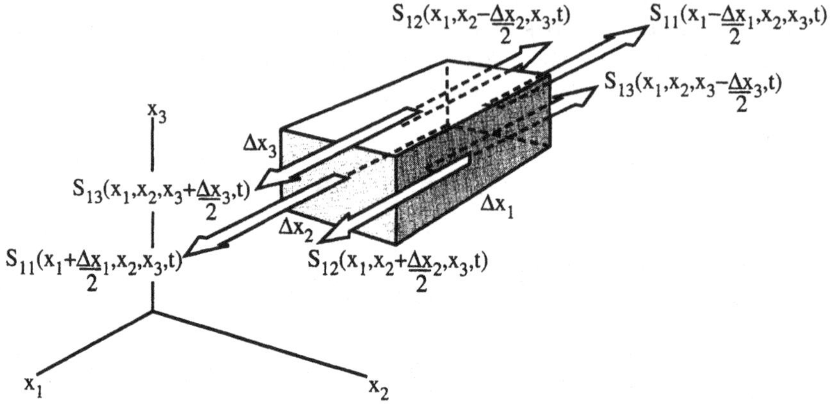

Figure 7.6. The x_1 components of the traction vectors that could act on the faces of a small rectangular parallelepiped within a continuum.

x_1 direction equals the mass $\rho\Delta x_1\Delta x_2\Delta x_3$ of the element times its acceleration $\frac{\partial^2 u_1}{\partial t^2}$ in the x_1 direction:

$$\begin{aligned}
&\left[S_{11}\left(x_1+\frac{\Delta x_1}{2},x_2,x_3,t\right)-S_{11}\left(x_1-\frac{\Delta x_1}{2},x_2,x_3,t\right)\right]\Delta x_2\Delta x_3\\
&+\left[S_{12}\left(x_1,x_2+\frac{\Delta x_2}{2},x_3,t\right)-S_{12}\left(x_1,x_2-\frac{\Delta x_2}{2},x_3,t\right)\right]\Delta x_1\Delta x_3\\
&+\left[S_{13}\left(x_1,x_2,x_3+\frac{\Delta x_3}{2},t\right)-S_{13}\left(x_1,x_2,x_3-\frac{\Delta x_3}{2},t\right)\right]\Delta x_1\Delta x_2\\
&=\rho\frac{\partial^2 u_1}{\partial t^2}(x_1,x_2,x_3,t)\Delta x_1\Delta x_2\Delta x_3.
\end{aligned}$$

Dividing by $\Delta x_1\Delta x_2\Delta x_3$, and proceeding to the limit by shrinking the element down to the point (x_1,x_2,x_3), we obtain

$$\frac{\partial S_{1j}}{\partial x_j}(x_1,x_2,x_3,t)=\rho\frac{\partial^2 u_1}{\partial t^2}(x_1,x_2,x_3,t).$$

By applying Newton's law in the x_2 and x_3 directions, we can obtain two more similar equations. By requiring the resultant moment about (x_1,x_2,x_3) of all the forces acting on the element to vanish, we can show that the stress tensor is symmetric: $S_{ij}=S_{ji}$. The final equations of equilibrium can thus be written as

$$\nabla\cdot\mathbf{S}=\rho\frac{\partial^2\mathbf{u}}{\partial t^2} \tag{7.17a}$$

or

$$\frac{\partial S_{ij}}{\partial x_j} = \rho \frac{\partial^2 u_i}{\partial t^2}. \tag{7.17b}$$

The theory is made complete by specifying a stress-strain law. We suppose the material is linear isotropic and the temperature effects are negligible, so the constitutive relations are, from (4.22),

$$\mathbf{S} = 2\mu\mathbf{\Gamma} + \lambda(\text{tr}\mathbf{\Gamma})\mathbf{1} \tag{7.18a}$$

or

$$\mathbf{\Gamma} = \frac{(1+\nu)}{Y}\mathbf{S} - \frac{\nu}{Y}(\text{tr }\mathbf{S})\mathbf{1}. \tag{7.18b}$$

We also suppose the material is *homogeneous*, meaning the elastic constants λ, μ, Y, ν and the density ρ are the same at every point within the material.

The linear field equations for an isotropic elastic solid consist of the strain-displacement relations (7.16), the equilibrium equations (7.17), and the stress-strain relations (7.18). Substituting (7.16) and (7.18) into (7.17), and using vector identities, we can obtain the so-called *Navier equations* for the displacement field $\mathbf{u}(x_1, x_2, x_3, t)$:

$$\rho \frac{\partial^2 \mathbf{u}}{\partial t^2} = (\lambda + \mu)\nabla\nabla \cdot \mathbf{u} + \mu\nabla^2\mathbf{u} \tag{7.19a}$$

or

$$\rho \frac{\partial^2 u_i}{\partial t^2} = (\lambda + \mu)\frac{\partial^2 u_j}{\partial x_i \partial x_j} + \mu \frac{\partial^2 u_i}{\partial x_j \partial x_j}. \tag{7.19b}$$

Appropriate boundary conditions are to specify the value of the displacement field $\mathbf{u}$ or the traction field $\mathbf{t}$ on the bounding surface of the body.

A useful consequence of these equations is the *energy-work theorem* for an arbitrary volume V enclosed by a surface S within the solid:

$$\frac{d}{dt}\int\!\!\int\!\!\int_V \left(\frac{\rho}{2}\frac{\partial u_i}{\partial t}\frac{\partial u_i}{\partial t} + \frac{1}{2}S_{ij}\gamma_{ij}\right) dV = \oint\!\!\!\oint_S S_{ij}\frac{\partial u_i}{\partial t} n_j \, dS. \tag{7.20}$$

The positive definite quantity $\frac{1}{2}S_{ij}\gamma_{ij}$ is called the *strain energy density* and represents the elastic energy stored reversibly within the body. Equation (7.20) says that the rate of change of kinetic energy plus strain energy of the material within V equals the rate at which work is being done by the tractions on S.

We can obtain exact solutions to these equations by guessing a solution based on physical intuition, then substituting it into the equations to verify that they are all satisfied.

Torsion of a circular cylindrical rod. Let us reconsider the problem discussed in Section 4.2 of the twisting of a rod with a circular cross section of radius a (Fig. 7.7a and b). Establish a Cartesian coordinate system (x_1, x_2, x_3) with origin at the center of the rod and x_1 axis along the centerline of the rod. From symmetry considerations, we guess that a circular cross section remains circular when the

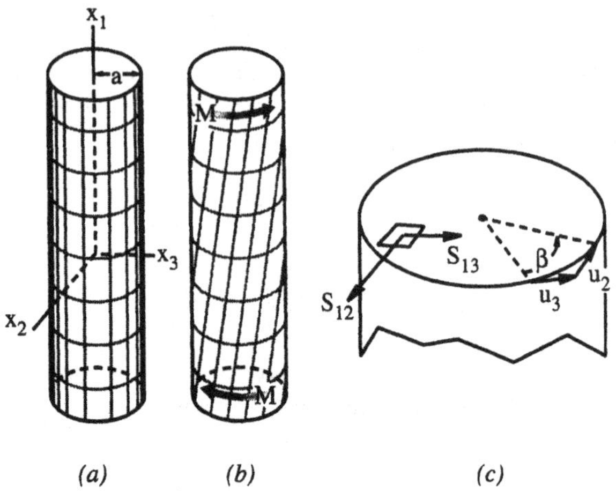

Figure 7.7. Torsion of a circular cylindrical rod.

rod is twisted. A solid particle at position (x_1, x_2, x_3) would then simply rotate to the position $(x_1, x_2 - \beta x_3,\ x_3 + \beta x_2)$, where the angle of rotation $\beta(x_1)$ varies with the axial distance x_1. The displacement of a particle is then (Fig. 7.7c)

$$u_1 = 0, \qquad u_2 = -\beta(x_1)x_3, \qquad u_3 = \beta(x_1)x_2.$$

Substituting this displacement field into the Navier equations (7.19b), we find that they reduce to $\dfrac{d^2\beta}{dx_1^2} = 0$. Hence the rotation per unit length is $\dfrac{d\beta}{dx_1} =$ constant. The nonvanishing components of the stress tensor $\mathbf{S}$, obtained from the strain-displacement relations (7.16b) and constitutive law (7.18a), are

$$S_{12} = -\mu\frac{d\beta}{dx_1}x_3, \qquad S_{13} = \mu\frac{d\beta}{dx_1}x_2. \tag{7.21}$$

The stress components acting on a small area of the surface of a circular cross section are sketched in Fig. 7.7c. Since

$$\int\!\!\int_A S_{12}dx_2dx_3 = \int\!\!\int_A S_{13}dx_2dx_3 = 0,$$

where the integrals are taken over the entire area A of a cross section, the net force acting on any cross section is zero. The resultant torque about the x_1 axis must be equal to the twisting moment M applied to each end of the rod:

$$\int\!\!\int_A (x_2S_{13} - x_3S_{12})\,dx_2dx_3 = \mu\frac{d\beta}{dx_1}J = M. \tag{7.22}$$

Here

$$J = \int\!\!\int (x_2^2 + x_3^2)\,dx_2dx_3 = \frac{\pi a^4}{2}$$

is the polar moment of inertia of the cross section. This determines $\frac{d\beta}{dx_1} = \frac{M}{\mu J}$.
To recapitulate, the displacement field

$$u_1 = 0, \qquad u_2 = \frac{-Mx_1x_3}{\mu J}, \qquad u_3 = \frac{Mx_1x_2}{\mu J}$$

is the exact solution to the linear elasticity equations (7.19) which produces no stresses on a circular cylindrical surface and no net force but a resultant twisting moment M on circular cross sections. When the ends of a real shaft are twisted in some mechanical manner (e.g., by wrenches), the shear stress distribution near the ends will not usually obey the equations (7.21). However, the local variations in the stress field usually die out quickly away from the points of application of the external load, so our solution should be accurate away from the ends of the shaft. This is known as *Saint-Venant's principle* in the engineering literature. Also, it is possible to obtain the exact solution to the problem of the torsion of a cylindrical rod of *arbitrary* cross-section. In this general case, the twisting moment M is related to the rotation per unit axial length $\frac{d\beta}{dx_1}$ by the same equation (7.22), where the *torsion constant* J for each cross-sectional shape must be obtained by solving an auxiliary boundary value problem.

Bending of a prismatic beam. Consider the problem of the pure bending of a prismatic beam (Fig. 7.8a and b). Establish a Cartesian coordinate system (x_1, x_2, x_3) with the x_1-axis along the length of the beam and the x_2 and x_3 axes in the plane of a cross section. We guess that the only nonvanishing stress component is (Fig. 7.8c)

$$S_{11} = -Cx_2,$$

where C is a constant. We require the net axial force acting on any cross section to vanish:

$$\int\int_A S_{11}dx_2dx_3 = -C\int\int_A x_2dx_2dx_3 = 0.$$

This equation can be satisfied by taking the x_1-axis through the *centroid* of the cross section. We also require the resultant torque about the x_2-axis to vanish:

$$\int\int_A x_3S_{11}dx_2dx_3 = -C\int\int_A x_2x_3dx_2dx_3 = 0.$$

This equation implies that the product of inertia of a cross-section vanishes (see Problem 3 in Section 4.2), so the x_2 and x_3 axes must be the *principal axes of inertia* of a cross section. We set the resultant torque about the x_3-axis equal to the bending moment M applied to each end of the beam:

$$-\int\int_A x_2S_{11}dx_2dx_3 = CI = M.$$

Here $I = I_{33} = \int\int_A x_2^2dx_2dx_3$ is the x_3 component of the moment of inertia of the cross section. This determines the constant $C = \frac{M}{I}$. The nonvanishing

(a)

(b)

(c)

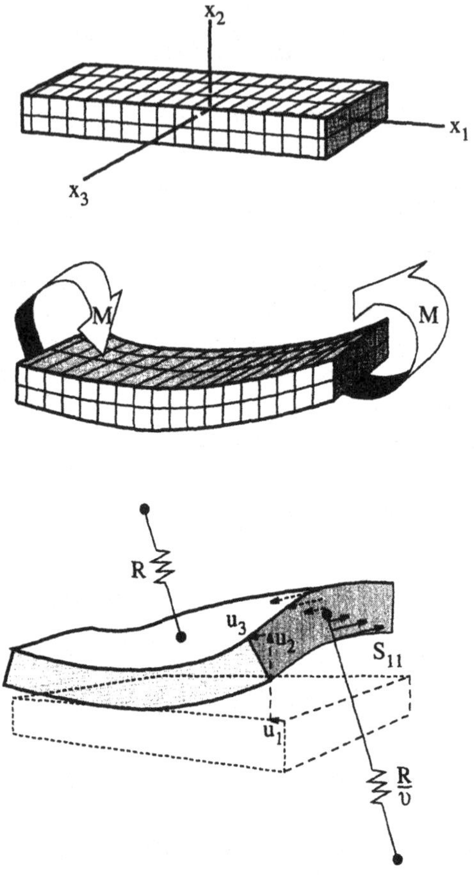

Figure 7.8. Bending of a prismatic beam.

components of the strain tensor $\mathbf{\Gamma}$, obtained from the constitutive law (7.18b), a

$$\gamma_{11} = -\frac{Mx_2}{YI}, \qquad \gamma_{22} = \gamma_{33} = \frac{\nu M x_2}{YI}.$$

The plane $x_2 = 0$ is called the ***neutral plane*** of the beam, since material elemen within it undergo no strain. Integrating the equations (7.16b),

$$\frac{\partial u_1}{\partial x_1} = \frac{-Mx_2}{YI}, \qquad \frac{\partial u_2}{\partial x_2} = \frac{\nu M x_2}{YI}, \qquad \frac{\partial u_3}{\partial x_3} = \frac{\nu M x_2}{YI}, \tag{7.2}$$

$$\frac{\partial u_1}{\partial x_2} + \frac{\partial u_2}{\partial x_1} = 0, \qquad \frac{\partial u_1}{\partial x_3} + \frac{\partial u_3}{\partial x_1} = 0, \qquad \frac{\partial u_2}{\partial x_3} + \frac{\partial u_3}{\partial x_2} = 0, \tag{7.2}$$

we obtain the displacement components[1]

$$u_1 = \frac{-Mx_1x_2}{YI}, \qquad u_2 = \frac{M(x_1^2 + \nu x_2^2 - \nu x_3^2)}{2YI},$$
$$u_3 = \frac{\nu M x_2 x_3}{YI}. \tag{7.25}$$

Hence the centroidal axis bends into an arc (Fig. 7.8c) of small curvature

$$\frac{1}{R} = \frac{\partial^2 u_2}{\partial x_1^2} = \frac{M}{YI}. \tag{7.26}$$

A rectangular cross section of the beam will, after the deformation, lie in a plane normal to the deformed centroidal axis and have the shape shown in Fig. 7.8c. Thus the neutral plane bends into an anticlastic or saddle-shaped surface. To recapitulate, the displacement field (7.25) is the exact solution to the linear elasticity equations (7.19) which produces no stresses on a prismatic surface and no net force but a resultant bending moment M on cross sections. *Euler-Bernoulli* beam theory is a good approximation for the bending of *slender* beams under other loadings, such as that shown in Fig. 1.14. In this simple theory, the normal deflection $u_2(x_1, 0, 0)$ of the centroidal axis is assumed to be governed by the same equation (7.26), where the bending moment $M(x_1)$ is obtained by considering equilibrium of a beam segment.

Waves in solids. According to the fundamental theorem (6.53), we can decompose the displacement field $\mathbf{u}(x_1, x_2, x_3, t)$ into the sum of an irrotational vector field $\mathbf{u}_\ell(x_1, x_2, x_3, t)$ and a solenoidal vector field $\mathbf{u}_t(x_1, x_2, x_3, t)$:

$$\mathbf{u} = \mathbf{u}_\ell + \mathbf{u}_t,$$

where

$$\nabla \times \mathbf{u}_\ell = 0, \qquad \nabla \cdot \mathbf{u}_t = 0. \tag{7.27}$$

Substituting this representation into (7.19a), and using the vector identity (6.15), we obtain

$$\rho \frac{\partial^2 \mathbf{u}_\ell}{\partial t^2} + \rho \frac{\partial^2 \mathbf{u}_t}{\partial t^2} = (\lambda + 2\mu)\nabla^2 \mathbf{u}_\ell + \mu \nabla^2 \mathbf{u}_t.$$

---

[1] Integrate (7.23) to obtain

$$u_1 = \frac{-Mx_1x_2}{YI} + f(x_2, x_3), \qquad u_2 = \frac{\nu M x_2^2}{2YI} + g(x_1, x_3), \qquad u_3 = \frac{\nu M x_2 x_3}{YI} + h(x_1, x_2),$$

where f, g, and h are arbitrary functions of the indicated arguments. Then substitute in (7.24) to obtain

$$-\frac{Mx_1}{YI} + \frac{\partial f}{\partial x_2} + \frac{\partial g}{\partial x_1} = 0, \qquad \frac{\partial f}{\partial x_3} + \frac{\partial h}{\partial x_1} = 0, \qquad \frac{\nu M x_3}{YI} + \frac{\partial g}{\partial x_3} + \frac{\partial h}{\partial x_2} = 0.$$

The simplest solution of these equations is $f = h = 0$, $g = \frac{M}{2YI}(x_1^2 - \nu x_3^2)$.

This equation will be satisfied if

$$\frac{\partial^2 \mathbf{u}_\ell}{\partial t^2} = c_\ell^2 \nabla^2 \mathbf{u}_\ell, \qquad c_\ell = \sqrt{\frac{\lambda + 2\mu}{\rho}}, \tag{7.28}$$

$$\frac{\partial^2 \mathbf{u}_t}{\partial t^2} = c_t^2 \nabla^2 \mathbf{u}_t, \qquad c_t = \sqrt{\frac{\mu}{\rho}}. \tag{7.29}$$

These are *wave equations*; c_ℓ is the *longitudinal speed of sound* in solids, and c_t is the *transverse speed of sound* in solids. Any twice differentiable function of the argument $(x_1 - c_\ell t)$ or $(x_1 - c_t t)$ is a solution of (7.28) or (7.29), respectively:

$$\mathbf{u}_\ell = \mathbf{u}_\ell(x_1 - c_\ell t), \qquad \mathbf{u}_t = \mathbf{u}_t(x_1 - c_t t). \tag{7.30}$$

These solutions are *plane waves* propagating in the x_1-direction. The displacement $\mathbf{u}_\ell$ of all the points on a plane $(x_1 - c_\ell t)$ = constant is the same, and the displacement $\mathbf{u}_t$ of all the points on a plane $(x_1 - c_t t)$ = constant is the same. Although the individual solid particles do not displace far from their initial positions, the entire deformation pattern travels in the x_1 direction with a speed c_ℓ or c_t. Substituting (7.30) into (7.27), we find that the components of $\mathbf{u}_\ell$ in the x_2 and x_3 directions vanish and the component of $\mathbf{u}_t$ in the x_1 direction vanishes (deleting rigid body displacements). Hence the displacement $\mathbf{u}_\ell$ is parallel to the x_1-axis, and the displacement $\mathbf{u}_t$ is perpendicular to the x_1-axis. The $\mathbf{u}_\ell$ waves are referred to as *longitudinal, irrotational, dilational, pressure*, or *P-waves*, while the $\mathbf{u}_t$ waves are referred to as *transverse, rotational, equivoluminal, shear*, or *S-waves*. If you suddenly strike the end of a rod, you will start a plane longitudinal wave moving down the rod (Fig. 7.5), whereas if you suddenly twist the end, you will start a plane transverse wave. A longitudinal wave is accompanied by changes in the density of the medium, whereas a transverse wave causes no changes in the density. To recapitulate, the plane longitudinal and plane transverse waves

$$\mathbf{u}_\ell = f_1(x_1 - c_\ell t)\mathbf{e}_1, \qquad \mathbf{u}_t = f_2(x_1 - c_t t)\mathbf{e}_2 + f_3(x_1 - c_t t)\mathbf{e}_3,$$

where f_1, f_2, f_3 are arbitrary functions of the indicated arguments, are exact solutions to the dynamic elasticity equations (7.19) in an unbounded solid. In a finite solid, a plane wave will generally be reflected and refracted at a boundary between different media. Geologists place seismographs at various locations to record the amplitudes and arrival times of the P, S, and other waves caused by a distant earthquake. Comparing the measurements with data from previous earthquakes of known location and magnitude, and using mathematical models of the earth, they can compute the magnitude and location of the earthquake.

PROBLEMS 7.2

1. Verify the energy-work theorem (7.20).

2. Assuming chalk is a linear isotropic material, describe the deformation of the small cube in Fig. 4.5(a). Show that the deformed shape may be obtained from the cube by a rigid body rotation plus a symmetrical deformation.

3. An exact solution for the torsion of a cylindrical bar of arbitrary cross section can be obtained by assuming a displacement field

$$u_1 = \frac{M}{\mu J}\phi(x_2, x_3), \qquad u_2 = -\frac{M x_1 x_3}{\mu J}, \qquad u_3 = \frac{M x_1 x_2}{\mu J}.$$

   Determine the equation satisfied by the *warping function* $\phi(x_2, x_3)$ and express the torsion constant J in terms of ϕ.

4. Find the location and magnitude of the maximum and minimum stress components in a bent beam with a rectangular cross section.

5. Verify that a cross section of the beam has the deformed shape sketched in Fig. 7.8c.

6. A linear isotropic prismatic bar is pulled on each of its ends with a uniform axial tension f. Calculate the displacement field in the bar.

7. A circular cylindrical rod is subjected to a twisting moment, bending moment, and tension on each of its ends. Determine the displacement field of the solid.

8. The most general solution of the one-dimensional wave equation

$$\frac{\partial^2 u}{\partial t^2} = c^2 \frac{\partial^2 u}{\partial x_1^2}$$

   is $u(x_1, t) = f(x_1 - ct) + g(x_1 + ct)$, where f and g are twice differentiable functions of the indicated arguments. Show that this is the solution by directly substituting into the differential equation.

9. Determine the nonvanishing stress components in an unbounded elastic medium with either a plane longitudinal wave or a plane transverse wave propagating in the x_1-direction. Show that

$$c_\ell = \sqrt{\frac{3(1-\nu)B}{(1+\nu)\rho}},$$

   where B is the bulk modulus of the solid. (See Problem 7 in Section 4.4.)

10. Obtain the appropriate equation governing the motion of torsional waves in a circular cylindrical shaft.

11. Find the stress in a helicopter rotor blade of length ℓ rotating about a vertical axis with constant angular velocity ω. Account only for the effect of the centrifugal force and assume the density ρ of the blade is constant.

12. Explain why trees, telephone poles, airplane wings, and so on increase in cross-sectional area from their free ends to their rooted ends.

13. Use Euler-Bernoulli beam theory to obtain the deflection $u_2\ (x_1,\ 0,\ 0)$ of a cantilevered beam, whose end $x_1 = 0$ satisfies the boundary conditions $u_2(0,0,0) = \frac{\partial u_2}{\partial x_1}(0,0,0) = 0$, and whose end $x_1 = \ell$ has a weight W hanging from it. (The solution could model the deflection of the crane shown in Fig. 1.14, with ρ neglected.)

7.3 FLUID MECHANICS

The primary field quantity in fluid mechanics is the *velocity* $\mathbf{v}\ (x_1,\ x_2,\ x_3,\ t)$ of the fluid particle passing through the fixed point (x_1, x_2, x_3) at the time t. If $\mathbf{r}(t) = x_i(t)\mathbf{e}_i$ denotes the position vector of this particle, then

$$\frac{d\mathbf{r}}{dt} = \mathbf{v} \tag{7.31a}$$

or

$$\frac{dx_i}{dt} = v_i. \tag{7.31b}$$

Knowing the velocity field $\mathbf{v}(x_1, x_2, x_3, t)$, we integrate (7.31) to obtain the trajectories of the fluid particles or *pathlines*. If the flow is *steady*, meaning $\mathbf{v}(x_1, x_2, x_3)$ does not explicitly contain the time t, the pathlines are identical to the field lines of $\mathbf{v}$ (obtained from (6.20)).

A fluid particle that is at the position $\mathbf{r} = x_i\mathbf{e}_i$ at time t will be at the position $\mathbf{r} + \mathbf{v}(x_1, x_2, x_3, t)\Delta t$ after a small time Δt (Fig. 7.9). By the same considerations as in the preceding section, we find that a vector $\delta\mathbf{r}$ joining two adjacent particles will become the vector $\mathbf{T}\cdot\delta\mathbf{r}$, where

$$\mathbf{T} = \mathbf{1} + (\nabla\mathbf{v})^t\Delta t.$$

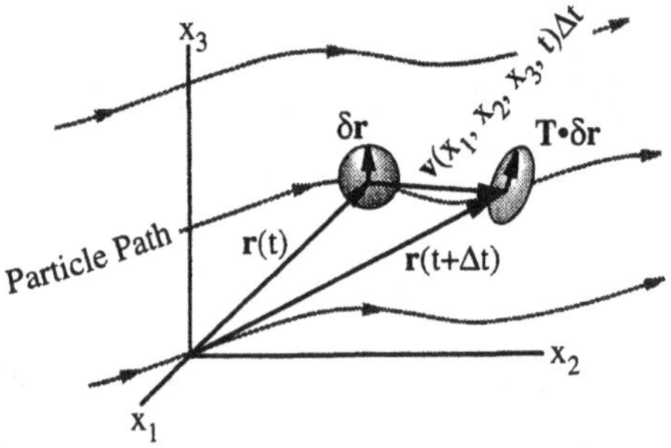

Figure 7.9. Kinematics of fluid flow.

It follows that $\nabla \cdot \mathbf{v}$ is the rate of volume expansion per unit volume and $\frac{1}{2}\nabla \times \mathbf{v}$ is the local angular velocity or *vorticity* of the fluid. It also follows directly that the rate of strain within the fluid is

$$\frac{d\Gamma}{dt} = \frac{1}{2}\left[\nabla\mathbf{v} + (\nabla\mathbf{v})^t\right] \tag{7.32a}$$

or

$$\frac{d\gamma_{ij}}{dt} = \frac{1}{2}\left(\frac{\partial v_i}{\partial x_j} + \frac{\partial v_j}{\partial x_i}\right). \tag{7.32b}$$

Next let us apply Newton's law to the fluid that at time t occupies the small rectangular parallelepiped shown in Fig. 7.6. In addition to the stress $\mathbf{S}$ that acts on the surface of the element, we also now include the gravitational force per unit volume $\rho\mathbf{g}$ that acts upon the entire fluid within the element, where $\rho(x_1, x_2, x_3, t)$ is the density of the fluid at the fixed point (x_1, x_2, x_3) and time t. Proceeding as in the preceding section, we obtain the equilibrium equations

$$\nabla \cdot \mathbf{S} + \rho\mathbf{g} = \rho\frac{d\mathbf{v}}{dt}. \tag{7.33}$$

Here $\frac{d\mathbf{v}}{dt}$ is the acceleration of a fluid particle, which from the chain rule, (6.10), and (7.31) is given by

$$\begin{aligned}\frac{d\mathbf{v}}{dt} &= \frac{\partial\mathbf{v}}{\partial t} + \frac{\partial\mathbf{v}}{\partial x_i}\frac{dx_i}{dt} = \frac{\partial\mathbf{v}}{\partial t} + \mathbf{v}\cdot\nabla\mathbf{v} \\ &= \frac{\partial\mathbf{v}}{\partial t} + \nabla\left(\frac{v^2}{2}\right) + (\nabla\times\mathbf{v})\times\mathbf{v}.\end{aligned} \tag{7.34}$$

The term $\frac{\partial\mathbf{v}}{\partial t}$ is called the *local acceleration* and arises when the flow is unsteady, while the term $\mathbf{v}\cdot\nabla\mathbf{v}$ is called the *convective acceleration* and arises when the particle moves through regions of differing speeds.

We can derive the *continuity equation* of fluid mechanics by enforcing the conservation of mass for an arbitrary control volume V bounded by a surface S within the fluid and fixed in space. The rate of increase of mass within V plus the rate of mass flow out of V across S must equal zero, if there are no sources of fluid within V:

$$\frac{d}{dt}\iiint_V \rho\, dV + \oint\!\!\oint_S \rho\mathbf{v}\cdot d\mathbf{S} = 0. \tag{7.35a}$$

Application of the divergence theorem yields the differential form of the continuity equation

$$\frac{\partial\rho}{\partial t} + \nabla\cdot(\rho\mathbf{v}) = 0. \tag{7.35b}$$

It is usually assumed that the density ρ, pressure p, and temperature T are related by an *equation of state* $\rho = \rho(p, T)$. In many fluid problems the density changes are insignificant so the equation of state may be approximated by

$$\rho = \text{ constant}.$$

Limiting our study to these *incompressible* flows, we conclude from (7.35b) that the rate of dilatation is zero,

$$\nabla \cdot \mathbf{v} = 0, \tag{7.36}$$

and therefore the fluid volume does not change.

The theory is made complete by specifying a constitutive law. We suppose that the stress $\mathbf{S}$ is linearly related to $\frac{d\mathbf{\Gamma}}{dt}$ and given by (2.10) when the fluid is at rest. We also suppose the fluid is isotropic, so by analogy with (4.22)

$$\mathbf{S} = 2\mu \frac{d\mathbf{\Gamma}}{dt} + \lambda \operatorname{tr}\left(\frac{d\mathbf{\Gamma}}{dt}\right)\mathbf{1} - p\mathbf{1}. \tag{7.37}$$

For an incompressible flow the λ term drops out, since $\operatorname{tr}\left(\frac{d\mathbf{\Gamma}}{dt}\right) = \nabla \cdot \mathbf{v} = 0$:

$$\mathbf{S} = 2\mu \frac{d\mathbf{\Gamma}}{dt} - p\mathbf{1}. \tag{7.38}$$

Here the scalar μ is the *viscosity coefficient*, which we suppose to be a constant. A fluid such as water or air that obeys the constitutive law (7.37) or (7.38) is sometimes called *Newtonian*.

Substituting (7.32), (7.36), and (7.38) into (7.33), we obtain the *Navier-Stokes equations* for an incompressible homogeneous fluid:

$$\rho \frac{d\mathbf{v}}{dt} = \rho \mathbf{g} - \nabla p + \mu \nabla^2 \mathbf{v} \tag{7.39a}$$

or

$$\rho \frac{\partial v_i}{\partial t} + \rho \frac{\partial v_i}{\partial x_j} v_j = \rho g_i - \frac{\partial p}{\partial x_i} + \mu \frac{\partial^2 v_i}{\partial x_j \partial x_j}. \tag{7.39b}$$

The equations (7.36) and (7.39) comprise four scalar equations for the velocity field $\mathbf{v}(x_1, x_2, x_3, t)$ and pressure field $p(x_1, x_2, x_3, t)$. Typical boundary conditions are to require the fluid to have the same velocity as the solid at a fluid-solid interface, and to prescribe the volume flux and the pressure at an inlet or outlet section of the flow.

A useful consequence of these equations is the *energy-work theorem* for an arbitrary volume V enclosed by a surface S within the fluid:

$$\frac{d}{dt}\iiint_V \left(\frac{\rho v^2}{2} + \rho\Phi\right) dV + 2\mu \iiint_V \frac{d\gamma_{ij}}{dt}\frac{d\gamma_{ij}}{dt} dV = \oiint_S S_{ij} v_j n_i dS. \tag{7.40}$$

Since a gravitational field is conservative (as shown in the next section), we have here set $\mathbf{g} = -\nabla\Phi$. The positive definite quantity $2\mu \frac{d\gamma_{ij}}{dt}\frac{d\gamma_{ij}}{dt}$ is called the *rate of dissipation of mechanical energy density* and represents the irreversible transfer of mechanical energy to thermal energy due to friction within the fluid. Equation (7.40) says that the rate of change of kinetic energy and gravitational potential energy plus the rate of dissipation of mechanical energy of the fluid within V equals the rate at which work is being done by the tractions on S.

We can define a dimensionless velocity field, spatial coordinates, time, gravitational acceleration, and pressure by

$$\mathbf{v}^* = \frac{\mathbf{v}}{v}, \quad x_i^* = \frac{x_i}{\ell}, \quad t^* = \frac{tv}{\ell}, \quad \mathbf{g}^* = \frac{\mathbf{g}\ell}{v^2}, \quad p^* = \frac{p}{\rho v^2}.$$

Here v denotes a speed and ℓ a length characteristic of the particular fluid problem. The differential equations (7.36) and (7.39) then become

$$\nabla^* \cdot \mathbf{v}^* = 0, \tag{7.41}$$

$$\frac{d\mathbf{v}^*}{dt^*} = \mathbf{g}^* - \nabla^* p^* + \frac{1}{\mathrm{Re}} \nabla^{*2} \mathbf{v}^*, \tag{7.42}$$

where Re is the dimensionless *Reynolds number*

$$\mathrm{Re} = \frac{\rho v \ell}{\mu} = \frac{v\ell}{\nu},$$

and the ratio $\nu = \frac{\mu}{\rho}$ is called the *kinematic viscosity*. It can be seen from (7.42) that the Reynolds number measures the relative magnitude of the inertial forces compared to the viscous forces acting on a fluid particle. The solutions $\mathbf{v}^*(x_1^*, x_2^*, x_3^*, t^*, \mathrm{Re})$ and $p^*(x_1^*, x_2^*, x_3^*, t^*, \mathrm{Re})$ to the equations (7.41) and (7.42) with nondimensionalized boundary conditions depend only upon the dimensionless variables and the Reynolds number. If we actually measure the flow around a scale model of an airplane in a low-speed wind tunnel, or around a submerged submarine in a towing tank, this theory predicts that the velocity field in the nondimensionalized variables will be identical to that around the prototype at the same Reynolds number. To model accurately the flow around jets flying above the speed of sound, or around ships moving on the ocean surface, additional terms must be added to the equations herein; these lead to additional dimensionless parameters.

In the following we study some special problems for which analytical solutions to these equations can be obtained.

Hydrostatics. If a fluid is motionless, the Navier-Stokes equations (7.39) reduce simply to

$$\nabla p = \rho \mathbf{g} = -\rho \nabla \Phi. \tag{7.43}$$

It follows that, in a fluid at rest, the *isobaric surfaces* $p(x_1, x_2, x_3) =$ constant coincide with the *gravitational equipotential surfaces* $\Phi(x_1, x_2, x_3) =$ constant. If the density ρ is constant, we can solve for the pressure:

$$p = p_0 - \rho \Phi. \tag{7.44}$$

If the gravitational field $\mathbf{g} = -g\mathbf{e}_1$ is also constant, then

$$p = p_0 - \rho g x_1. \tag{7.45}$$

In accordance with this law, for every 10 meters of depth the pressure p in water increases by roughly one atmosphere. The linear pressure law is not as good an approximation in the atmosphere because the air density ρ decreases significantly with height. Note that the total force $\mathbf{f}$ of a fluid on the surface S of a submerged body is

$$\mathbf{f} = \oiint_S \mathbf{S} \cdot \mathbf{n} dS = -\oiint_S p\mathbf{n} dS.$$

Using the linear pressure law (7.45) and the divergence theorem, we find that

$$\mathbf{f} = \rho g V \mathbf{e}_1,$$

where V is the volume of the body. Thus the effect of gravity is to exert a vertical *buoyant force* equal to the weight of the fluid displaced by the body — this is *Archimedes' principle.* The centroid of the displaced volume is called the *center of buoyancy.* For a ship the buoyancy force is all important, while for an airplane the buoyancy force is negligible, since the density of water is roughly 10^3 times the density of air. For a floating ship the buoyancy force $\rho g V$ equals the ship weight mg and the center of buoyancy lies on the same vertical line as the center of mass, because for static equilibrium there can be no net forces or moments acting on the ship. When a symmetrical floating ship is tilted through a small angle, a vertical line drawn through the new center of buoyancy will intersect the line of symmetry at a point called the *metacenter* (Fig. 7.10). If the metacenter is above the center of mass, a restoring couple is present and the upright position is

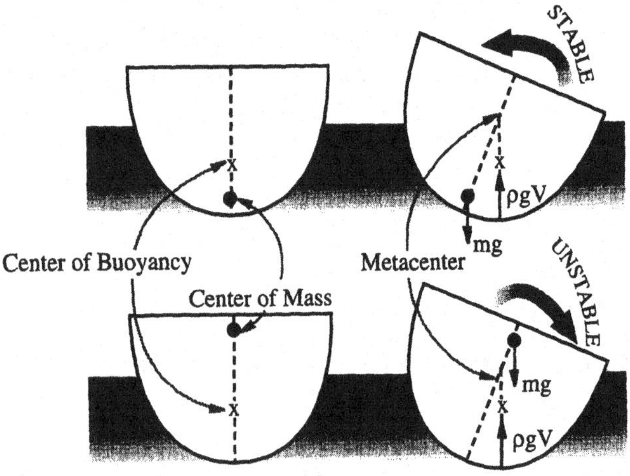

Figure 7.10. The ship is stable if the metacenter is above the center of mass but unstable if the metacenter is below the center of mass.

stable, whereas if the metacenter is below the center of mass, the upright position is unstable and the ship will overturn when disturbed.

Laminar flow in a circular cylindrical pipe. Suppose a horizontal pipe of circular cross section carries a steady volume flux I (fluid volume per unit time passing any cross section). We guess (a derivation is given in Section 8.2) that the velocity field is

$$v_1 = C(a^2 - x_2^2 - x_3^2), \qquad v_2 = 0, \qquad v_3 = 0,$$

where x_1 is measured along the central axis of the pipe (Fig. 7.11) and C is a constant. This field, commonly known as *Hagen-Poiseuille flow*, satisfies the continuity equation (7.36) identically, and the no slip condition $v_1 = 0$ at the wall of the pipe of radius a. Since the volume flux is

$$I = \int\int_{\substack{\text{cross}\\\text{section}}} \mathbf{v} \cdot d\mathbf{S} = C \int_0^{2\pi} \int_0^a (a^2 - r^2) r\,dr\,d\theta = \frac{\pi a^4 C}{2},$$

the constant $C = \dfrac{2I}{\pi a^4}$. Substituting this velocity field into the Navier-Stokes' equations (7.39b), we obtain

$$\frac{\partial p}{\partial x_1} = -\frac{8\mu I}{\pi a^4}, \qquad \frac{\partial p}{\partial x_2} = 0, \qquad \frac{\partial p}{\partial x_3} = 0.$$

Here we have not included the effect of any gravitational field, but gravity can always be accounted for simply by adding the term $-\rho\Phi$ on the right side of (7.44) to the pressure field. It follows that the pressure is the same on any cross section and drops linearly along the pipe:

$$p = p_1 - \left(\frac{p_1 - p_2}{\ell}\right) x_1.$$

The pressure drop is related to the volume flux by

$$p_1 - p_2 = IR, \tag{7.46}$$

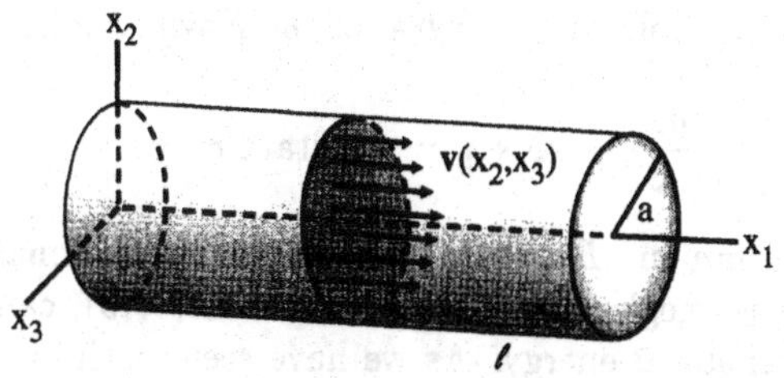

Figure 7.11. Laminar flow in a pipe of circular cross section.

where $R = \dfrac{8\mu\ell}{\pi a^4}$ is the *flow resistance* of a length ℓ of the pipe. The coordinates of a fluid particle, obtained by solving (7.31b), are

$$x_1(t) = x_1(0) + \frac{2It}{\pi a^4}\left[a^2 - x_2^2(0) - x_3^2(0)\right],$$

$$x_2(t) = x_2(0), \quad x_3(t) = x_3(0),$$

where $(x_1(0), x_2(0), x_3(0))$ are the coordinates of the particle at time $t = 0$. To recapitulate, the parabolic velocity distribution

$$\mathbf{v} = \frac{2I}{\pi a^4}(a^2 - x_2^2 - x_3^2)\,\mathbf{e}_1$$

and pressure field

$$p = p_1 - \frac{8\mu I}{\pi a^4}x_1$$

is the exact solution to the incompressible homogeneous fluid equations (7.36) and (7.39), which vanishes on a circular cylindrical surface. This solution is accurate for Reynolds numbers up to roughly 2000, away from the entrance to the pipe. If the Reynolds number is much larger, a pipe flow may become turbulent.

Irrotational steady incompressible flows. If we restrict our study to irrotational ($\nabla \times \mathbf{v} = 0$) and steady $\left(\dfrac{\partial \mathbf{v}}{\partial t} = 0\right)$ flows, we can introduce a *velocity potential* ϕ such that

$$\mathbf{v}(x_1, x_2, x_3) = \nabla\phi(x_1, x_2, x_3).$$

If also the flow is incompressible ($\nabla \cdot \mathbf{v} = 0$), the velocity potential must satisfy Laplace's equation

$$\nabla^2\phi = 0.$$

A complete boundary value problem for ϕ is obtained by prescribing boundary conditions, such as $\mathbf{v} \cdot \mathbf{n} = \dfrac{\partial \phi}{\partial n} = 0$ at a solid-fluid interface, where $\mathbf{n}$ is normal to the interface. Since Laplace's equation and the boundary conditions are linear in ϕ, we can add together irrotational steady incompressible velocity fields to model a variety of flow phenomena. Knowing the velocity field, we can immediately obtain the pressure field from the following integral of the Navier-Stokes equation:

$$\frac{\rho v^2}{2} + \rho\Phi + p = \text{ constant } c. \tag{7.47}$$

Equation (7.47) is the famous *Bernoulli equation* for irrotational steady incompressible flows and, if we neglect the viscous terms in (7.40), can be interpreted as conservation of mechanical energy. As we have seen, each of the vector fields in the examples of Sections 6.3 through 6.5 is the velocity field of an irrotational steady incompressible flow. An interesting velocity field is obtained by adding together a *line vortex* of strength K_1 and a *line source* of strength K_2 (refer to the last example plus Problem 7 in Section 6.4):

$$\mathbf{v} = \frac{K_1(-x_2\mathbf{e}_1 + x_1\mathbf{e}_2)}{x_1^2 + x_2^2} + \frac{K_2(x_1\mathbf{e}_1 + x_2\mathbf{e}_2)}{x_1^2 + x_2^2}. \tag{7.48}$$

The stream function for $\mathbf{v}(x_1, x_2)$ is a linear combination of our previous results:

$$\psi = -\frac{K_1}{2}\ln(x_1^2 + x_2^2) + K_2 \tan^{-1}\left(\frac{x_2}{x_1}\right).$$

Since the flow is steady, the trajectories of the fluid particles are identical to the streamlines $\psi(x_1, x_2) = \text{constant}$. Introducing polar coordinates (r, θ), we can write the equations of the pathlines/streamlines in the form

$$r = Ce^{K_2\theta/K_1},$$

where C is a constant. These are logarithmic or equiangular spirals, similar to the streamlines sketched in Fig. 5.6. Substituting the solution

$$x_1 = r\cos\theta = C\cos\theta e^{K_2\theta/K_1}, \qquad x_2 = r\sin\theta = C\sin\theta e^{K_2\theta/K_1}$$

into the differential equations (7.31b), we find that the angular speed of a fluid particle is $\dfrac{d\theta}{dt} = \dfrac{K_1}{r^2}$. Since the linear speed is

$$v = \frac{\sqrt{K_1^2 + K_2^2}}{r}$$

we obtain from Bernoulli's equation (7.47) the pressure in the absence of gravity:

$$p = c - \frac{\rho(K_1^2 + K_2^2)}{2r^2}. \tag{7.49}$$

In the case $K_2 < 0$ (a *line sink*), the speed of a fluid particle increases and the pressure drops as it spirals towards the origin. To recapitulate, the velocity distribution (7.48) and pressure distribution (7.49) of spiral flow are an exact solution to the incompressible homogeneous fluid equations (7.36) and (7.39) in all of space, excluding the x_3-axis. Spiral flow is a simple model for the flow of tornados, hurricanes, and water down the bathtub drain. Of course, near the vortex center the three-dimensional nature of such real flows must be accounted for. Irrotational theory may also be used to calculate the *lift* (component of aerodynamic force normal to the air velocity) on an airplane wing, because the effects of vorticity will normally be negligible outside of a thin boundary layer near the solid surface and a thin wake behind the body. However, to account for *drag* (component of aerodynamic force parallel to the air velocity) we must return to the full equations (7.39), because we cannot satisfy the no-slip condition at the fluid-solid interface within the confines of irrotational theory.

PROBLEMS 7.3

1. Verify Newton's law for an arbitrary control volume V enclosed by a surface S within the fluid:

$$\frac{d}{dt}\int\int\int_V \rho \mathbf{v} dV = \int\int\int_V \rho \mathbf{g} dV + \oint\!\!\oint_S (\mathbf{S} - \rho \mathbf{v}\mathbf{v}) \cdot d\mathbf{S}.$$

2. Estimate the order of magnitude of the Reynolds number for

   (a) flow around a submarine of length 100 m cruising at 30 $\frac{\text{km}}{\text{hr}}$,

   (b) flow around an airplane with a wing chord length of 3 m flying at 800 $\frac{\text{km}}{\text{hr}}$, and

   (c) flow through the hose of problem 9 in this section.

   (*Note:* $\nu = 10^{-6}\ \frac{\text{m}^2}{\text{s}}$ for water at 20°C, $\nu = 1.5 \times 10^{-5}\ \frac{\text{m}^2}{\text{s}}$ for air at one atmosphere and 20°C.)

3. Why is it so much more difficult to move through water than through the air?

4. Why does a bird fly forward when flapping its wings?

5. Approximate the hull of a submarine by a long circular cylindrical shell. For stability, when completely submerged or when floating on the surface, the center of gravity must be kept below what height? If the fluid forces exert a rolling moment M about the axis of the cylinder, what is the angle ϕ of roll?

6. Derive a better approximation to the variation of atmospheric pressure with height by assuming that air obeys the perfect gas law $\rho = \frac{Mp}{RT}$, where M is the molecular mass, R is the universal gas constant, and T is the absolute temperature (assume constant).

7. For Hagen-Poiseuille flow, find the strain rate and stress fields, the net force exerted by the fluid upon the pipe, and verify (7.40).

8. Find the velocity profile and pressure drop in a vertical pipe flow that is retarded by a uniform gravitational field.

9. A long hose with inside diameter 1 centimeter delivers 1 liter of water per minute. How many liters per minute will a similar hose of the same length but inside diameter 9 millimeters deliver? Assume the inlet pressure remains the same.

10. Determine the pressure distribution and find the coordinates of a fluid particle as a function of time for the velocity field $\mathbf{v} = Kx_1\mathbf{e}_1 - Kx_2\mathbf{e}_2$.

11. A fluid is constrained between infinite parallel plates a distance ℓ apart, the bottom being held stationary while the top plate moves with constant speed V. Determine the steady velocity and pressure fields, known as *plane Couette* or *linear shear flow.* Find the coordinates of a fluid particle and the strain and stress fields as a function of time.

12. A long circular cylindrical tube of radius 1 meter is immersed in a fluid and rotating about its axis with constant angular velocity 1 radian per second. Determine the steady velocity and pressure fields and find the coordinates of a fluid particle as a function of time.

13. A circular cylindrical glass of radius a and the fluid within are rotating steadily about the central axis with constant angular velocity $\boldsymbol{\omega} = \omega \mathbf{e}_3$. Assuming a constant gravitational field, find the pressure in the fluid and the equation of its free surface.

14. A *geostrophic* current or wind is a flow parallel to the earth's surface in which the Coriolis force balances the horizontal pressure gradient force and the inertial force is negligible. Determine the velocity of a geostrophic flow and show it is parallel to the isobars.

15. Derive the acoustic wave equation for the propagation of small disturbances in fluids: $\dfrac{\partial^2 p}{\partial t^2} = c^2 \nabla^2 p$, where $c = \sqrt{\frac{B}{\rho}}$ is the speed of sound in fluids and $B = \rho \dfrac{dp}{d\rho}$ is the bulk modulus of the fluid (cf. Problem 9 in the last section). Neglect the effects of gravitation and viscosity.

7.4 NEWTONIAN ORBITAL MECHANICS

The subject of orbital mechanics deals with the movements of celestial bodies about one another. Here we focus on the problem of predicting the orbit of a point mass m moving under the influence of the gravitational field of a finite body, which is at rest with respect to an inertial coordinate system (x_1, x_2, x_3).

We can think of a finite body as consisting of small particles of mass $\rho \Delta V$, where $\rho(x_1, x_2, x_3)$ is the density of the body and ΔV is a small volume element. We assume that the gravitational force $m\mathbf{g}$ exerted on the point mass m by a particle within the finite body is given by Newton's inverse square law, and that the total gravitational field $\mathbf{g}$ is the sum of the fields due to each of the individual particles:

$$\mathbf{g}(x_1, x_2, x_3) = -G \int\!\!\int\!\!\int_V \frac{\rho(x_1', x_2', x_3')(\mathbf{r} - \mathbf{r}')}{|\mathbf{r} - \mathbf{r}'|^3} dV'. \tag{7.50}$$

Here G is the *gravitational constant*, $\mathbf{r} = x_i \mathbf{e}_i$ is the location of the point mass, $\mathbf{r}' = x_i' \mathbf{e}_i$ is the location of a particle within the finite body, and the integration proceeds over the entire volume V of the finite body (Fig. 7.12). Since (7.50) has

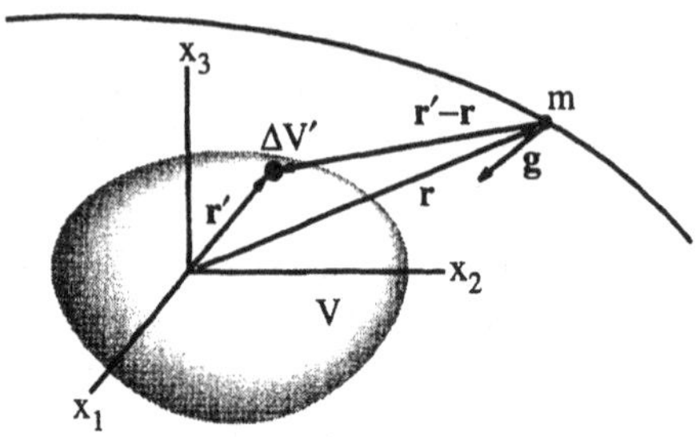

Figure 7.12. The gravitational force exerted on the point mass m by the particle in the finite body is $-\dfrac{mG\rho(x'_1, x'_2, x'_3)(\mathbf{r}-\mathbf{r}')\Delta V'}{|\mathbf{r}-\mathbf{r}'|^3}$.

the same form as (6.48), we conclude from (6.51) that

$$\mathbf{g} = -\boldsymbol{\nabla}\Phi,$$

where the *gravitational potential* Φ is given by

$$\Phi(x_1, x_2, x_3) = -G\int\int\int_V \frac{\rho(x'_1, x'_2, x'_3)}{|\mathbf{r}-\mathbf{r}'|}dV'.$$

From (6.52), for points (x_1, x_2, x_3) *outside* the finite body, the gravitational potential obeys Laplace's equation

$$\nabla^2\Phi(x_1, x_2, x_3) = 0.$$

Knowing the gravitational potential Φ, we can find the orbit by solving Newton's law

$$\frac{d^2\mathbf{r}}{dt^2} = -\boldsymbol{\nabla}\Phi \tag{7.51}$$

for the position vector $\mathbf{r}(x_1, x_2, x_3)$ of the point mass. An immediate integral follows from the principle of conservation of mechanical energy (5.15):

$$\frac{1}{2}v^2 + \Phi = \text{constant.} \tag{7.52}$$

From this we see that Φ is simply the *specific potential energy* of the particle. Unfortunately, for a general potential no other exact integrals of (7.51) are known, so one must in general resort to approximate means to obtain a solution.

So that we can obtain an analytical formula for the orbit, in the remainder of this section we suppose that the finite body is spherical and has a radially symmetric density distribution $\rho(x_1, x_2, x_3)$ relative to the origin of an inertial coordinate system (x_1, x_2, x_3). We can easily find the gravitational field $\mathbf{g}$ of such a body by calculating the flux of $\mathbf{g}$ out of a spherical surface S surrounding the body. The total flux of **g** out of S is the sum of the fluxes due to each individual particle within the finite body (see (6.49)):

$$\oiint_S \mathbf{g} \cdot d\mathbf{S} = -4\pi G \iiint_V \rho(x_1, x_2, x_3) dV = -4\pi GM. \qquad (7.53)$$

Here M is the total mass of the finite body. By symmetry the **g** field must be pointing directly towards the center of the finite body, so the surface integral in (7.53) has the value $-4\pi g r^2$ and

$$\mathbf{g} = -\frac{GM\mathbf{r}}{r^3}. \qquad (7.54)$$

The gravitational potential corresponding to (7.54) is (dropping a possible additive constant)

$$\Phi = -\frac{GM}{r}.$$

Thus the external gravitational field of a spherically symmetric body is the same as if the entire mass were concentrated at its center.

The task now before us is to solve Newton's law

$$\frac{d^2\mathbf{r}}{dt^2} = -\frac{GM\mathbf{r}}{r^3}. \qquad (7.55)$$

First, cross multiply (7.55) by **r** to obtain

$$\mathbf{r} \times \frac{d^2\mathbf{r}}{dt^2} = \frac{d}{dt}\left(\mathbf{r} \times \frac{d\mathbf{r}}{dt}\right) = 0.$$

Thus the ***specific angular momentum*** **h** of the point mass about the center of the spherical body is conserved:

$$\mathbf{h} = \mathbf{r} \times \mathbf{v} = \text{ constant}.$$

We also conclude that the motion of the point mass must be confined to a plane with normal **h**.

Next, cross multiply (7.55) by **h** and use vector identities to obtain

$$\frac{d}{dt}\left(\frac{d\mathbf{r}}{dt} \times \mathbf{h}\right) = GM\frac{d}{dt}\left(\frac{\mathbf{r}}{r}\right).$$

Integrating yields

$$\frac{d\mathbf{r}}{dt} \times \mathbf{h} = GM\left(\frac{\mathbf{r}}{r} + \mathbf{c}\right), \qquad (7.56)$$

where **c** is a constant vector. If we now dot multiply (7.56) by **r**, we can solve for r to obtain

$$r = \frac{a(1-e^2)}{1 + e\cos\theta}. \qquad (7.57)$$

Here the *true anomaly* θ is the angle between $\mathbf{c}$ and $\mathbf{r}$, the *eccentricity* $e = |\mathbf{c}|$, and the *semi-major axis*

$$a = \frac{h^2}{GM(1-e^2)}.$$

You may recognize (7.57) as the equation of a *conic section* in polar coordinates (r, θ), with r being measured from a *focus*, and with θ being measured from the line joining the focus to the *perigee* or *perihelion*, which is the point on the conic closest to the focus. The type of conic section is determined by the eccentricity e:

$$\begin{aligned} e = 0 &: \text{circle } (a > 0), \\ 0 < e < 1 &: \text{ellipse } (a > 0), \\ e = 1 &: \text{parabola } (a = \infty), \\ e > 1 &: \text{hyperbola } (a < 0). \end{aligned}$$

We see from the geometry shown in Fig. 7.13 that the time Δt it takes the point mass to move through a small angle $\Delta\theta$ is

$$\Delta t \approx \frac{r^2}{h}\Delta\theta.$$

The time it takes to travel through a finite angle θ is thus

$$t - t_0 = \frac{[a(1-e^2)]^{3/2}}{\sqrt{GM}} \int_0^\theta \frac{d\theta}{(1 + e\cos\theta)^2}, \tag{7.58}$$

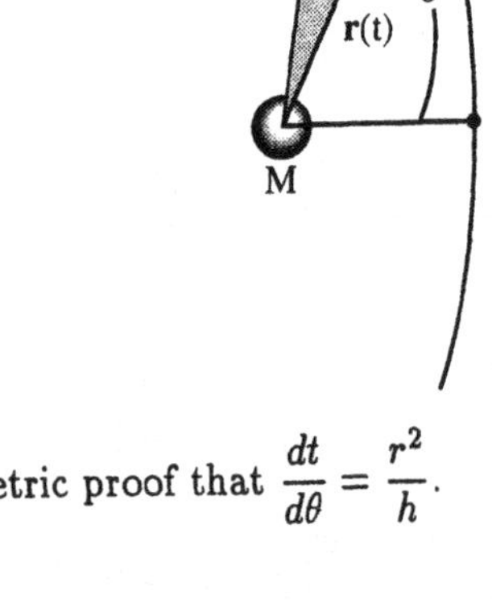

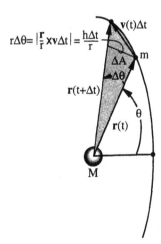

Figure 7.13. Geometric proof that $\dfrac{dt}{d\theta} = \dfrac{r^2}{h}$.

where t_0 is the *time of perigee passage.* Since the area of a sector of a circle of radius r and central angle $\Delta\theta$ is

$$\frac{1}{2}r^2\Delta\theta \approx \frac{h}{2}\Delta t \approx \Delta A,$$

we can also express the time of travel beyond perigee in terms of an integral over the area A swept out by the radius vector:

$$t - t_0 = \frac{2}{h}\int\int_A dA. \tag{7.59}$$

Equation (7.59) is commonly known as *Kepler's second law.*

We also see from the geometry shown in Fig. 7.13 that the speed of the point mass is given by

$$v = \sqrt{\left(\frac{dr}{dt}\right)^2 + r^2\left(\frac{d\theta}{dt}\right)^2} = \sqrt{\left(\frac{dr}{d\theta}\right)^2 + r^2}\,\frac{d\theta}{dt}.$$

Using (7.57) and (7.58) produces

$$v = \sqrt{GM\left(\frac{2}{r} - \frac{1}{a}\right)}. \tag{7.60}$$

It follows immediately that

$$\frac{1}{2}v^2 + \Phi = -\frac{GM}{2a}$$

so the specific mechanical energy (7.52) is indeed conserved.

The orbits of many celestial objects are approximately elliptical in shape. Fig. 7.14 illustrates for a representative elliptical orbit the quantities we have defined. During one revolution the radius vector sweeps out the ellipse's entire area of $\pi a^2\sqrt{1-e^2}$, so from (7.59) the period of revolution is

$$T = \frac{2\pi a^{3/2}}{\sqrt{GM}}.$$

Since we can arbitrarily prescribe the initial position $\mathbf{r}(0)$ and velocity $\mathbf{v}(0)$ vectors of a point mass, it takes six independent scalar quantities to uniquely specify an orbit in space. The six orbital elements $(i, \Omega, \omega, a, e, t_0)$ are commonly used to characterize the orbit of an artificial earth satellite. Here the *inclination* i is the angle between the earth's equatorial plane and the orbital plane, the *right ascension of the ascending node* Ω is the angle between $O\Upsilon$ and the line of intersection of the orbital and equatorial planes, and the *argument of perigee* ω is the angle in the orbital plane from the ascending node to perigee (Fig. 7.15). Knowing these six orbital elements and the value of GM, we can predict the position vector $\mathbf{r}(t)$ and velocity vector $\mathbf{v}(t)$ of the satellite at any time t.

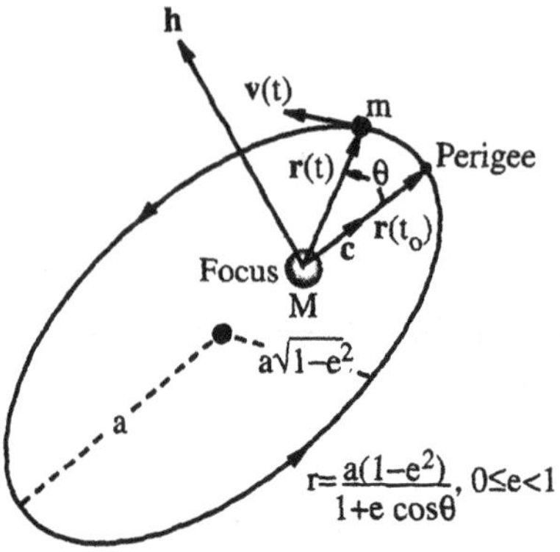

Figure 7.14. Newtonian theory predicts that the orbit of a point mass m ab
a spherical body M is a conic section.

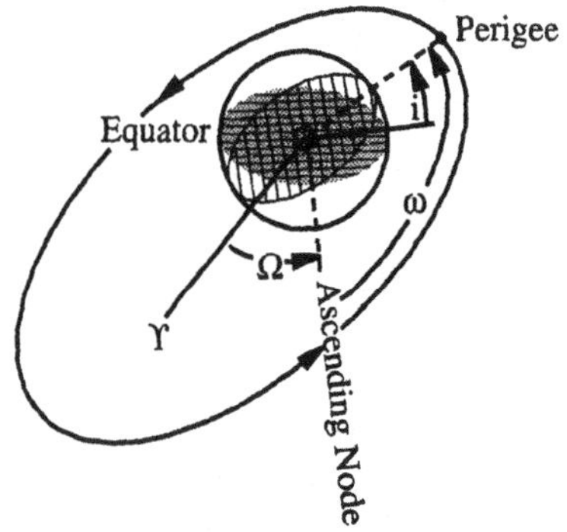

Figure 7.15. The angles (i, Ω, ω) specify the location of a conic section w
respect to a set of inertial axes fixed in space.

Actual celestial orbits may deviate from the two-body solution due to effects of small perturbation forces. The main perturbations of artificial sa lite orbits are the nonspherical components of the earth's gravitational field, drag of the earth's atmosphere, and the gravitational fields of the moon and sun. Over a time period of many revolutions these perturbation forces may ca significant changes in a satellite's orbital elements.

PROBLEMS 7.4

1. A better model for the two-body problem would allow both bodies to m through space. Consider the problem of two spherically symmetric bo of total mass M and m, respectively, which are orbiting about each ot Show that the solution for the position vector $\mathbf{r}(t)$ from the center of

body to the center of the other is the same as that in the text with M replaced by $M+m$.

2. The ellipse $\frac{x_1^2}{a^2}+\frac{x_2^2}{b^2}=1$ can be written in the parametric form $x_1 = a\cos E$, $x_2 = b\sin E$, where the angle E is called the *eccentric anomaly*. Show that the elliptical orbit can be written as $r = a(1-e\cos E)$, where E is the solution to *Kepler's equation* $t - t_0 = \frac{a^{3/2}}{\sqrt{GM}}(E - e\sin E)$.

3. Show that the orbit of a point mass under the influence of an arbitrary central force $f\mathbf{r}$ lies in a plane and obeys Kepler's second law. Here f could be any scalar function of position, velocity, time, and so on.

4. Compute the radius of the orbit and the speed of a *geosynchronous* satellite which moves around the earth in a circular orbit in the equatorial plane so that the period of revolution of the satellite is equal to the period of rotation of the earth about its axis. (*Note*: $G = 6.67 \times 10^{-11}\,\frac{\mathrm{m}^3}{\mathrm{kg}\cdot\mathrm{s}^2}$, $M_{\mathrm{Earth}} = 5.97 \times 10^{24}$ kg.)

5. Calculate the minimum speed with which it is required to launch a satellite from the surface of the earth in order to escape the earth's gravitational field (neglect air resistance).

6. Using the data for the Voyager 1 flyby of Jupiter given in Section 1.5, compute the distance of the spacecraft from Jupiter at its point of closest approach and the speed of the spacecraft relative to the sun. Assume the spacecraft is at infinity when it enters and leaves the gravitational field of Jupiter. (*Note*: $M_{\mathrm{Jupiter}} = 1.9 \times 10^{27}$ kg.)

7. Given the six orbital elements and GM, show how to calculate the position and velocity vectors of a satellite in an inertial coordinate system.

8. A radar tracking station determines that the position and velocity of a satellite in an inertial geocentric coordinate system at a given instant are $\mathbf{r}(0) = r_0\mathbf{e}_1, \mathbf{v}(0) = v_0\mathbf{e}_3$. Determine the position $\mathbf{r}(t)$ and velocity $\mathbf{v}(t)$ at any time t.

9. A drag force per unit mass f acts in a direction opposite to the velocity $\mathbf{v}$ of a satellite. Show that the drag force causes a decrease in the semi-major axis of the orbit given by
$$\frac{da}{dt} = -\frac{2va^2 f}{GM}.$$
(There may also be changes in e, ω, and t_0.)

10. Show that, if we are far enough away from a finite body, the gravitational field is approximately given by the inverse square law (7.54).

11. Show that the gravitational potential inside a finite body is governed by Poisson's equation $\nabla^2\Phi = 4\pi G\rho$. Find the gravitational and pressure fields inside a spherical body of constant density.

7.5 ELECTROMAGNETIC THEORY

The entire edifice of electromagnetism is built upon *Maxwell's equations*:

$$\nabla \cdot \mathcal{D} = \rho, \tag{7.61a}$$

$$\nabla \cdot \mathcal{B} = 0, \tag{7.62a}$$

$$\nabla \times \mathcal{E} = -\frac{\partial \mathcal{B}}{\partial t}, \tag{7.63a}$$

$$\nabla \times \mathcal{H} = \mathcal{J} + \frac{\partial \mathcal{D}}{\partial t}. \tag{7.64a}$$

Here $\mathcal{D}(x_1, x_2, x_3, t)$ is the *electric displacement*, $\mathcal{H}(x_1, x_2, x_3, t)$ is the *magnetic intensity*, $\rho(x_1, x_2, x_3, t)$ is the *charge density* (charge per unit volume), and $\mathcal{J}(x_1, x_2, x_3, t)$ is the electric *current density* (charge per unit time and area).

By integrating (7.61a) and (7.62a) within an arbitrary volume V of the field and (7.63a) and (7.64a) within an arbitrary surface S_1 in the field, and using the divergence and Stokes' theorems to transform the left sides into integrals over the surface S enclosing V or around the curve C_1 bounding S_1, we can obtain the integral form of Maxwell's equations:

$$\oiint_S \mathcal{D} \cdot d\mathbf{S} = Q, \tag{7.61b}$$

$$\oiint_S \mathcal{B} \cdot d\mathbf{S} = 0, \tag{7.62b}$$

$$\oint_{C_1} \mathcal{E} \cdot d\mathbf{r} = -\frac{d}{dt}\int\int_{S_1} \mathcal{B} \cdot d\mathbf{S}, \tag{7.63b}$$

$$\oint_{C_1} \mathcal{H} \cdot d\mathbf{r} = I + \frac{d}{dt}\int\int_{S_1} \mathcal{D} \cdot d\mathbf{S}. \tag{7.64b}$$

Here

$$Q = \int\int\int_V \rho dV \tag{7.65}$$

is the total charge within V and

$$I = \int\int_{S_1} \mathcal{J} \cdot d\mathbf{S} \tag{7.66}$$

is the total electric current (charge per unit time) passing through the surface S_1. Equations (7.61) are sometimes called *Gauss's laws for electricity*, (7.62) an called *Gauss's laws for magnetism*, (7.63) are called *Faraday's laws of induction*, and (7.64) are called *generalized Ampere's laws*.

If material bodies are present, one must determine experimentally the constitutive relations $\mathcal{D} = \mathcal{D}(\mathcal{E})$, $\mathcal{H} = \mathcal{H}(\mathcal{B})$, $\mathcal{J} = \mathcal{J}(\mathcal{E})$. In a vacuum, or in a homogeneous linear isotropic medium,

$$\mathcal{D} = \epsilon\mathcal{E}, \tag{7.67}$$

$$\mathcal{H} = \frac{\mathcal{B}}{\mu}, \tag{7.68}$$

where ϵ and μ are constants called the *permittivity* and *permeability*, respectively. In a homogeneous linear isotropic conducting medium, *Ohm's law* applies:

$$\mathcal{J} = \sigma\mathcal{E}, \tag{7.69}$$

where σ is a constant called the *conductivity*.

In summary, Maxwell's equations (7.61)–(7.64) and the constitutive relations (7.67)–(7.69) form a complete system of linear differential equations for the electric field $\mathcal{E}(x_1, x_2, x_3, t)$ and magnetic field $\mathcal{B}(x_1, x_2, x_3, t)$. Appropriate boundary conditions for the electric and magnetic fields at the interface between two media may be deduced by applying (7.61) and (7.62) to a small cylindrical volume and (7.63) and (7.64) to a small rectangular loop.

An immediate consequence of Maxwell's equations (7.61) and (7.64) is the *continuity equation* (cf. 7.3 and 7.35):

$$\frac{\partial \rho}{\partial t} + \nabla \cdot \mathcal{J} = 0 \tag{7.70a}$$

or

$$\frac{dQ}{dt} + I = 0. \tag{7.70b}$$

In (7.70b) Q is the net charge contained in an arbitrary volume and I is the net current flowing out of the surface enclosing the volume.

The total force exerted by the electromagnetic field on the charges and currents within a volume V is

$$\mathbf{f} = \int\int\int_V (\rho\mathcal{E} + \mathcal{J} \times \mathcal{B})dV. \tag{7.71}$$

This follows from the Lorentz force law upon noting that, if the charge ρ is in motion with velocity $\mathbf{v}$, the resulting current density is

$$\mathcal{J} = \rho\mathbf{v}. \tag{7.72}$$

Using (7.61a)–(7.64a), (7.67)–(7.69), the vector identities (6.11) and (6.14), and the divergence theorem, we can transform (7.71) into

$$\int\int\int_V (\rho\mathcal{E} + \mathcal{J} \times \mathcal{B})dV = -\frac{d}{dt}\int\int\int \mathcal{D} \times \mathcal{B}dV + \oiint_S \mathcal{T} \cdot d\mathbf{S}. \tag{7.73}$$

The quantity $\mathcal{D} \times \mathcal{B}$ may be identified as the *momentum density* (momentum per unit volume) stored in the electromagnetic field. *Maxwell's stress tensor*

$$\mathcal{T} = \mathcal{E}\mathcal{D} + \mathcal{B}\mathcal{H} - \frac{1}{2}(\mathcal{E}\mathcal{D} + \mathcal{B}\mathcal{H})\mathbf{1} \tag{7.74}$$

may be interpreted as *momentum flux density*; $-\mathcal{T} \cdot \Delta\mathbf{S}$ is the momentum tranferred per unit of time out of V across a surface element ΔS. Equation (7.73) says that the force exerted by the electromagnetic field on the material within V is equal to the rate of decrease of electromagnetic momentum within V plus the rate at which electromagnetic momentum is transferred into V across S.

The rate at which the electromagnetic field does work on the moving charges within V is, from (7.71) and (7.72),

$$\int\int\int_V (\rho\mathcal{E} + \mathcal{J} \times \mathcal{B}) \cdot \mathbf{v} dV = \int\int\int_V \mathcal{E} \cdot \mathcal{J} dV. \tag{7.75}$$

If the charges flowing within a conductor obey Ohm's law (7.69), then (7.75) is the *Joule heating rate* and represents the irreversible transfer of electrical energy to thermal energy within the conductor. We can transform (7.75) into

$$\int\int\int_V \mathcal{E} \cdot \mathcal{J} dV = -\frac{d}{dt}\int\int\int_V \left(\frac{\epsilon\mathcal{E}^2}{2} + \frac{\mathcal{B}^2}{2\mu}\right) dV - \oiint_S \mathcal{E} \times \mathcal{H} \cdot d\mathbf{S}. \tag{7.76}$$

The quantities $\frac{\epsilon\mathcal{E}^2}{2}$ and $\frac{\mathcal{B}^2}{2\mu}$ may be identified as the *energy densities* (energies per unit volume) stored in the electric and magnetic fields, respectively. The *Poynting vector* $\mathcal{E} \times \mathcal{H}$ may be interpreted as *energy flux density*; $\mathcal{E} \times \mathcal{H} \cdot \Delta\mathbf{S}$ is the energy transferred per unit of time out of V across a surface element ΔS. Equation (7.76) says that the power delivered by the electromagnetic field to the charges within V equals the rate of decrease of the electromagnetic energy within V plus the rate at which electromagnetic energy is transferred into V across S.

In the following we construct some exact solutions to the electromagnetic field equations.

Electrostatics. If the charge density $\rho(x_1, x_2, x_3)$ and the current density $\mathcal{J}(x_1, x_2, x_3)$ are independent of the time t, then the electric and magnetic fields are also independent of t and the equations for the electric field $\mathcal{E}(x_1, x_2, x_3)$ can be solved independently of the equations for the magnetic field $\mathcal{B}(x_1, x_2, x_3)$. Consider first the electrostatic field equations obtained from (7.61), (7.63) with the time derivative dropped, and (7.67):

$$\nabla \cdot \mathcal{E} = \frac{\rho}{\epsilon}, \tag{7.77a}$$

$$\nabla \times \mathcal{E} = 0, \tag{7.78a}$$

$$\oiint_S \mathcal{E} \cdot d\mathbf{S} = \frac{Q}{\epsilon}, \tag{7.77b}$$

$$\oint_C \mathcal{E} \cdot d\mathbf{r} = 0. \tag{7.78b}$$

From (7.78a), $\mathcal{E}(x_1, x_2, x_3)$ is a conservative vector field so an *electrostatic potential* $\phi(x_1, x_2, x_3)$ can be defined such that

$$\mathcal{E} = -\nabla\phi.$$

Then, from (7.77a), we find that ϕ obeys Poisson's equation

$$\nabla^2\phi = -\frac{\rho}{\epsilon}.$$

The electrostatic potential is obtained by comparison with (6.48) and (6.49):

$$\phi(x_1, x_2, x_3) = \frac{1}{4\pi\epsilon}\int\int\int_V \frac{\rho(x_1', x_2', x_3')}{|\mathbf{r} - \mathbf{r}'|}dV'. \tag{7.79}$$

Coulomb's law for the electrostatic field follows upon taking the negative gradient of (7.79):

$$\mathcal{E}(x_1, x_2, x_3) = \frac{1}{4\pi\epsilon}\int\int\int_V \frac{\rho(x_1', x_2', x_3')}{|\mathbf{r} - \mathbf{r}'|^3}(\mathbf{r} - \mathbf{r}')dV'. \tag{7.80}$$

Here the integration is over the volume V occupied by *all* the charges throughout three-dimensional space. If the charge distribution is all concentrated in a finite region near the origin, we can approximate the electric field at distances r far from the charge distribution by expanding the quantity $|\mathbf{r} - \mathbf{r}'|^{-3}$ in a series of ascending powers of $\frac{\mathbf{r}'}{r}$:

$$\frac{1}{|\mathbf{r} - \mathbf{r}'|^3} = \frac{1}{[r^2 - 2\mathbf{r}\cdot\mathbf{r}' + (r')^2]^{3/2}} = \frac{1}{r^3}\left(1 + \frac{3\mathbf{r}\cdot\mathbf{r}'}{r^2} + \cdots\right). \tag{7.81}$$

Insertion of this expansion into (7.80) leads to

$$\mathcal{E} = \frac{Q\mathbf{r}}{4\pi\epsilon r^3} + \frac{(3\mathbf{r}\mathbf{r} - r^2\mathbf{1})\cdot\mathbf{p}}{4\pi\epsilon r^5} + \cdots, \tag{7.82}$$

where $\mathbf{p}$ is the *dipole moment* of the charge distribution:

$$\mathbf{p} = \int\int\int_V \mathbf{r}'\rho(x_1', x_2', x_3')dV'.$$

The first term on the right side of (7.82) is the electric field that would result if the total charge Q were concentrated at the origin (Fig. 6.6), while the second term is the electric field of a *point dipole* (Fig. 7.16). Thus, if we are far enough away from a charge distribution, the electric field approximates that of a point charge when $Q \neq 0$ or that of a point dipole when $Q = 0$ and $\mathbf{p} \neq 0$. For instance, the electric field far away from some neutral molecules, such as the water molecule H_2O, is approximately that of a point dipole. It can be seen that the solution of

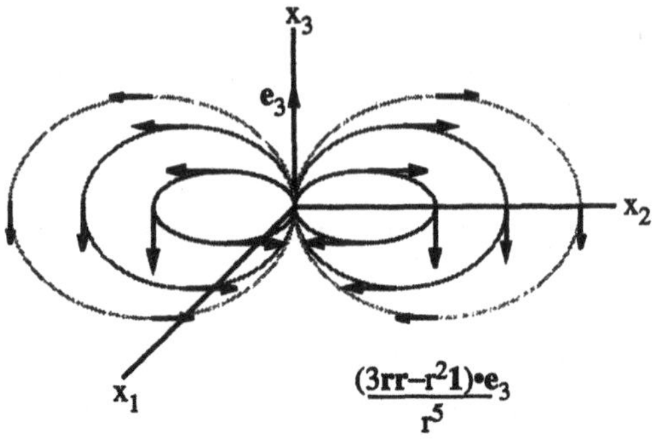

Figure 7.16. Field lines of an electric or magnetic point dipole oriented along the x_3-axis.

electrostatic field problems is straightforward when the entire charge distribution is known. If the charge distribution is not known in advance, it may be necessary to determine the electric field before the charge distribution can be calculated. An example of this kind of problem is the determination of the electrostatic field caused by conductors at fixed potentials. The charges within each conductor must distribute themselves on the surface — otherwise by (7.77) there would be an electric field inside the conductor. We can obtain a formula relating the surface charge density q_S (charge per unit area) to the electric field $\mathcal{E}$ just outside the surface of the conductor by applying Gauss's law (7.77b) to a small cylindrical volume half inside and half outside the surface, as sketched in Fig. 7.17. Since $\mathcal{E}$ is perpendicular to the surface of the conductor, there is a contribution to the

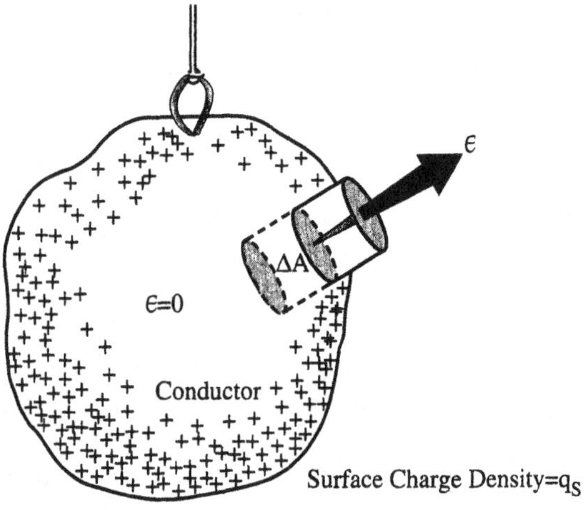

Figure 7.17. The excess charges of an insulated conductor reside entirely on its outer surface.

total flux of $\mathcal{E}$ only through the surface of the box outside of the conductor, and the total charge enclosed by the box is $q_S \Delta A$, where ΔA is the surface area of the ends of the box:

$$\mathcal{E}\Delta A = \frac{q_S \Delta A}{\epsilon}.$$

Thus

$$q_S = \epsilon \mathcal{E}.$$

Any irrotational solenoidal vector field can be regarded as the electric field outside of charged conductors whose surfaces coincide with the equipotential lines of the vector field. Thus, each of the vector fields in the examples of Sections 6.3 through 6.5 corresponds to a possible electrostatic field. For example, the electric field of Fig. 7.18 (cf. Fig. 6.15) may be created by using four hyperbola-shaped electrodes held at the potentials indicated. This field is used in devices called *quadrupole lenses* to focus particle beams.

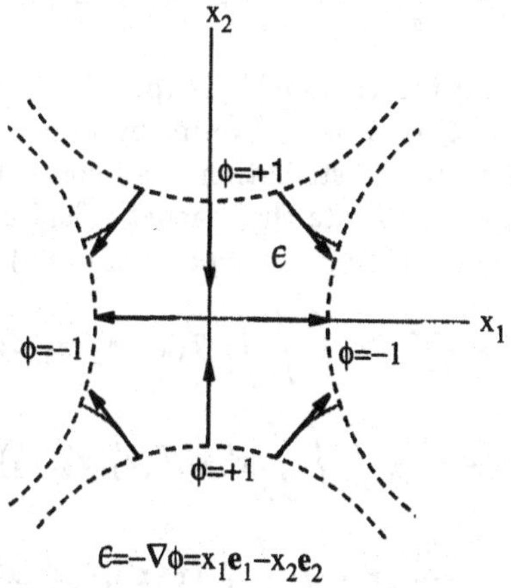

Figure 7.18. Electric field of a quadrupole lense.

Magnetostatics. Continuing our study of the case when $\rho(x_1, x_2, x_3)$ and $\mathcal{J}(x_1, x_2, x_3)$ are independent of t, we now consider the magnetostatic field equations obtained from (7.62), (7.64) with the time derivative dropped, and (7.68):

$$\nabla \cdot \boldsymbol{B} = 0, \tag{7.83a}$$

$$\nabla \times \boldsymbol{B} = \mu \mathcal{J}, \tag{7.84a}$$

or

$$\oiint_S \boldsymbol{B} \cdot d\mathbf{S} = 0, \tag{7.83b}$$

$$\oint_C \boldsymbol{B} \cdot d\mathbf{r} = \mu I. \tag{7.84b}$$

From (7.83a), $\boldsymbol{B}(x_1, x_2, x_3)$ is a solenoidal vector field so a *magnetostatic potential* $\mathbf{A}(x_1, x_2, x_3)$ can be defined such that

$$\boldsymbol{B} = \nabla \times \mathbf{A}.$$

Assuming for the moment that we can choose $\nabla \cdot \mathbf{A} = 0$, we find from (7.83a) and the vector identity (6.18) that $\mathbf{A}$ obeys the vector Poisson equation

$$\nabla^2 \mathbf{A} = -\mu \boldsymbol{\mathcal{J}}.$$

A magnetostatic potential is obtained by comparison with (6.54) and (6.55):

$$\mathbf{A}(x_1, x_2, x_3) = \frac{\mu}{4\pi} \iiint_V \frac{\boldsymbol{\mathcal{J}}(x_1', x_2', x_3')}{|\mathbf{r} - \mathbf{r}'|} dV'. \tag{7.85}$$

The *Biot-Savart* law for the magnetostatic field follows upon taking the curl of (7.85) and using the vector identity (6.10):

$$\boldsymbol{B}(x_1, x_2, x_3) = \frac{\mu}{4\pi} \iiint_V \frac{\boldsymbol{\mathcal{J}}(x_1', x_2', x_3') \times (\mathbf{r} - \mathbf{r}')}{|\mathbf{r} - \mathbf{r}'|^3} dV'. \tag{7.86}$$

Here the integration is over the volume V occupied by *all* the currents throughout three-dimensional space. Now you can verify by direct differentiation of (7.85) that $\nabla \cdot \mathbf{A} = 0$. If the current distribution is all concentrated in a finite region near the origin, we can approximate the magnetic field at distances far from the current distribution by substituting the expansion (7.81) into (7.86) to obtain

$$\begin{aligned} \boldsymbol{B}(x_1, x_2, x_3) = &-\frac{\mu \mathbf{r}}{4\pi r^3} \times \iiint_V \boldsymbol{\mathcal{J}}(x_1', x_2', x_3') dV' \\ &+ \frac{\mu}{4\pi r^3} \iiint_V \mathbf{r}' \times \boldsymbol{\mathcal{J}}(x_1', x_2', x_3') dV' \\ &- \frac{3\mu}{4\pi r^5} \mathbf{r} \times \iiint_V (\mathbf{r} \cdot \mathbf{r}') \boldsymbol{\mathcal{J}}(x_1', x_2', x_3') dV' + \cdots. \end{aligned} \tag{7.87}$$

The first integral on the right of (7.87) must vanish to ensure the satisfaction of (7.83b) for large surfaces enclosing the origin:

$$\iiint_V \boldsymbol{\mathcal{J}}(x_1', x_2', x_3') dV' = 0.$$

Thus there can be no magnetic point sources. One-half of the second integral on the right of (7.87) is the *magnetic moment* of the current distribution:

$$\mathbf{m} = \frac{1}{2} \iiint_V \mathbf{r}' \times \boldsymbol{\mathcal{J}}(x_1', x_2', x_3') dV'.$$

The last integral in (7.87) can be manipulated into a more convenient form using the triple vector product:

$$\mathbf{r} \cdot \iiint_V \mathbf{r}' \boldsymbol{\mathcal{J}}(x_1', x_2', x_3') dV' = \mathbf{r} \cdot \iiint_V \boldsymbol{\mathcal{J}}(x_1', x_2', x_3') \mathbf{r}' dV' - 2\mathbf{r} \times \mathbf{m}. \tag{7.88}$$

The integral on the right side of (7.88) can be shown to be the negative of the integral on the left of (7.88) by using (7.70a), the divergence theorem, and the fact that no current flows through the surface S bounding the current distribution:

$$\begin{aligned}\iiint_V \mathcal{J}_i x_j' dV' &= \iiint_V \frac{\partial}{\partial x_k'}(x_i'\mathcal{J}_k)x_j' dV' \\ &= \oiint_S x_i'x_j'\boldsymbol{\mathcal{J}}\cdot d\mathbf{S}' - \iiint_V x_i'\mathcal{J}_k \frac{\partial x_j'}{\partial x_k'} dV \\ &= -\iiint x_i'\mathcal{J}_j dV'.\end{aligned}$$

The expression (7.87) finally reduces to

$$\boldsymbol{B} = \frac{\mu(3\mathbf{rr} - r^2\mathbf{1})\cdot\mathbf{m}}{4\pi r^5} + \cdots.$$

Thus, if we are far enough away from a current distribution, the magnetic field approximates that of a *point dipole* when $\mathbf{m} \neq 0$ (Fig. 7.16). The magnetic field far from a permanent magnet also may be approximated by a dipole field. It can be seen that the solution of magnetostatic field problems is straightforward when the entire current distribution is known. If the current distribution is not known in advance, it may be necessary to determine the electric field first; the current density can then be calculated from Ohm's law (7.69). A simple example is a long straight wire conducting a steady current I (Fig. 7.19). There can be no component of $\boldsymbol{\mathcal{E}}$ within the wire at right angles to its axis since by (7.69) this would produce a continual charging of the wire's surface. Then setting $\boldsymbol{\mathcal{E}} = \mathcal{E}\mathbf{e}_1$ in $\nabla\times\boldsymbol{\mathcal{E}} = 0$ and $\nabla\cdot\boldsymbol{\mathcal{E}} = 0$ yields

$$\frac{\partial\mathcal{E}}{\partial x_1} = \frac{\partial\mathcal{E}}{\partial x_2} = \frac{\partial\mathcal{E}}{\partial x_3} = 0.$$

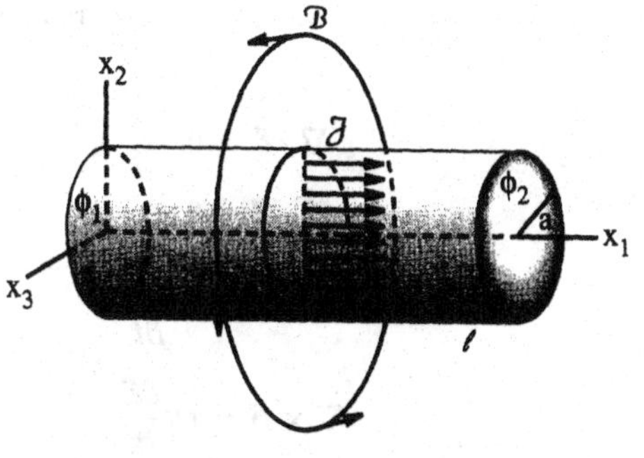

Figure 7.19. Steady current in a wire of circular cross section.

Thus the $\mathcal{E}$ field within the wire is a constant and, from (7.66) and (7.69),

$$\mathcal{E} = \frac{I}{\sigma A}\mathbf{e}_1 \quad \text{for} \quad r \leq a,$$

where A is the cross-sectional area of the wire. The electrostatic potential drop linearly along the wire:

$$\phi = \phi_1 - \left(\frac{\phi_1 - \phi_2}{\ell}\right) x_1.$$

The potential drop is related to the current by (cf. (7.13) and (7.46))

$$\phi_1 - \phi_2 = IR,$$

where $R = \dfrac{\ell}{A\sigma}$ is the *electrical resistance* of a length ℓ of the wire. If we suppos the wire has circular cross section of radius a, then from symmetry the field line of $\boldsymbol{B}$ go around the axis of the wire in closed circles and the magnitude B has a constant value on each field line (Fig. 7.19). We can easily find the magnitude B by applying Ampere's law (7.84b) to a circular field line:

$$2\pi\sqrt{x_2^2 + x_3^2}B = \begin{cases} \dfrac{\mu I \pi(x_2^2 + x_3^2)}{\pi a^2} & \text{for } x_2^2 + x_3^2 \leq a^2, \\ \mu I & \text{for } x_2^2 + x_3^2 \geq a^2. \end{cases}$$

Hence

$$B = \begin{cases} \dfrac{\mu I\sqrt{x_2^2 + x_3^2}}{2\pi a^2} & \text{for } x_2^2 + x_3^2 \leq a^2, \\ \dfrac{\mu I}{2\pi\sqrt{x_2^2 + x_3^2}} & \text{for } x_2^2 + x_3^2 \geq a^2. \end{cases}$$

Electromagnetic waves. Finally, we look for solutions to the complete Maxwe equations in regions of space that are free of charges and currents. The fiel equations (7.61a)–(7.64a) and (7.67) and (7.68) reduce in the case $\rho = 0$ an $\mathcal{J} = 0$ to

$$\nabla \cdot \mathcal{E} = 0, \tag{7.89}$$

$$\nabla \cdot \boldsymbol{B} = 0, \tag{7.90}$$

$$\nabla \times \mathcal{E} = -\frac{\partial \boldsymbol{B}}{\partial t}, \tag{7.91}$$

$$\nabla \times \boldsymbol{B} = \epsilon\mu\frac{\partial \mathcal{E}}{\partial t}. \tag{7.92}$$

By taking the curl of both sides of (7.91) and using (6.18), (7.89), and (7.92), w can obtain

$$\frac{\partial^2 \boldsymbol{\mathcal{E}}}{\partial t^2} = c^2 \nabla^2 \boldsymbol{\mathcal{E}}, \qquad c = \frac{1}{\sqrt{\epsilon\mu}}. \tag{7.93}$$

Similarly, by taking the curl of both sides of (7.92) and using (6.18), (7.90), and (7.91), we can obtain

$$\frac{\partial^2 \boldsymbol{B}}{\partial t^2} = c^2 \nabla^2 \boldsymbol{B}. \tag{7.94}$$

Thus the electric and magnetic fields each obey the same wave equation, with c being the *speed of electromagnetic radiation.* Any twice differentiable function of the argument $(x_1 - ct)$ is a solution of (7.93) and (7.94):

$$\boldsymbol{\mathcal{E}} = \boldsymbol{\mathcal{E}}(x_1 - ct)\,, \qquad \boldsymbol{B} = \boldsymbol{B}(x_1 - ct). \tag{7.95}$$

Substituting (7.95) into (7.89)–(7.92), we find that the components of $\boldsymbol{\mathcal{E}}$ and $\boldsymbol{B}$ in the x_1 direction vanish and $\boldsymbol{\mathcal{E}}$ and $\boldsymbol{B}$ are related by

$$\boldsymbol{B} = \frac{\mathrm{e}_1 \times \boldsymbol{\mathcal{E}}}{c}. \tag{7.96}$$

Hence $(\mathrm{e}_1, \boldsymbol{\mathcal{E}}, \boldsymbol{B})$ form a right-handed set of orthogonal vectors and the ratio $\frac{\mathcal{E}}{B}$ is the constant c. The most general plane wave solutions of (7.89)–(7.92) propagating in the x_1 direction are thus

$$\begin{aligned} \boldsymbol{\mathcal{E}} &= f(x_1 - ct)\mathrm{e}_2 + g(x_1 - ct)\mathrm{e}_3, \\ \boldsymbol{B} &= \frac{-g(x_1 - ct)\mathrm{e}_2 + f(x_1 - ct)\mathrm{e}_3}{c}, \end{aligned} \tag{7.97}$$

where f and g are arbitrary functions of the indicated arguments. An important special case of (7.97) are the plane *monochromatic* (single frequency) waves

$$\boldsymbol{\mathcal{E}} = a\sin(kx_1 - \omega t + \alpha)\mathrm{e}_2 + b\sin(kx_1 - \omega t + \beta)\mathrm{e}_3, \tag{7.98}$$

$$\boldsymbol{B} = \frac{-b\sin(kx_1 - \omega t + \beta)\mathrm{e}_2 + a\sin(kx_1 - \omega t + \alpha)\mathrm{e}_3}{c}. \tag{7.99}$$

Here ω is the *frequency*, $k = \frac{\omega}{c}$ is the *wave number*, a and b are *amplitudes*, and α and β are *phase constants*. At a point fixed in space, the $\boldsymbol{\mathcal{E}}$ and $\boldsymbol{B}$ fields are the same at any two instants differing by the *period* $\frac{2\pi}{\omega}$; at an instant of time, the $\boldsymbol{\mathcal{E}}$ and $\boldsymbol{B}$ fields are the same on any two planes perpendicular to the x_1-axis and spaced the *wavelength* $\frac{2\pi}{k}$ apart. The waves (7.98)–(7.99) are said to be *elliptically polarized* because, at a point fixed in space, the tips of the $\boldsymbol{\mathcal{E}}$

and $\boldsymbol{B}$ vectors trace out elliptical paths, as shown in Fig. 7.20. Special cases of (7.98)–(7.99) are *circularly polarized* waves with $\boldsymbol{\mathcal{E}}$ and $\boldsymbol{B}$ vectors tracing out a circular path ($a = b$ and $\alpha = \beta \pm \frac{\pi}{2}$), and *linearly polarized* waves with $\boldsymbol{\mathcal{E}}$ and $\boldsymbol{B}$ vectors tracing out a line segment ($a = 0$ or $b = 0$ or $\alpha = \beta$). It can be shown that *every* solution of the wave equation is a linear superposition of (possibly an infinite number of) plane monochromatic waves traveling in various directions and having various frequencies.

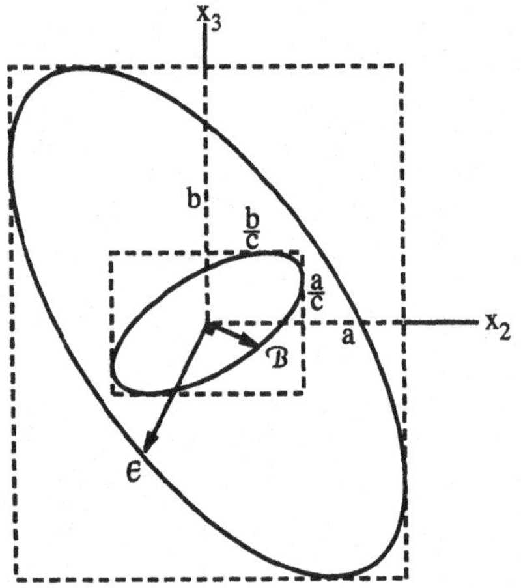

Figure 7.20. At a given point in space the $\boldsymbol{\mathcal{E}}$ and $\boldsymbol{B}$ vectors of a plane monochromatic wave trace out ellipses.

PROBLEMS 7.5

1. Show that the dipole moment of a charge distribution is independent of the origin of coordinates if and only if the total charge is zero.

2. If charge distribution is all concentrated in a finite region near the origin, show that the electrostatic potential at distances r far from the charge distribution can be approximated by

$$\phi = \frac{Q}{4\pi\epsilon r} + \frac{\mathbf{r}\cdot\mathbf{p}}{4\pi\epsilon r^3} + \frac{\mathbf{r}\cdot\mathbf{Y}\cdot\mathbf{r}}{8\pi\epsilon r^5} + \cdots,$$

where the *quadrupole moment tensor*

$$\mathbf{Y} = \int\int\int_V (3\mathbf{r}'\mathbf{r}' - r'^2\mathbf{1})\rho(x_1', x_2', x_3')dV'.$$

3. A conducting plate at potential ϕ_1 is a relatively small distance d from a parallel conducting plate at potential ϕ_2. Neglecting the fringing field at the

edges of the plates, determine the *capacitance* $C = \dfrac{Q}{\phi_1 - \phi_2}$, where $\pm Q$ are the total charges stored on each plate.

4. A conducting medium has a volume density of prescribed charge $\rho(x_1, x_2, x_3)$ at time $t = 0$. Determine the time it takes the charge to decay to $\dfrac{1}{e}$ of its original value.

5. If the only sources are steady currents I in circuits C of small wires, show that the magnetic field is given by

$$\boldsymbol{B} = \frac{\mu}{4\pi} \int_C \frac{I d\mathbf{r}' \times (\mathbf{r} - \mathbf{r}')}{|\mathbf{r} - \mathbf{r}'|^3}.$$

Show that the total force exerted on a current loop C_1 carrying a current I_1 by a current loop C_2 carrying a current I_2 is

$$\mathbf{f} = \frac{\mu I_1 I_2}{4\pi} \oint_{C_2} \oint_{C_1} \frac{d\mathbf{r}_1 \times [d\mathbf{r}_2 \times (\mathbf{r}_1 - \mathbf{r}_2)]}{|\mathbf{r}_1 - \mathbf{r}_2|^3}.$$

6. Find the magnetic moment of a circuit consisting of a thin wire loop lying in the x_1-x_2 plane and carrying a steady current I.

7. Verify (7.76) for the steady flow of current in a straight wire (cf. Problem 7 in Section 7.3).

8. A planar wire coil C is rotating with angular velocity ω about a diameter perpendicular to a uniform magnetic field $\boldsymbol{B}$. Find the *induced electromotive force* $\oint_C \boldsymbol{\mathcal{E}} \cdot d\mathbf{r}$ in the coil. This is the prototype of an alternating current generator.

9. Show that the area swept out per unit of time by the $\boldsymbol{\mathcal{E}}(t)$ vectors (7.98) is a constant.

10. Show that the wave (7.98) may be written as the real or imaginary part of $\mathbf{v}e^{i(kx-\omega t)}$, where $i = \sqrt{-1}$ and the amplitude $\mathbf{v}$ is a constant *complex vector*.

11. Find the rate of momentum and energy transferred per unit area by a plane electromagnetic wave across a surface normal to the direction of propagation.

12. Radiation from the sun striking the earth has an average intensity of 1400 $\frac{\mathrm{W}}{\mathrm{m}^2}$. What is the radiation pressure exerted on a surface normal to the direction of radiation (such as a portion of a satellite skin)? (*Note*: The speed of light in free space is $c = 3 \times 10^8 \frac{\mathrm{m}}{\mathrm{s}}$.)

13. Show that the time-varying $\mathcal{E}$ and $\mathcal{B}$ fields are derivable from potential functions such that $\mathcal{E} = -\nabla\phi - \frac{\partial \mathbf{A}}{\partial t}$ and $\mathcal{B} = \nabla \times \mathbf{A}$. Find the equations satisfied by ϕ and $\mathbf{A}$ if the *Lorentz gauge condition*

$$\nabla \cdot \mathbf{A} + \frac{1}{c^2}\frac{\partial \phi}{\partial t} = 0$$

is imposed.

7.6 NOTATION OF OTHERS

Table 7

OUR NOTATION	OUR NAME	OTHER NOTATIONS	OTHER NAMES
$R = \frac{L}{k}$	Thermal resistance.	$R = \frac{L}{Ak}$	Same name and symbol may be ours divided by area A of cross section normal to direction of heat transfer.
$\frac{1}{h}$	Surface resistance.	$\frac{1}{Ah}$	
$\mathbf{q}$	Heat flux density.		Heat flux.
$\frac{d\mathbf{v}}{dt} = \frac{\partial \mathbf{v}}{\partial t} + \mathbf{v} \cdot \nabla\mathbf{v}$	Acceleration of a fluid particle.	$\frac{D\mathbf{v}}{Dt}$	Substantial derivative or material derivative of $\mathbf{v}$.

Chapter 8

GENERAL COORDINATES

8.1 GENERAL CURVILINEAR COORDINATES

Until now we have always labeled the points in space with a set of right-handed Cartesian coordinates. However, there is no need to make this restriction. Suppose that the Cartesian coordinates (x_1, x_2, x_3) are expressed in terms of new *curvilinear coordinates* (x^1, x^2, x^3) by the transformation

$$x_i = x_i(x^1, x^2, x^3). \tag{8.1}$$

If we vary x^1 while holding x^2 and x^3 constant, a space curve is generated called an x^1 *coordinate curve.* Similarly, x^2 and x^3 coordinate curves may be generated (Fig. 8.1). The position vector $\mathbf{r}(x^1, x^2, x^3) = x_i\mathbf{e}_i$, joining the origin O of the Cartesian system to a point P, can now be regarded as a function of the curvilinear coordinates (x^1, x^2, x^3). Then a triad of vectors tangent to the coordinate curves may be defined by

$$\boxed{\mathbf{g}_i = \frac{\partial \mathbf{r}}{\partial x^i}.} \tag{8.2}$$

The base vectors $(\mathbf{g}_1, \mathbf{g}_2, \mathbf{g}_3)$ are the ***covariant*** base vectors associated with the curvilinear coordinates (x^1, x^2, x^3). If the condition

$$\mathbf{g}_1 \times \mathbf{g}_2 \cdot \mathbf{g}_3 \neq 0$$

is satisfied in a region of space (the base vectors are linearly independent), then a single-valued inverse of the transformation (8.1) exists in that region:

$$x^1 = x^1(x_1, x_2, x_3), \qquad x^2 = x^2(x_1, x_2, x_3), \qquad x^3 = x^3(x_1, x_2, x_3). \tag{8.3}$$

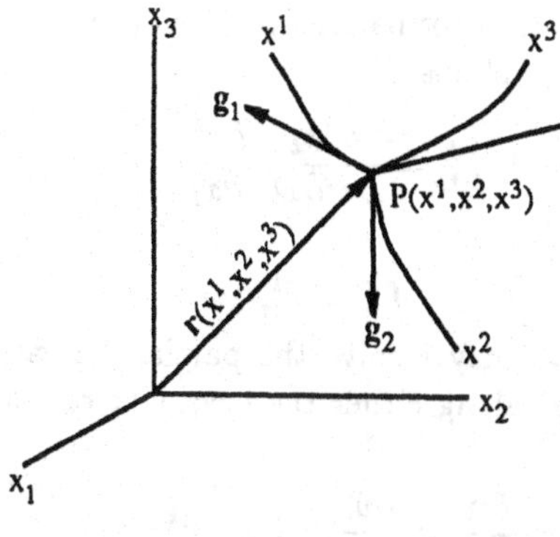

Figure 8.1. Arbitrary curvilinear coordinates (x^1, x^2, x^3).

Contravariant base vectors $(\mathbf{g}^1, \mathbf{g}^2, \mathbf{g}^3)$ may be defined by (4.23), and tensors can be resolved along either the covariant or contravariant base vectors using the formalism established in Section 4.5. Remember that if the coordinate system is not Cartesian, we must be careful with the position of the indices on tensor quantities. Note that the index i on the right of (8.2) is a superscript of x but may be regarded as being in the denominator of a quotient and hence as a subscript to the entire symbol; that is, the reciprocal of a superscript is a subscript.

The calculus of multivariable tensor fields whose spatial points are labeled with curvilinear coordinates (x^1, x^2, x^3) is a generalization of that developed in Chapter 6 for Cartesian coordinates. Now we have to take cognizance of the possibility that the covariant and contravariant base vectors may not be constant but may vary from point to point in space. The vectors $\dfrac{\partial \mathbf{g}_i}{\partial x^j}$ may be decomposed into a linear combination of the covariant base vectors in the form

$$\boxed{\frac{\partial \mathbf{g}_i}{\partial x^j} = \Gamma^k_{ij}\mathbf{g}_k.} \tag{8.4}$$

The quantity Γ^k_{ij} is called a *Christoffel symbol* and represents the kth contravariant component of the vector $\dfrac{\partial \mathbf{g}_i}{\partial x^j}$. Similarly, the decomposition of the vectors $\dfrac{\partial \mathbf{g}^i}{\partial x^j}$ along the contravariant base vectors is

$$\boxed{\frac{\partial \mathbf{g}^i}{\partial x^j} = -\Gamma^i_{kj}\mathbf{g}^k.} \tag{8.5}$$

The Christoffel symbols may be expressed solely in terms of components of the metric tensor and their derivatives:

$$\boxed{\Gamma^k_{ij} = \frac{1}{2}g^{k\ell}\left(\frac{\partial g_{i\ell}}{\partial x^j} + \frac{\partial g_{j\ell}}{\partial x^i} - \frac{\partial g_{ij}}{\partial x^\ell}\right).} \tag{8.6}$$

You can verify (8.5) and (8.6) by first partial differentiating (4.23) and (4.25) and then using the definition (8.4). The Christoffel symbols may also be expressed in terms of derivatives of the transformations (8.1) and (8.3) relating the Cartesian and curvilinear coordinate systems:

$$\Gamma^k_{ij} = \frac{\partial^2 x_\ell}{\partial x^i \partial x^j}\,\frac{\partial x^k}{\partial x_\ell}. \tag{8.7}$$

Note also that

$$\Gamma^k_{ij} = \Gamma^k_{ji}.$$

Using (8.4) or (8.5), we may resolve the partial derivatives of a vector field $\mathbf{u}(x^1, x^2, x^3) = u^i\mathbf{g}_i = u_i\mathbf{g}^i$ along either the covariant or contravariant base vectors:

$$\begin{aligned}\frac{\partial \mathbf{u}}{\partial x^j} &= \frac{\partial u^i}{\partial x^j}\mathbf{g}_i + \Gamma^i_{kj}u^k \\ &= \frac{\partial u_i}{\partial x^j}\mathbf{g}^i - \Gamma^k_{ij}u_k\mathbf{g}^i.\end{aligned}$$

We write these equations in the shorthand notation

$$\boxed{\begin{aligned}\frac{\partial \mathbf{u}}{\partial x^j} &= u^i{}_{,j}\mathbf{g}_i = u_{i,j}\mathbf{g}^i, \\ u^i{}_{,j} &= \frac{\partial u^i}{\partial x^j} + \Gamma^i_{kj}u^k, &(8.8)\\ u_{i,j} &= \frac{\partial u_i}{\partial x^j} - \Gamma^k_{ij}u_k. &(8.9)\end{aligned}}$$

The quantities $u^i{}_{,j}$ and $u_{i,j}$ are called *covariant derivatives* of u^i or u_i and represent the ith contravariant or covariant components, respectively, of the vector $\frac{\partial \mathbf{u}}{\partial x^j}$. The covariant derivatives have been defined so that the component form of $\frac{\partial \mathbf{u}}{\partial x^j}$ is the same as in a Cartesian coordinate system *except that the partial derivative sign is replaced by a comma.*

The covariant derivatives of other objects may also be computed. We define the covariant derivative of a scalar, vector, or higher-order tensor to be the same as the ordinary derivative; for example, we write

$$\mathbf{u}_{,i} = \frac{\partial \mathbf{u}}{\partial x^i}.$$

The various covariant derivatives of a second-order tensor $\mathbf{T} = T^{ij}\mathbf{g}_i\mathbf{g}_j = T_{ij}\mathbf{g}^i\mathbf{g}^j = T_i{}^j\mathbf{g}^i\mathbf{g}_j = T^i{}_j\mathbf{g}_i\mathbf{g}^j$ are defined so that the curvilinear components of $\frac{\partial \mathbf{T}}{\partial x^k}$ are analogous to the Cartesian components:

$$\mathbf{T}_{,k} = T^{ij}{}_{,k}\mathbf{g}_i\mathbf{g}_j = T_{ij,k}\mathbf{g}^i\mathbf{g}^j = T_i{}^j{}_{,k}\mathbf{g}^i\mathbf{g}_j = T^i{}_{j,k}\mathbf{g}_i\mathbf{g}^j,$$

where

$$\begin{aligned}T^{ij}{}_{,k} &= \frac{\partial T^{ij}}{\partial x^k} + \Gamma^i_{\ell k}T^{\ell j} + \Gamma^j_{\ell k}T^{i\ell}, \\ T_{ij,k} &= \frac{\partial T_{ij}}{\partial x^k} - \Gamma^\ell_{ik}T_{\ell j} - \Gamma^\ell_{jk}T_{i\ell}, &(8.10)\\ T_i{}^j{}_{,k} &= \frac{\partial T_i{}^j}{\partial x^k} + \Gamma^j_{\ell k}T_i{}^\ell - \Gamma^\ell_{ik}T_\ell{}^j, \\ T^i{}_{j,k} &= \frac{\partial T^i{}_j}{\partial x^k} + \Gamma^i_{\ell k}T^\ell{}_j - \Gamma^\ell_{jk}T^i{}_\ell.\end{aligned}$$

The covariant derivative of a product obeys the same rule as the ordinary derivative; for example,

$$(u_iT^{jk})_{,\ell} = u_{i,\ell}T^{jk} + u_iT^{jk}{}_{,\ell}. \qquad (8.11)$$

The covariant and contravariant base vectors and various components of the metric and permutation tensors are *covariantly constant*:

$$\mathbf{g}_{i,j} = \frac{\partial \mathbf{g}_i}{\partial x^j} - \Gamma^k_{ij}\mathbf{g}_k = 0,$$

$$g_{ij,k} = 0, \qquad e_{ijk,\ell} = 0. \tag{8.12}$$

The curvilinear components of the del operator may be obtained from (6.3) using the chain rule:

$$\mathbf{g}_i \cdot \boldsymbol{\nabla} = \frac{\partial \mathbf{r}}{\partial x^i} \cdot \mathbf{e}_j \frac{\partial}{\partial x_j} = \frac{\partial x_j}{\partial x^i} \frac{\partial}{\partial x_j} = \frac{\partial}{\partial x^i}.$$

Hence

$$\boldsymbol{\nabla} = \mathbf{g}^i \frac{\partial}{\partial x^i}. \tag{8.13}$$

Our formulas (6.4)–(6.8) for the gradient, divergence, and curl of tensor fields may be generalized to curvilinear coordinates simply by replacing the partial derivative signs by commas, $\mathbf{e}_i$ by $\mathbf{g}_i$, ϵ_{ijk} by e_{ijk}, and raising the appropriate indices:

$$\boldsymbol{\nabla} f = f_{,i}\mathbf{g}^i,$$

$$\boldsymbol{\nabla} \cdot \mathbf{u} = u^i{}_{,i}, \tag{8.14}$$

$$\boldsymbol{\nabla} \times \mathbf{u} = e^{ijk} u_{j,i}\mathbf{g}_k, \tag{8.15}$$

$$\boldsymbol{\nabla}\mathbf{u} = u_{j,i}\mathbf{g}^i\mathbf{g}^j, \tag{8.16}$$

$$\boldsymbol{\nabla} \cdot \mathbf{T} = T^{ij}{}_{,i}\mathbf{g}_j.$$

Note from equation (8.16) that the covariant derivatives $u_{j,i}$ are the covariant components of the second-order tensor $\boldsymbol{\nabla}\mathbf{u}$.

The covariant derivative may be applied more than once; successive indices following a comma indicate successive covariant differentiation. The formulas (6.15) and (6.19) for the Laplacian and Hessian are readily generalized; for example,

$$\nabla^2 f = g^{ij} f_{,ij}, \tag{8.17}$$

$$\boldsymbol{\nabla\nabla}\mathbf{u} = u_{i,jk}\mathbf{g}^k\mathbf{g}^j\mathbf{g}^i. \tag{8.18}$$

Note from equation (8.18) that the second covariant derivatives $u_{i,jk}$ are the covariant components of the third-order tensor $\boldsymbol{\nabla\nabla}\mathbf{u}$. Covariantly differentiating (8.9), we obtain

$$u_{i,jk} - u_{i,kj} = R^\ell{}_{ijk} u_\ell, \tag{8.19}$$

$$R^\ell{}_{ijk} = \frac{\partial \Gamma^\ell_{ik}}{\partial x^j} - \frac{\partial \Gamma^\ell_{ij}}{\partial x^k} + \Gamma^m_{ik}\Gamma^\ell_{mj} - \Gamma^m_{ij}\Gamma^\ell_{mk}. \tag{8.20}$$

Since u_i are the components of a vector and $u_{i,jk}$ and $u_{i,kj}$ are the components of third-order tensors, $R_{ijk\ell}$ are the components of a fourth-order tensor called the *Riemann curvature tensor* or *Riemann-Christoffel tensor*. Using (8.6) and (8.20), you can verify that

$$R_{ijk\ell} = -R_{jik\ell} = -R_{ij\ell k} = R_{k\ell ij} \qquad \text{and} \qquad R_{ijk\ell} + R_{ik\ell j} + R_{i\ell jk} = 0$$

so most of the components of the Riemann curvature tensor are zero and only six of the remaining components are distinct. The covariant derivatives of the components of the Riemann curvature tensor are related by the *Bianchi identities*

$$R_{ijk\ell,m} + R_{ij\ell m,k} + R_{ijmk,\ell} = 0. \tag{8.21}$$

The Riemann curvature tensor will turn out to be important in our subsequent work.

Line, surface, and volume integrals can also be performed in curvilinear coordinates. In order to derive formulas for the length, area, and volume of elemental regions, let us form an elemental parallelepiped with edges $(\mathbf{g}_1\Delta x^1, \mathbf{g}_2\Delta x^2, \mathbf{g}_3\Delta x^3)$ as sketched in Fig. 8.2. Then the square of the arc length Δs of the line joining the two opposite corners $P(s)$ and $P(s+\Delta s)$ of the elemental parallelepiped is

$$\boxed{\Delta s^2 = \Delta\mathbf{r}\cdot\Delta\mathbf{r} = \mathbf{g}_i\cdot\mathbf{g}_j\Delta x^i\Delta x^j = g_{ij}\Delta x^i\Delta x^j.} \tag{8.22}$$

This formula explains why $\mathbf{1} = g_{ij}\mathbf{g}^i\mathbf{g}^j$ is called the *metric* tensor; the g_{ij}'s are a set of scale factors for converting increments in x^i to changes in length. The surface area ΔS_1 of a face of the elemental parallelepiped on which x^1 = constant is

$$\boxed{\Delta S_1 = |\mathbf{g}_2\times\mathbf{g}_3|\Delta x^2\Delta x^3 = |g_{22}g_{33} - g_{23}^2|^{1/2}\Delta x^2\Delta x^3} \tag{8.23}$$

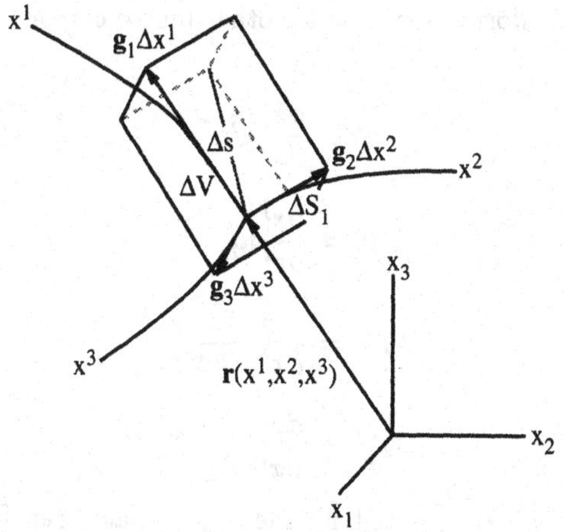

Figure 8.2. Elemental parallelepiped used for deriving formulas for arc length, surface area, and volume in curvilinear coordinates.

with similar formulas holding for the other two faces. The volume of the elemental parallelepiped is, from (4.28):

$$\boxed{\Delta V = |\mathbf{g}_1 \times \mathbf{g}_2 \cdot \mathbf{g}_3| \Delta x^1 \Delta x^2 \Delta x^3 = \sqrt{g}\Delta x^1 \Delta x^2 \Delta x^3.} \tag{8.24}$$

Now suppose we introduce a new set of curvilinear coordinates $(\bar{x}^1, \bar{x}^2, \bar{x}^3)$ with the understanding that the relations between the new and the old curvilinear coordinates (x^1, x^2, x^3) are known. By the chain rule, the new base vectors are a linear combination of the old base vectors with coefficients that are elements of the Jacobian matrix characterizing the transformation:

$$\bar{\mathbf{g}}_i = \frac{\partial \mathbf{r}}{\partial \bar{x}^i} = \frac{\partial x^j}{\partial \bar{x}^i}\frac{\partial \mathbf{r}}{\partial x^j} = \frac{\partial x^j}{\partial \bar{x}^i}\mathbf{g}_j.$$

Resolving a vector $\mathbf{u}$ along either set of base vectors:

$$\mathbf{u} = u^i \mathbf{g}_i = \bar{u}^i \bar{\mathbf{g}}_i,$$

we can write the transformation laws between components as

$$\boxed{\bar{u}^i = \frac{\partial \bar{x}^i}{\partial x^j} u^j, \quad u^i = \frac{\partial x^i}{\partial \bar{x}^j}\bar{u}^j.}$$

These formulas have a simple structure that makes it easy to figure out the correct location of the indices and overbars. Understanding this structure, we can quickly write down the transformation laws for other tensor components; for example,

$$\begin{aligned}
\bar{\mathbf{g}}^i &= \frac{\partial \bar{x}^i}{\partial x^j}\mathbf{g}^j, \\
\bar{u}_i &= \frac{\partial x^j}{\partial \bar{x}^i}u_j, \\
\bar{T}^{ij} &= \frac{\partial \bar{x}^i}{\partial x^k}\frac{\partial \bar{x}^j}{\partial x^\ell}T^{k\ell}, \\
\bar{T}_i{}^j &= \frac{\partial x^k}{\partial \bar{x}^i}\frac{\partial \bar{x}^j}{\partial x^\ell}T_k{}^\ell.
\end{aligned} \tag{8.25}$$

If indexed quantities are related by the appropriate transformation laws for change of curvilinear coordinates, they can be regarded as the components of tensor. Conversely, some indexed quantities such as the partial derivatives $\frac{\partial u_i}{\partial x^j}$ and the Christoffel symbols Γ^i_{jk} do not obey the correct transformation laws and hence are not the components of any one tensor in every coordinate system.

Finally, we note that there is a very simple method of proving the validity of tensor formulas. If we can show that the components of a tensor are zero in an particular coordinate system, then the components must be zero in all coordinate systems. To illustrate, consider the Riemann curvature tensor, whose components in an arbitrary system are defined by (8.20). In a Euclidean space, we can always select a Cartesian coordinate system in which the components g_{ij} of the metric tensor are equal to δ_{ij}. Since the right side of (8.20) vanishes in a Cartesian coordinate system, the Riemann curvature tensor must be identically zero:

$$\boxed{R_{ijkl} = 0.} \tag{8.26}$$

The identities (8.26) are equivalent to a set of six second-order partial differential equations that must be obeyed by the components $g_{ij}(x^1, x^2, x^3)$ of the metric tensor. Conversely, it can be shown that if the six components $g_{ij}(x^1, x^2, x^3)$ of a positive definite symmetric tensor satisfy (8.26), then a curvilinear coordinate system (x^1, x^2, x^3) with corresponding g_{ij} can be embedded in the three-dimensional Euclidean space, and a coordinate transformation exists which reduces the components of the metric tensor to δ_{ij}.

Armed with the formulas in this section, we can write the equations of engineering and physics in terms of arbitrary curvilinear coordinates.

Newtonian mechanics of a particle in curvilinear coordinates. As an example, let us figure out the form of Newton's law (5.12) in a curvilinear coordinate system (x^1, x^2, x^3) fixed in space (an inertial system). The position vector $\mathbf{r}(x^1(t), x^2(t), x^3(t))$ of a particle is now regarded as being a function of its curvilinear coordinates, which are in turn functions of the time t. The contravariant components of the velocity vector may be obtained from the chain rule and the definition (8.2):

$$\frac{d\mathbf{r}}{dt} = \frac{dx^i}{dt}\frac{\partial \mathbf{r}}{\partial x^i} = \frac{dx^i}{dt}\mathbf{g}_i.$$

The contravariant components of the acceleration vector may be obtained from the chain rule and (8.4):

$$\frac{d^2\mathbf{r}}{dt^2} = \frac{d^2x^i}{dt^2}\mathbf{g}_i + \frac{dx^j}{dt}\frac{dx^k}{dt}\frac{\partial \mathbf{g}_j}{\partial x^k} = \left(\frac{d^2x^i}{dt^2} + \Gamma^i_{jk}\frac{dx^j}{dt}\frac{dx^k}{dt}\right)\mathbf{g}_i.$$

Newton's law (5.12a) thus transforms into

$$m\left(\frac{d^2x^i}{dt^2} + \Gamma^i_{jk}\frac{dx^j}{dt}\frac{dx^k}{dt}\right)\mathbf{g}_i = f^i\mathbf{g}_i.$$

Equating the contravariant components, we arrive at the following component form of Newton's law in arbitrary curvilinear coordinates:

$$m\left(\frac{d^2x^i}{dt^2} + \Gamma^i_{jk}\frac{dx^j}{dt}\frac{dx^k}{dt}\right) = f^i. \tag{8.27}$$

PROBLEMS 8.1

1. The *Jacobian determinant* of the transformation (8.1) is defined by

$$J = \begin{vmatrix} \dfrac{\partial x_1}{\partial x^1} & \dfrac{\partial x_1}{\partial x^2} & \dfrac{\partial x_1}{\partial x^3} \\ \dfrac{\partial x_2}{\partial x^1} & \dfrac{\partial x_2}{\partial x^2} & \dfrac{\partial x_2}{\partial x^3} \\ \dfrac{\partial x_3}{\partial x^1} & \dfrac{\partial x_3}{\partial x^2} & \dfrac{\partial x_3}{\partial x^3} \end{vmatrix}.$$

   Show that $J^2 = (\mathbf{g}_1 \times \mathbf{g}_2 \cdot \mathbf{g}_3)^2 = g$.

2. Verify the identity $\Gamma^j_{ij} = \dfrac{1}{2g}\dfrac{\partial g}{\partial x^i}$.

3. Show that

$$\nabla \cdot \mathbf{u} = \frac{1}{\sqrt{g}}\frac{\partial}{\partial x^i}(\sqrt{g}u^i) \quad \text{and} \quad \nabla^2 f = \frac{1}{\sqrt{g}}\frac{\partial}{\partial x^i}\left(\sqrt{g}g^{ij}\frac{\partial f}{\partial x^j}\right).$$

4. List six distinct nonzero components of the Riemann curvature tensor in three dimensions.

5. Write down the component form of Stokes' theorem and the divergence theorem in arbitrary curvilinear coordinates.

6. Show that the partial derivatives $\dfrac{\partial u_i}{\partial x^j}$ and the Christoffel symbols Γ^k_{ij} do not obey the correct laws for transformation of coordinates but the covariant derivatives $u_{i,j}$ do.

7. Verify (8.11) and (8.12) and show that $f_{,ij} = f_{,ji}$.

8. Write down the equation of a straight line in arbitrary curvilinear coordinates (x^1, x^2, x^3).

9. Figure out a component form of the Navier-Stokes equations (7.39) in arbitrary curvilinear coordinates.

8.2 ORTHOGONAL CURVILINEAR COORDINATES

In this section we specialize the general formulas in the preceding section to the case when the curvilinear coordinate curves intersect at right angles. The covariant base vectors defined by (8.2) are then orthogonal but not necessarily of unit length. If we denote the lengths of the covariant base vectors by (h_1, h_2, h_3), an orthonormal curvilinear triad $(\bar{\mathbf{e}}_1(x^1, x^2, x^3), \bar{\mathbf{e}}_2(x^1, x^2, x^3), \bar{\mathbf{e}}_3(x^1, x^2, x^3))$ may be defined by (4.29). Here we use a bar over the base vectors to emphasize that they may be functions of the curvilinear coordinates (x^1, x^2, x^3). We suppose that the coordinate curves are labeled in such a way that the curvilinear triads of base vectors $(\mathbf{g}_1, \mathbf{g}_2, \mathbf{g}_3)$, $(\mathbf{g}^1, \mathbf{g}^2, \mathbf{g}^3)$, and $(\bar{\mathbf{e}}_1, \bar{\mathbf{e}}_2, \bar{\mathbf{e}}_3)$ are right-handed.

The contravariant base vectors and the various components of the metric tensor are given by formulas at the end of Section 4.5. Some of the Christoffel symbols, obtained with the use of (8.6), are:

$$\Gamma^1_{11} = \frac{1}{h_1}\frac{\partial h_1}{\partial x^1}, \qquad \Gamma^1_{12} = \frac{1}{h_1}\frac{\partial h_1}{\partial x^2}, \qquad \Gamma^2_{11} = -\frac{h_1}{h_2^2}\frac{\partial h_1}{\partial x^2}, \qquad \Gamma^3_{12} = 0. \tag{8.28}$$

The rest of the Christoffel symbols can be obtained by changing the indices in these results. The analogs of (8.4) for the derivatives of the curvilinear unit base vectors are

$$\frac{\partial \bar{\mathbf{e}}_1}{\partial x^1} = -\frac{1}{h_2}\frac{\partial h_1}{\partial x^2}\bar{\mathbf{e}}_2 - \frac{1}{h_3}\frac{\partial h_1}{\partial x^3}\bar{\mathbf{e}}_3,$$

$$\frac{\partial \bar{\mathbf{e}}_1}{\partial x^2} = \frac{1}{h_1}\frac{\partial h_2}{\partial x^1}\bar{\mathbf{e}}_2, \qquad \frac{\partial \bar{\mathbf{e}}_1}{\partial x^3} = \frac{1}{h_1}\frac{\partial h_3}{\partial x^1}\bar{\mathbf{e}}_3,$$

with similar results for $\bar{\mathbf{e}}_2$ and $\bar{\mathbf{e}}_3$.

We now resolve tensors into their physical components along the triad $(\bar{\mathbf{e}}_1, \bar{\mathbf{e}}_2, \bar{\mathbf{e}}_3)$; for example,

$$\mathbf{u}(x^1, x^2, x^3) = \bar{u}_i(x^1, x^2, x^3)\bar{\mathbf{e}}_i(x^1, x^2, x^3).$$

The del operator (8.13) becomes

$$\nabla = \frac{\bar{\mathbf{e}}_1}{h_1}\frac{\partial}{\partial x^1} + \frac{\bar{\mathbf{e}}_2}{h_2}\frac{\partial}{\partial x^2} + \frac{\bar{\mathbf{e}}_3}{h_3}\frac{\partial}{\partial x^3}. \tag{8.29}$$

The analogs of (8.14), (8.15), and (8.17) may be obtained with the help of the results of Problem 3 in Section 4.5 and Problem 3 in Section 8.1:

$$\nabla \cdot \mathbf{u} = \frac{1}{h_1 h_2 h_3}\left[\frac{\partial}{\partial x^1}(h_2 h_3 \bar{u}_1) + \frac{\partial}{\partial x^2}(h_1 h_3 \bar{u}_2) + \frac{\partial}{\partial x^3}(h_1 h_2 \bar{u}_3)\right], \tag{8.30}$$

$$\nabla \times \mathbf{u} = \frac{1}{h_1 h_2 h_3}\begin{vmatrix} h_1\bar{\mathbf{e}}_1 & h_2\bar{\mathbf{e}}_2 & h_3\bar{\mathbf{e}}_3 \\ \dfrac{\partial}{\partial x^1} & \dfrac{\partial}{\partial x^2} & \dfrac{\partial}{\partial x^3} \\ h_1\bar{u}_1 & h_2\bar{u}_2 & h_3\bar{u}_3 \end{vmatrix}, \tag{8.31}$$

$$\nabla^2 f = \frac{1}{h_1 h_2 h_3}\left[\frac{\partial}{\partial x^1}\left(\frac{h_2 h_3}{h_1}\frac{\partial f}{\partial x^1}\right) + \frac{\partial}{\partial x^2}\left(\frac{h_1 h_3}{h_2}\frac{\partial f}{\partial x^2}\right) + \frac{\partial}{\partial x^3}\left(\frac{h_1 h_2}{h_3}\frac{\partial f}{\partial x^3}\right)\right]. \tag{8.32}$$

The formulas (8.22)–(8.24) for the length, area, and volume of elemental regions become simply

$$\Delta s^2 = h_1^2(\Delta x^1)^2 + h_2^2(\Delta x^2)^2 + h_3^2(\Delta x^3)^2, \tag{8.33}$$

$$\Delta S_1 = h_2 h_3 \Delta x^2 \Delta x^3, \tag{8.34}$$

$$\Delta V = h_1 h_2 h_3 \Delta x^1 \Delta x^2 \Delta x^3. \tag{8.35}$$

Below we summarize some of these formulas for the two most commonly used curvilinear coordinate systems.

Circular cylindrical coordinates. When there is an axis of symmetry in a problem, it is convenient to use circular cylindrical coordinates (ρ, θ, z) to locate points in space. Here (ρ, θ) are just the polar coordinates in a plane perpendicular to the axis of symmetry and z is measured along the axis of symmetry. If we establish a set of Cartesian axes as shown in Fig. 8.3, then the Cartesian and cylindrical coordinates are related by the familiar transformations

$$x_1 = \rho\cos\theta, \qquad x_2 = \rho\sin\theta, \qquad x_3 = z. \tag{8.36}$$

Since the position vector from the origin O of the Cartesian coordinate system to a point P is

$$\mathbf{r} = \rho\cos\theta\mathbf{e}_1 + \rho\sin\theta\mathbf{e}_2 + z\mathbf{e}_3$$

the covariant base vectors (8.2) are

$$\mathbf{g}_1 = \cos\theta\mathbf{e}_1 + \sin\theta\mathbf{e}_2, \qquad \mathbf{g}_2 = -\rho\sin\theta\mathbf{e}_1 + \rho\cos\theta\mathbf{e}_2, \qquad \mathbf{g}_3 = \mathbf{e}_3.$$

The scale factors (h_1, h_2, h_3) are just the magnitudes of $(\mathbf{g}_1, \mathbf{g}_2, \mathbf{g}_3)$:

$$h_1 = 1, \qquad h_2 = \rho, \qquad h_3 = 1.$$

The unit curvilinear base vectors $(\bar{\mathbf{e}}_1, \bar{\mathbf{e}}_2, \bar{\mathbf{e}}_3)$ are now denoted by $(\mathbf{e}_\rho, \mathbf{e}_\theta, \mathbf{e}_z)$ and given by

$$\mathbf{e}_\rho = \cos\theta\mathbf{e}_1 + \sin\theta\mathbf{e}_2, \qquad \mathbf{e}_\theta = -\sin\theta\mathbf{e}_1 + \cos\theta\mathbf{e}_2, \qquad \mathbf{e}_z = \mathbf{e}_3.$$

We also denote the physical components of tensors by letter subscripts; for example,

$$\mathbf{u} = u_\rho\mathbf{e}_\rho + u_\theta\mathbf{e}_\theta + u_z\mathbf{e}_z.$$

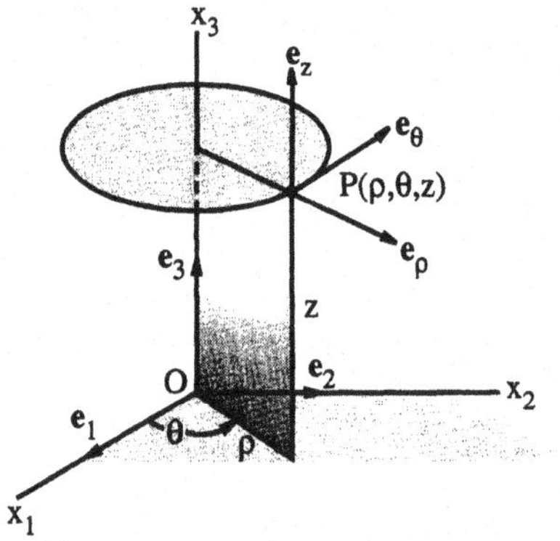

Figure 8.3. Circular cylindrical coordinates.

The formulas (8.29)–(8.35) imply

$$\nabla f = \frac{\partial f}{\partial \rho}\mathbf{e}_\rho + \frac{1}{\rho}\frac{\partial f}{\partial \theta}\mathbf{e}_\theta + \frac{\partial f}{\partial z}\mathbf{e}_z, \tag{8.37}$$

$$\nabla \cdot \mathbf{u} = \frac{1}{\rho}\frac{\partial}{\partial \rho}(\rho u_\rho) + \frac{1}{\rho}\frac{\partial u_\theta}{\partial \theta} + \frac{\partial u_z}{\partial z}, \tag{8.38}$$

$$\nabla \times \mathbf{u} = \frac{1}{\rho}\begin{vmatrix} \mathbf{e}_\rho & \rho \mathbf{e}_\theta & \mathbf{e}_z \\ \dfrac{\partial}{\partial \rho} & \dfrac{\partial}{\partial \theta} & \dfrac{\partial}{\partial z} \\ u_\rho & \rho u_\theta & u_z \end{vmatrix},$$

$$\nabla^2 f = \frac{1}{\rho}\frac{\partial}{\partial \rho}\left(\rho\frac{\partial f}{\partial \rho}\right) + \frac{1}{\rho^2}\frac{\partial^2 f}{\partial \theta^2} + \frac{\partial^2 f}{\partial z^2}, \tag{8.39}$$

$$\Delta s^2 = \Delta \rho^2 + \rho^2 \Delta\theta^2 + \Delta z^2,$$

$$\Delta S_\rho = \rho \Delta\theta \Delta z,$$

$$\Delta V = \rho \Delta\rho \Delta\theta \Delta z.$$

Spherical coordinates. When there is a center of symmetry in a problem, it is convenient to use spherical coordinates (r, ϕ, θ) to locate points in space. Here r is the distance from the center of symmetry O to a point P, ϕ is the angle measured down from the polar axis to the line OP, and θ is the polar angle measured around the polar axis. If we establish a set of Cartesian axes as shown in Fig. 8.4, then the Cartesian and spherical coordinates are related by the transformations

$$x_1 = r \sin\phi \cos\theta, \qquad x_2 = r\sin\phi\sin\theta, \qquad x_3 = r\cos\phi. \tag{8.40}$$

The scale factors (h_1, h_2, h_3) turn out to be

$$h_1 = 1, \qquad h_2 = r, \qquad h_3 = r\sin\phi.$$

The unit curvilinear base vectors $(\bar{\mathbf{e}}_1, \bar{\mathbf{e}}_2, \bar{\mathbf{e}}_3)$ are now

$$\mathbf{e}_r = \sin\phi\cos\theta\,\mathbf{e}_1 + \sin\phi\sin\theta\,\mathbf{e}_2 + \cos\phi\,\mathbf{e}_3,$$

$$\mathbf{e}_\phi = \cos\phi\cos\theta\,\mathbf{e}_1 + \cos\phi\sin\theta\,\mathbf{e}_2 - \sin\phi\,\mathbf{e}_3,$$

$$\mathbf{e}_\theta = -\sin\theta\,\mathbf{e}_1 + \cos\theta\,\mathbf{e}_2.$$

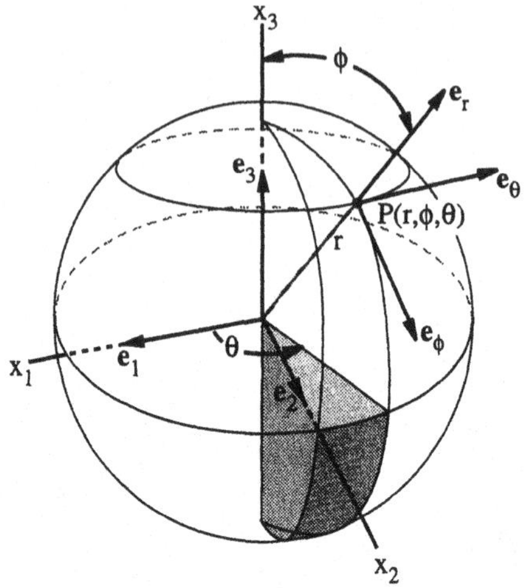

Figure 8.4. Spherical coordinates.

The formulas (8.29)–(8.35) imply

$$\nabla f = \frac{\partial f}{\partial r}\mathbf{e}_r + \frac{1}{r}\frac{\partial f}{\partial \phi}\mathbf{e}_\phi + \frac{1}{r\sin\phi}\frac{\partial f}{\partial \theta}\mathbf{e}_\theta,$$

$$\nabla\cdot\mathbf{u} = \frac{1}{r^2}\frac{\partial}{\partial r}(r^2 u_r) + \frac{1}{r\sin\phi}\frac{\partial}{\partial\phi}(\sin\phi\, u_\phi) + \frac{1}{r\sin\phi}\frac{\partial u_\theta}{\partial\theta},$$

$$\nabla\times\mathbf{u} = \frac{1}{r^2\sin\phi}\begin{vmatrix} \mathbf{e}_r & r\mathbf{e}_\phi & r\sin\phi\,\mathbf{e}_\theta \\ \dfrac{\partial}{\partial r} & \dfrac{\partial}{\partial\phi} & \dfrac{\partial}{\partial\theta} \\ u_r & ru_\phi & r\sin\phi\, u_\theta \end{vmatrix},$$

$$\nabla^2 f = \frac{1}{r^2}\frac{\partial}{\partial r}\left(r^2\frac{\partial f}{\partial r}\right) + \frac{1}{r^2\sin\phi}\frac{\partial}{\partial\phi}\left(\sin\phi\frac{\partial f}{\partial\phi}\right) + \frac{1}{r^2\sin^2\phi}\frac{\partial^2 f}{\partial\theta^2},$$

$$\Delta s^2 = \Delta r^2 + r^2\Delta\phi^2 + r^2\sin^2\phi\,\Delta\theta^2,$$

$$\Delta S_r = r^2\sin\phi\,\Delta\phi\Delta\theta,$$

$$\Delta V = r^2\sin\phi\,\Delta r\Delta\phi\,\Delta\theta.$$

Laminar flow in a circular pipe. As an example of the utility of curvilinear coordinate systems, let us reconsider the problem of finding the velocity and pressure fields that satisfy the continuity equation (7.36) for incompressible flows

$$\nabla\cdot\mathbf{v} = 0,$$

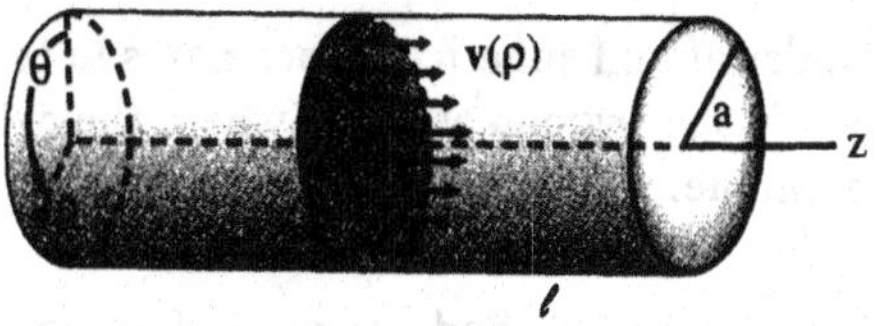

Figure 8.5. Laminar flow in a pipe of circular cross section: circular cylindrical coordinates.

the Navier-Stokes equations (7.39a) for steady flows

$$\rho \mathbf{v} \cdot \nabla \mathbf{v} = -\nabla p + \mu \nabla^2 \mathbf{v},$$

and the no-slip boundary condition $\mathbf{v} = 0$ on the wall of a circular cylindrical pipe. From symmetry considerations, we now use cylindrical coordinates (ρ, θ, z) with the z-axis measured along the pipe axis (Fig. 8.5) and assume that

$$v_\rho = v_\theta = 0, \qquad v_z = v_z(\rho).$$

From the formulas (8.37)–(8.39), we find that the continuity equation is satisfied identically and the Navier-Stokes equations reduce to

$$\frac{\partial p}{\partial \rho} = \frac{\partial p}{\partial \theta} = 0, \tag{8.41}$$

$$\frac{\mu}{\rho} \frac{d}{d\rho}\left(\rho \frac{dv_z}{d\rho}\right) = \frac{\partial p}{\partial z}. \tag{8.42}$$

Equation (8.41) implies that the pressure p cannot depend upon the coordinates ρ or θ. Since the left side of (8.42) is a function of ρ only and the right side of (8.42) is a function of z only, each side must be equal to a constant. Equation (8.42) can then be readily integrated to yield

$$v_z = \frac{1}{4\mu} \frac{dp}{dz} \rho^2 + c_1 \ln \rho + c_2.$$

We choose the constant $c_1 = 0$ to eliminate the singularity at $\rho = 0$, and determine the constant c_2 from the no-slip condition $v_z(a) = 0$. Thus

$$v_z = -\frac{1}{4\mu} \frac{dp}{dz} (a^2 - \rho^2).$$

Of course, this solution is equivalent to the one given in terms of Cartesian coordinates in Section 7.3.

PROBLEMS 8.2

1. Express the cylindrical and spherical coordinates in terms of Cartesian coordinates, and specify the regions of space where the cylindrical and spherical coordinates are unique.

2. For Problem 7 of Section 5.2, find the path of an ant.

3. A heat-conducting spherical shell has inner radius a_1 and outer radius a_2. If the inner surface is held at temperature T_1 and the outer surface at temperature T_2, determine the steady-state temperature distribution in the shell.

4. From the solution in Section 8.2 for the torsion of a circular cylindrical rod, determine the physical components of the displacement and stress fields in cylindrical coordinates.

5. Find a spherically symmetric solution, representing a wave propagating outward from the origin, to the acoustic wave equation derived in Problem 15 of Section 7.3.

6. By solving Newton's law (7.55) in spherical coordinates, rederive the solution (7.57) and (7.58) for the orbit of a point mass about a spherical body.

7. Resolve the dipole field

$$\boldsymbol{\mathcal{E}} = \frac{(3\mathbf{rr} - r^2\mathbf{1}) \cdot \mathbf{e}_3}{r^5}$$

into spherical components, then find and sketch the equations of the field lines in spherical coordinates.

8. Determine the electrostatic field around a long straight wire of circular cross section having a constant charge density λ (charge per unit length).

9. Use the Biot-Savart law (Problem 5 in Section 7.5) in cylindrical coordinates to find the magnetic field $\boldsymbol{B}$ around a long straight wire carrying a current I.

10. *Parabolic cylindrical coordinates* (x, y, z) are related to Cartesian coordinates (x_1, x_2, x_3) by the transformations $x_1 = \frac{1}{2}(x^2 - y^2)$, $x_2 = xy$, $x_3 = z$. Sketch the (x, y) coordinate curves, determine the scale factors (h_1, h_2, h_3), and find the unit vectors $(\mathbf{e}_x, \mathbf{e}_y, \mathbf{e}_z)$.

8.3 SURFACE COORDINATES

We can specify a surface in three-dimensional Euclidean space by expressing its Cartesian coordinates (x_1, x_2, x_3) as functions of *two* curvilinear coordinates (x^1, x^2):

$$x_1 = x_1(x^1, x^2), \qquad x_2 = x_2(x^1, x^2), \qquad x_3 = x_3(x^1, x^2). \tag{8.43}$$

If we vary x^1 while holding x^2 constant, an x^1 coordinate curve lying on the surface is generated; an x^2 coordinate curve may be similarly generated (Fig. 8.6). A pair of vectors tangent to the coordinate curves on the surface may be defined by

$$\boxed{\mathbf{g}_\alpha = \frac{\partial \mathbf{r}}{\partial x^\alpha}, \qquad \alpha = 1 \text{ or } 2.} \tag{8.44}$$

Here and throughout this section, we adopt the convention that *Greek indices can have only the values 1 or 2.* The vectors $(\mathbf{g}_1, \mathbf{g}_2)$ are the covariant base vectors associated with the surface coordinates (x^1, x^2). If the condition

$$\mathbf{g}_1 \times \mathbf{g}_2 \neq 0$$

is satisfied in a region on the surface, then a single-valued inverse of the transformation (8.43) exists in that region.

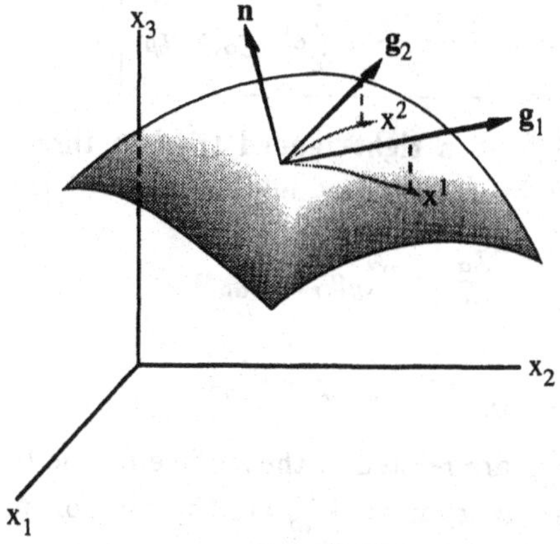

Figure 8.6. Surface curvilinear coordinates (x^1, x^2).

Contravariant base vectors $(\mathbf{g}^1, \mathbf{g}^2)$, which along with the covariant base vectors $(\mathbf{g}_1, \mathbf{g}_2)$ lie in a plane tangent to the surface, may be defined by

$$\mathbf{g}^\alpha \cdot \mathbf{g}_\beta = \delta^\alpha_\beta.$$

A *surface tensor* is a tensor that can be resolved along $(\mathbf{g}_1, \mathbf{g}_2)$ or $(\mathbf{g}^1, \mathbf{g}^2)$; for example,

$$\mathbf{v} = v^\alpha \mathbf{g}_\alpha, \qquad \mathbf{T} = T^\alpha_{\ \beta} \mathbf{g}_\alpha \mathbf{g}^\beta.$$

Many of the mathematical formulas that we have developed for tensor fields of three variables in arbitrary curvilinear coordinates may be specialized to surface tensor fields of two variables simply by *replacing the Latin indices by Greek indices.* For instance, the replacement of Latin indices by Greek indices in (4.25) yields the following formulas for the covariant and contravariant components of the *surface metric tensor*:

$$g_{\alpha\beta} = \mathbf{g}_\alpha \cdot \mathbf{g}_\beta, \qquad g^{\alpha\beta} = \mathbf{g}^\alpha \cdot \mathbf{g}^\beta. \tag{8.45}$$

From now on we explicitly write down only formulas that are somewhat *different* from their three-variable counterparts.

The covariant components of the *surface permutation tensor* are defined by

$$e_{\alpha\beta} = \sqrt{g}\epsilon_{\alpha\beta},$$

where now $g = g_{11}g_{22} - g_{12}^2$ and $\epsilon_{\alpha\beta}$ is the two-dimensional permutation symbol with components

$$\epsilon_{11} = \epsilon_{22} = 0, \qquad \epsilon_{12} = -\epsilon_{21} = 1.$$

The surface permutation tensor may be used to calculate the cross product of two surface vectors $\mathbf{u} = u^\alpha \mathbf{g}_\alpha$ and $\mathbf{v} = v^\beta \mathbf{g}_\beta$:

$$\mathbf{u} \times \mathbf{v} = e_{\alpha\beta} u^\alpha v^\beta \mathbf{n}.$$

Here $\mathbf{n}$ is the unit normal to the surface defined by

$$\boxed{\mathbf{n} = \frac{\mathbf{g}_1 \times \mathbf{g}_2}{\sqrt{g}} = \frac{1}{2} e^{\alpha\beta} \mathbf{g}_\alpha \times \mathbf{g}_\beta.} \tag{8.46}$$

The vectors $(\mathbf{g}_1, \mathbf{g}_2, \mathbf{n})$ form a right-handed triad in three dimensions (Fig. 8.6). Their derivatives may be decomposed into the form (8.4) (with $\mathbf{g}_3 = \mathbf{n}$):

$$\frac{\partial \mathbf{g}_\alpha}{\partial x^\beta} = \Gamma^\gamma_{\alpha\beta} \mathbf{g}_\gamma + \Gamma^3_{\alpha\beta} \mathbf{n}, \tag{8.47}$$

$$\frac{\partial \mathbf{n}}{\partial x^\alpha} = \Gamma^\beta_{3\alpha} \mathbf{g}_\beta + \Gamma^3_{3\alpha} \mathbf{n}. \tag{8.48}$$

The Christoffel symbols $\Gamma^\gamma_{\alpha\beta}$ are related to the surface metric tensor by the usual formulas (8.6). The Christoffel symbols $\Gamma^3_{\alpha\beta}$ are the components of a symmetric surface tensor $\mathbf{K}$ called the *surface curvature tensor*:

$$\Gamma^3_{\alpha\beta} = K_{\alpha\beta}.$$

The following formulas for the components of the surface curvature tensor are a consequence of (8.47) and the fact that $\mathbf{g}_\alpha \cdot \mathbf{n} = 0$:

$$\begin{aligned} K_{\alpha\beta} &= \frac{\partial \mathbf{g}_\alpha}{\partial x^\beta} \cdot \mathbf{n} \\ &= \frac{\partial}{\partial x^\beta}(\mathbf{g}_\alpha \cdot \mathbf{n}) - \mathbf{g}_\alpha \cdot \frac{\partial \mathbf{n}}{\partial x^\beta} = -\mathbf{g}_\alpha \cdot \frac{\partial \mathbf{n}}{\partial x^\beta}. \end{aligned} \tag{8.49}$$

We can infer from this that $K_{\alpha\beta}$ is a measure of the change in the base vector $\mathbf{g}_\alpha$ in the direction of $\mathbf{n}$, and the change in the normal vector $\mathbf{n}$ in the direction of $-\mathbf{g}_\alpha$, along the x^β coordinate curve. The *mean curvature* K_M and the *Gaussian curvature* or *total curvature* K_G are scalar invariants of the surface curvature tensor defined by

$$K_M = \tfrac{1}{2}\,\mathrm{tr}\,\mathbf{K} = \tfrac{1}{2}K^\alpha_\alpha = \tfrac{1}{2}(K^1_1 + K^2_2), \tag{8.50}$$

$$K_G = \det\mathbf{K} = \tfrac{1}{2}(K^\alpha_\alpha K^\beta_\beta - K^\alpha_\beta K^\beta_\alpha) = K^1_1 K^2_2 - K^1_2 K^2_1. \tag{8.51}$$

Now using a comma to denote *surface* covariant differentiation, we can write (8.47) in the shorthand notation

$$\mathbf{g}_{\alpha,\beta} = \frac{\partial \mathbf{g}_\alpha}{\partial x^\beta} - \Gamma^\gamma_{\alpha\beta}\mathbf{g}_\gamma = K_{\alpha\beta}\mathbf{n}. \tag{8.52}$$

Using (8.48) and (8.49) to show that

$$\Gamma^\beta_{3\alpha} = -K^\beta_\alpha,$$

we can also obtain

$$\mathbf{g}^\alpha{}_{,\beta} = \frac{\partial \mathbf{g}^\alpha}{\partial x^\beta} + \Gamma^\alpha_{\gamma\beta}\mathbf{g}^\gamma = K^\alpha_\beta\mathbf{n}.$$

Differentiating $\mathbf{n}\cdot\mathbf{n} = 1$ to show that

$$\Gamma^3_{3\alpha} = 0,$$

we can reduce (8.48) to

$$\mathbf{n}_{,\alpha} = -K^\beta_\alpha\mathbf{g}_\beta. \tag{8.53}$$

Since the surface covariant derivatives of the base vectors $(\mathbf{g}_1, \mathbf{g}_2)$ and $(\mathbf{g}^2, \mathbf{g}^2)$ need not be zero, it may not be possible to resolve the derivatives of surface tensor fields solely along these vectors. For instance, if $\mathbf{u}(x^1, x^2)$ is a surface vector field and $\mathbf{T}(x^2, x^2)$ is a surface tensor field of order two:

$$\mathbf{u}_{,\alpha} = u^\beta{}_{,\alpha}\mathbf{g}_\beta + K_{\alpha\beta}u^\beta\mathbf{n},$$

$$\nabla\mathbf{u} = u_{\beta,\alpha}\mathbf{g}^\alpha\mathbf{g}^\beta + K^\beta_\alpha u_\beta\mathbf{g}^\alpha\mathbf{n}, \tag{8.54}$$

$$\nabla\cdot\mathbf{T} = T^{\alpha\beta}{}_{,\alpha}\mathbf{g}_\beta + K_{\alpha\beta}T^{\alpha\beta}\mathbf{n}. \tag{8.55}$$

The calculation of the second covariant derivatives of the triad $(\mathbf{g}_1, \mathbf{g}_2, \mathbf{n})$ will lead us to some fundamental equations. From (8.52) and (8.53),

$$\mathbf{n}_{,\alpha\gamma} = -K^\beta_{\alpha,\gamma}\mathbf{g}_\beta - K^\beta_\alpha K_{\beta\gamma}\mathbf{n}.$$

Since $\mathbf{n}_{,\alpha\gamma} = \mathbf{n}_{,\gamma\alpha}$, it follows that

$$K_{\beta\alpha,\gamma} = K_{\beta\gamma,\alpha}. \tag{8.56}$$

These are the *Codazzi equations*, in which there are only two independent nontrivial relations:

$$\boxed{K_{11,2} = K_{12,1}, \qquad K_{22,1} = K_{12,2}.} \tag{8.57}$$

Again from (8.52) and (8.53)

$$\mathbf{g}_{\alpha,\beta\gamma} = K_{\alpha\beta,\gamma}\mathbf{n} - K_{\alpha\beta}K^{\delta}_{\gamma}\mathbf{g}_{\delta}.$$

With the use of the definition (8.19) and the Codazzi equations (8.56), it follows that

$$\mathbf{g}_{\alpha,\beta\gamma} - \mathbf{g}_{\alpha,\gamma\beta} = R^{\delta}{}_{\alpha\beta\gamma}\mathbf{g}_{\delta} = (K_{\alpha\gamma}K^{\delta}_{\beta} - K_{\alpha\beta}K^{\delta}_{\gamma})\mathbf{g}_{\delta}.$$

Hence the Riemann curvature tensor is related to the surface curvature tensor by

$$R_{\delta\alpha\beta\gamma} = K_{\delta\beta}K_{\alpha\gamma} - K_{\delta\gamma}K_{\alpha\beta}.$$

These are the *Gauss equations*, in which there is only one independent nontrivial relation that can be written as

$$\boxed{K_G = \frac{R_{1212}}{g}.} \tag{8.58}$$

The Codazzi-Gauss equations (8.57)–(8.58) are equivalent to a set of two first-order and one second order differential equations that must be obeyed by the components $g_{\alpha\beta}(x^1, x^2)$ and $K_{\alpha\beta}(x^1, x^2)$ of the surface metric and curvature tensors. Conversely, it can be shown that if the three components $g_{\alpha\beta}(x^1, x^2)$ of a positive definite symmetric tensor and the three components $K_{\alpha\beta}(x^1, x^2)$ of a symmetric tensor satisfy (8.57)–(8.58), then a surface with corresponding $g_{\alpha\beta}$ and $K_{\alpha\beta}$ can be embedded in three-dimensional Euclidean space, and the surface is unique except for its position in space.

Any quantity that is unchanged when a surface is bent into another shape without stretching or shrinking is said to be an *intrinsic* property of the surface. Obvious examples of intrinsic properties are the length of a curve on a surface, the components of the surface metric tensor, and the components of the Riemann curvature tensor. The Gaussian curvature is also an intrinsic property, as can be seen from the Gauss equation (8.58). In particular, since the Gaussian curvature vanishes on planes, it (and the Riemann curvature tensor) vanishes for all *developable surfaces*, which are surfaces such as cylinders and cones that can be obtained by bending a plane. Thus, if you bend the page on which this is written, all the surfaces you can form will have $K_G = 0$.

We can analyze the geometry of an arbitrary curve C *lying on a surface* in a manner similar to our analysis of space curves in Section 5.4. Label the points on C by the arc length s measured along the curve from some fixed point on C, and suppose that the surface coordinates of C can be expressed parametrically in terms of s:

$$x^1 = x^1(s), \qquad x^2 = x^2(s) \quad \text{along } C.$$

Define a unit tangent vector $\bar{\mathbf{e}}_1$ to C by (5.23), and resolve $\bar{\mathbf{e}}_1$ into components along the covariant basis (8.44):

$$\bar{\mathbf{e}}_1 = \frac{dx^\alpha}{ds}\mathbf{g}_\alpha. \tag{8.59}$$

Differentiate (8.59) and use (8.52) to obtain

$$\frac{d\bar{\mathbf{e}}_1}{ds} = \left(\frac{d^2x^\gamma}{ds^2} + \Gamma^\gamma_{\alpha\beta}\frac{dx^\alpha}{ds}\frac{dx^\beta}{ds}\right)\mathbf{g}_\gamma + K_{\alpha\beta}\frac{dx^\alpha}{ds}\frac{dx^\beta}{ds}\mathbf{n}. \tag{8.60}$$

Now decompose $\frac{d\bar{\mathbf{e}}_1}{ds}$ into components tangential and normal to the surface:

$$\boxed{\frac{d\bar{\mathbf{e}}_1}{ds} = \kappa_g\bar{\mathbf{e}}_2 + \kappa_n\mathbf{n}.} \tag{8.61}$$

Unlike in our previous analysis of space curves, $\bar{\mathbf{e}}_2 = e_{\alpha\beta}\frac{dx^\alpha}{ds}\mathbf{g}^\beta$ is now a unit vector *in the plane tangent to the surface*, chosen so that $(\bar{\mathbf{e}}_1, \bar{\mathbf{e}}_2, \mathbf{n})$ form a right-handed orthonormal triad (Fig. 8.7). Equating the right sides of (8.60) and (8.61), we obtain formulas for the *geodesic curvature* κ_g and *normal curvature* κ_n of C:

$$\kappa_g = e_{\delta\gamma}\frac{dx^\delta}{ds}\left(\frac{d^2x^\gamma}{ds^2} + \Gamma^\gamma_{\alpha\beta}\frac{dx^\alpha}{ds}\frac{dx^\beta}{ds}\right), \tag{8.62}$$

$$\kappa_n = K_{\alpha\beta}\frac{dx^\alpha}{ds}\frac{dx^\beta}{ds}. \tag{8.63}$$

It follows from (8.62) that surface curves passing through a given point P and having the same tangent (8.59) may have different geodesic curvatures κ_g at P. The *normal section*, or curve that is the intersection of a plane containing $\bar{\mathbf{e}}_1$ and $\mathbf{n}$ with the surface, has zero geodesic curvature at P. A curve on the surface that has zero geodesic curvature at *every point* is called a *geodesic*. Through any point

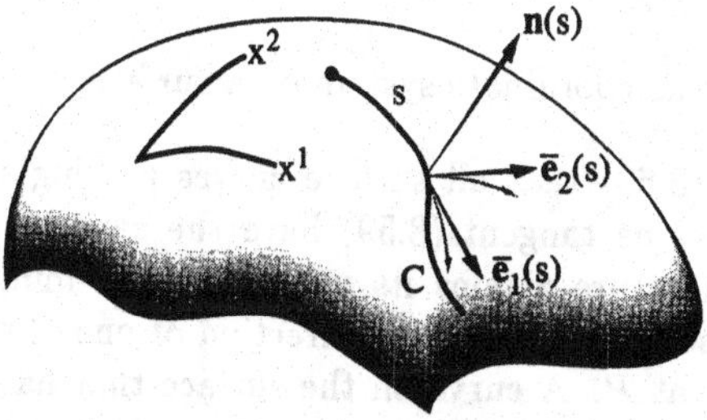

Figure 8.7. The moving trihedron $(\bar{\mathbf{e}}_1(s), \bar{\mathbf{e}}_2(s), \mathbf{n}(s))$ associated with a curve C lying on a surface.

of the surface there passes a unique geodesic in every direction. It can be shown that the curve of shortest distance joining two points on a surface is a geodesic. The parametric equations of the geodesics are, from (8.62),

$$\frac{d^2x^\alpha}{ds^2} + \Gamma^\alpha_{\beta\gamma}\frac{dx^\beta}{ds}\frac{dx^\gamma}{ds} = 0. \tag{8.64}$$

If the Gaussian curvature of a surface is not zero, then it is not possible to find a surface coordinate system in which $g_{\alpha\beta}$ equals $\delta_{\alpha\beta}$ everywhere. Geometry based on a metric tensor whose components may not be able to be transformed into the Kronecker delta is sometimes called *Riemannian*. However, it is always possible to construct a so-called *geodesic coordinate system* in which $g_{\alpha\beta}$ equals $\delta_{\alpha\beta}$ and the derivatives of $g_{\alpha\beta}$ are zero *at a particular point on the surface*. Namely, let (x^1, x^2) be arc lengths measured along two orthogonal geodesics passing through a point O, and construct an orthogonal net of surface coordinates as shown in Fig. 8.8. At the origin O of this geodesic coordinate system

$$g_{\alpha\beta}(0,0) = \delta_{\alpha\beta}, \qquad \frac{\partial g_{\alpha\beta}}{\partial x^\gamma}(0,0) = 0, \qquad \Gamma^\gamma_{\alpha\beta}(0,0) = 0. \tag{8.65}$$

We can prove the validity of formulas for the components of surface tensors by showing that they are valid in geodesic coordinates, and hence in all coordinate systems. However, note that the *second* derivatives of $g_{\alpha\beta}$ are not necessarily zero in geodesic coordinates so, for instance, we cannot conclude that the surface Riemann curvature tensor vanishes.

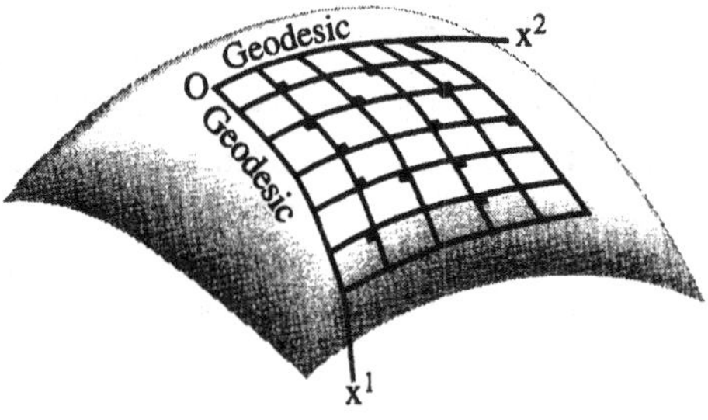

Figure 8.8. Geodesic coordinate system on a surface.

It follows from (8.63) that all surface curves passing through a given point P and having the same tangent (8.59) have the same normal curvature κ_n at P. The normal curvature attains its maximum and minimum values at P for surface curves whose tangent is in the direction of one of the eigenvectors of the curvature tensor $\mathbf{K}$ at P. A curve on the surface that has the property that an eigenvector of the surface curvature tensor is tangent to it at every point is called a *line of curvature*. The most convenient coordinate system to select on a given surface is usually one in which the coordinate curves are lines of curvature. If

(x^1, x^2) denote an orthogonal lines-of-curvature coordinate system in which the x^1 coordinate curve has the maximum possible values of κ_n, then

$$K_1^1 = (\kappa_n)_{\text{maximum}}, \qquad K_2^2 = (\kappa_n)_{\text{minimum}}, \qquad K_{12} = g_{12} = 0 \tag{8.66}$$

at all points of the surface.

Surface of revolution. Let us apply the foregoing theory to the surface of revolution generated by revolving about the x_3 axis a curve that lies in a plane perpendicular to the x_1-x_2 axis (Fig. 8.9). We select as surface coordinates (ℓ, θ), where ℓ denotes arc length measured along a meridian from the top of the surface, and where θ denotes the usual polar angle.[1] We suppose that the equations of the generating curve are given parametrically by

$$\rho = \rho(\ell), \qquad x_3 = x_3(\ell),$$

where ρ is the distance from the x_3-axis to a point on the surface. Since ℓ is arc length, ρ and x_3 are related by

$$\left(\frac{d\rho}{d\ell}\right)^2 + \left(\frac{dx_3}{d\ell}\right)^2 = 1. \tag{8.67}$$

The Cartesian coordinates of a point on the surface are

$$x_1 = \rho(\ell)\cos\theta, \qquad x_2 = \rho(\ell)\sin\theta, \qquad x_3 = x_3(\ell).$$

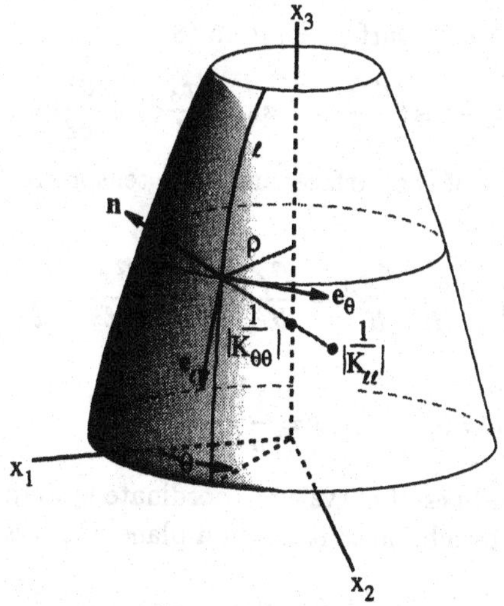

Figure 8.9. Geometry of a surface of revolution.

[1]This choice of coordinates produces negative values of components of the surface curvature tensor, but is consistent with our convention that the unit normal vector $\mathbf{n}$ points outward from an enclosed volume.

The coordinate curves of constant ℓ are the parallels, and the coordinate curves of constant θ are the meridians, of the surface. The covariant base vectors tangent to the coordinate curves are, from (8.44),

$$\mathbf{g}_1 = \frac{d\rho}{d\ell}\cos\theta\,\mathbf{e}_1 + \frac{d\rho}{d\ell}\sin\theta\,\mathbf{e}_2 + \frac{dx_3}{d\ell}\mathbf{e}_3,$$
$$\mathbf{g}_2 = -\rho\sin\theta\,\mathbf{e}_1 + \rho\cos\theta\,\mathbf{e}_2. \tag{8.68}$$

The covariant components of the surface metric tensor are, from (8.45) and (8.67)–(8.68):

$$g_{11} = (h_1)^2 = 1, \qquad g_{12} = 0, \qquad g_{22} = (h_2)^2 = \rho^2. \tag{8.69}$$

Since $g_{12} = 0$ this coordinate system is orthogonal, and we can resolve surface tensors into their physical components along the *unit* vectors $(\mathbf{e}_\ell, \mathbf{e}_\theta)$ obtained by dividing $(\mathbf{g}_1, \mathbf{g}_2)$ by their lengths:

$$\mathbf{e}_\ell = \frac{d\rho}{d\ell}\cos\theta\,\mathbf{e}_1 + \frac{d\rho}{d\ell}\sin\theta\,\mathbf{e}_2 + \frac{dx_3}{d\ell}\mathbf{e}_3.$$
$$\mathbf{e}_\theta = -\sin\theta\,\mathbf{e}_1 + \cos\theta\,\mathbf{e}_2.$$

The nonzero surface Christoffel symbols are, from (8.28),

$$\Gamma^1_{22} = -\rho\frac{d\rho}{d\ell}, \qquad \Gamma^2_{12} = \Gamma^2_{21} = \frac{1}{\rho}\,\frac{d\rho}{d\ell}. \tag{8.70}$$

The unit normal vector to the surface is, from (8.46)

$$\mathbf{n} = -\cos\theta\frac{dx_3}{d\ell}\mathbf{e}_1 - \sin\theta\frac{dx_3}{d\ell}\mathbf{e}_2 + \frac{d\rho}{d\ell}\mathbf{e}_3. \tag{8.71}$$

The physical components of the surface curvature tensor are, from (4.30), (8.49), (8.67)–(8.69), and (8.71):

$$K_{\ell\ell} = \frac{d\rho}{d\ell}\,\frac{d^2x_3}{d\ell^2} - \frac{d^2\rho}{d\ell^2}\,\frac{dx_3}{d\ell} = \frac{d^2x_3}{d\ell^2}\Big/\frac{d\rho}{d\ell},$$
$$K_{\ell\theta} = 0, \qquad K_{\theta\theta} = \frac{1}{\rho}\,\frac{dx_3}{d\ell}. \tag{8.72}$$

Since $K_{\ell\theta} = 0$, this is a lines-of-curvature coordinate system. The geodesic curvature of a meridian is zero because it lies in a plane which is everywhere normal to the surface:

$$\kappa_g = 0 \quad \text{on} \quad \theta = \text{constant}.$$

The geodesic curvature of a parallel may be obtained from (8.62) with (8.70) and $\dfrac{d\theta}{ds} = \dfrac{1}{\rho}$:

$$\kappa_g = \frac{1}{\rho}\,\frac{d\rho}{d\ell} \quad \text{on} \quad \ell = \text{constant}.$$

The normal curvature of a meridian or parallel is by (8.66) a physical component of the curvature tensor:

$$\kappa_n = K_{\ell\ell} \quad \text{on} \quad \theta = \text{constant},$$

$$\kappa_n = K_{\theta\theta} \quad \text{on} \quad \ell = \text{constant}.$$

Note that $\dfrac{1}{|K_{\ell\ell}|}$ is simply the radius of curvature of the osculating circle tangent to a meridian, and $\dfrac{1}{|K_{\theta\theta}|}$ is the distance measured along the normal line from the surface to the x_3-axis (Fig. 8.9).

PROBLEMS 8.3

1. Resolve the curl of a surface vector field $\mathbf{u}(x^1, x^2)$ along the base vectors $(\mathbf{g}_1, \mathbf{g}_2, \mathbf{n})$ and the derivatives of a surface tensor field $\mathbf{T}(x^1, x^2)$ along the dyads formed from $(\mathbf{g}_1, \mathbf{g}_2, \mathbf{n})$.

2. Write down the component form of Stokes' theorem and the divergence theorem for surface vectors.

3. List some intrinsic properties of a surface (other than the ones explicitly mentioned in the text).

4. Write down all the nonzero components of the Riemann curvature tensor for a surface.

5. Explain why an undistorted map of a spherical earth cannot be made upon a flat piece of paper.

6. Find simplified expressions for $K_{\ell\ell}$ and $K_{\theta\theta}$ at the apex of a closed surface of revolution.

7. Specialize the Codazzi-Gauss equations to an orthogonal lines-of-curvature coordinate system. Then show that they are satisfied by the metric and curvature tensors of the surface of revolution example in the text.

8. Show that the derivatives of the moving trihedron $(\bar{\mathbf{e}}_1, \bar{\mathbf{e}}_2, \mathbf{n})$ for a surface curve can be written as

$$\frac{d\bar{\mathbf{e}}_\alpha}{ds} = \boldsymbol{k} \times \bar{\mathbf{e}}_\alpha, \qquad \frac{d\mathbf{n}}{ds} = \boldsymbol{k} \times \mathbf{n}, \qquad \text{where} \qquad \boldsymbol{k} = \tau_g \bar{\mathbf{e}}_1 - \kappa_n \bar{\mathbf{e}}_2 + \kappa_g \mathbf{n}$$

and τ_g is the torsion of the geodesic curve with tangent $\bar{\mathbf{e}}_1$.

9. Show that the geodesics on a spherical surface are great circles and the geodesics on a circular cylindrical surface are helices.

10. Verify that the geodesic coordinate system shown in Fig. 8.8 has the properties (8.65).

11. Verify the formulas $g_{\alpha\beta,\gamma} = 0$, $e_{\alpha\beta,\gamma} = 0$, and $e_{\alpha\beta}e_{\gamma\delta} = g_{\alpha\gamma}g_{\beta\delta} - g_{\alpha\delta}g_{\beta\gamma}$.

12. Find the directions of the tangent vectors to the surface curves that have the maximum and minimum normal curvatures at a given point.

13. An *ellipsoidal surface of revolution* is determined by $x_1 = a\cos\theta\sin\delta$, $x_2 = a\sin\theta\sin\delta$, $x_3 = b\cos\delta$, where a and b are positive constants. Find the components of the surface metric and curvature tensors in the (δ, θ) system.

8.4 MECHANICS OF CURVED MEMBRANES

The use of surface coordinates is essential in the study of the mechanical behavior of curved membranes. Envision a thin membrane or film that is subjected to a pressure loading p normal to its lateral surface and edge tractions tangential to its lateral surface. We characterize the reactive forces per unit *length* within the membrane by the *membrane* stress tensor $\mathbf{T}$. Then the force acting on an elemental slice through the membrane having width Δs and unit normal vector $\mathbf{e}$ tangential to the lateral surface is $\mathbf{T}\cdot\mathbf{e}\Delta s$. If the membrane is at rest under the action of the applied loads, the sum of the forces acting on an arbitrary section having a lateral surface S bounded by a curve C must be zero (see Fig. 8.10):

$$\oint_C \mathbf{T}\cdot\mathbf{e}ds + \int\int_S p\mathbf{n}dS = 0.$$

Application of the surface divergence theorem yields

$$\int\int_S (\nabla\cdot\mathbf{T} + p\mathbf{n})dS = 0$$

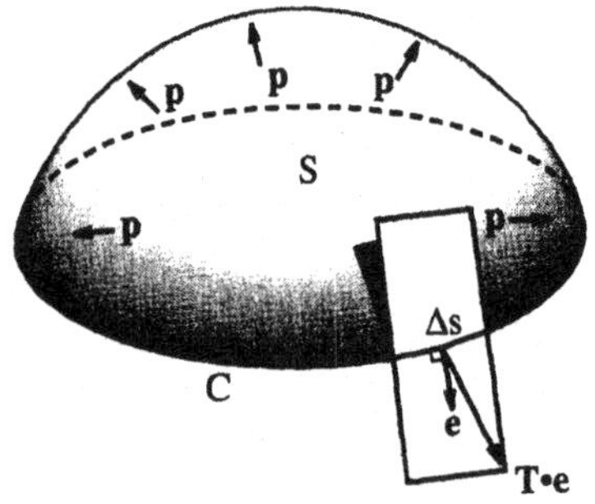

Figure 8.10. Thin membrane subjected to normal pressure loading p and edge forces per unit length $\mathbf{T}\cdot\mathbf{e}$.

whence the arbitrariness of the surface S implies the differential form of the force equilibrium equation:

$$\nabla \cdot \mathbf{T} + p\mathbf{n} = 0. \tag{8.73}$$

By enforcing moment equilibrium, we can show that the membrane stress tensor $\mathbf{T}$ is symmetric.

We assume that the membrane stress tensor depends only upon coordinates (x^1, x^2) on a lateral surface of the membrane and can be resolved solely along the dyads formed from the base vectors $(\mathbf{g}_1, \mathbf{g}_2)$ tangent to the surface coordinate curves:

$$\mathbf{T}(x^1, x^2) = T^{\alpha\beta}\mathbf{g}_\alpha \mathbf{g}_\beta.$$

Then, using (8.55), we can obtain the component form of the membrane equilibrium equations (8.73):

$$T^{\alpha\beta}{}_{,\beta} = 0, \qquad K_{\alpha\beta}T^{\alpha\beta} + p = 0. \tag{8.74}$$

In the following we apply these equations to three somewhat different physical situations. For each application it is necessary to make a somewhat different assumption about the stress distribution within the membrane.

Interface between two fluids. The interface between two motionless fluids — for example, between a liquid and a gas — is subject to a uniform *surface tension* T arising from molecular forces within the fluids. The membrane stress tensor characterizing the interface has components

$$T_{\alpha\beta} = T g_{\alpha\beta}. \tag{8.75}$$

Inserting (8.75) into (8.74), we find that the surface between the two fluids must have constant mean curvature:

$$K_M = -\frac{p}{2T}. \tag{8.76}$$

If the pressure difference p between the two fluids is zero, then the mean curvature K_M of the interface must be zero. Surfaces with zero mean curvature are called *minimal surfaces* and are among the best-studied surfaces in differential geometry. It can be shown that the surface of minimum area passing through a closed space curve is a minimal surface. Our analysis shows that the soap film obtained by dipping a wire in the form of a closed space curve into a soap solution must be a minimal surface.

Thin shell structure. A shell is a three-dimensional solid with one thin dimension. Even very thin curved shell structures can support large loads and hence are much utilized in nature and technology. Think of eggshells, peanut shells, concrete domes, metal cans, light bulb casings, the outer skin of aircraft and automobiles, and so on. If a shell is thin enough, the stress components $S_{\alpha\beta}$ (forces per unit area) will be approximately uniform through the thickness w, so we can write

$$T_{\alpha\beta} = wS_{\alpha\beta}.$$

If the equilibrium equations (8.74) can be solved for $T_{\alpha\beta}$, then the stresses
will be statically determined throughout the thin shell. For a linear isotropic s
material, the strain components $\gamma_{\alpha\beta}$ follow from the constitutive law (7.18b):

$$\gamma_{\alpha\beta} = \frac{(1+\nu)}{wY} T_{\alpha\beta} - \frac{\nu}{wY} g_{\alpha\beta} T^{\gamma}_{\gamma}. \qquad (8.$$

The displacements $u_i(x^1, x^2)$ of a point (x^1, x^2) on the shell surface are found
integrating the following components of (7.16a), obtained with the use of (8.
and (8.54):

$$\gamma_{\alpha\beta} = \frac{1}{2}(u_{\alpha,\beta} + u_{\beta,\alpha}) - K_{\alpha\beta} u_3. \qquad (8.$$

Many shells have the form of surfaces of revolution and are loaded symme
cally with respect to their axes, so their deflection field in the lines-of-curvat
coordinate system used in the preceding section satisfies

$$u_\ell = u_\ell(\ell), \qquad u_\theta = 0, \qquad u_3 = u_3(\ell).$$

When specialized to the axisymmetrical deformation of shells of revolution,
foregoing equations simplify to

$$\frac{d}{d\ell}(\rho T_{\ell\ell}) - \frac{d\rho}{d\ell} T_{\theta\theta} = 0, \qquad K_{\ell\ell} T_{\ell\ell} + K_{\theta\theta} T_{\theta\theta} + p = 0, \qquad (8.$$

$$\gamma_{\ell\ell} = \frac{T_{\ell\ell} - \nu T_{\theta\theta}}{wY} = \frac{du_\ell}{d\ell} - K_{\ell\ell} u_3, \qquad \gamma_{\theta\theta} = \frac{T_{\theta\theta} - \nu T_{\ell\ell}}{wY} = \frac{d\rho}{d\ell}\frac{u_\ell}{\rho} - K_{\theta\theta} u_3. \quad (8$$

Eliminating $T_{\theta\theta}$ from the equations (8.79) and substituting (8.72), we get a fi
order differential equation for $T_{\ell\ell}$:

$$\frac{d}{d\ell}\left(\rho \frac{dx_3}{d\ell} T_{\ell\ell}\right) = -\rho \frac{d\rho}{d\ell} p.$$

When p is uniform, this can be readily integrated to give

$$T_{\ell\ell} = -\frac{\rho p}{2\frac{dx_3}{d\ell}} + \frac{C}{\rho \frac{dx_3}{d\ell}},$$

$$(8$$

$$T_{\theta\theta} = -\frac{\rho p}{\frac{dx_3}{d\ell}} + \frac{\rho^2 p \frac{d^2 x_3}{d\ell^2}}{2\frac{d\rho}{d\ell}\left(\frac{dx_3}{d\ell}\right)^2} - \frac{C\frac{d^2 x_3}{d\ell^2}}{\frac{d\rho}{d\ell}\left(\frac{dx_3}{d\ell}\right)^2},$$

where C is an arbitrary constant. Knowing the membrane stresses, we can d
mine the displacement field by eliminating u_3 in (8.80) and solving the resu
differential equation for u_ℓ.

Tension field. If you grasp a pinch of skin between your fingers, you will note that the skin slides over its foundation in such a way that points remain on nearly the same surface. We can describe the mechanical behavior of a membrane constrained to slide over a rigid mandrel by the equations of thin shell theory, with the metric $g_{\alpha\beta}$ being that of the foundation and the normal displacement $u_3 = 0$. In some such problems, one of the components of the membrane stress tensor in its principal axes is positive and the other is negative. In these problems, since a very thin shell cannot support a large amount of negative stress without buckling or wrinkling, we replace the shell equations by *tension field theory*. In this theory the only nonzero stress component is the positive stress along a *tension ray*. You can easily see the tension rays by grasping a pinch of skin on the back of your hand; note how the force is carried by rays of stretched skin. We choose the surface coordinates (x^1, x^2) so the tangents to the coordinate curves coincide with the principal axes of the membrane stress tensor, and take T_{11} to be the one nonvanishing stress component so

$$T_{12} = T_{22} = 0.$$

The membrane equilibrium equations (8.73) then reduce to

$$\frac{\partial T^{11}}{\partial x^1} + 2T^{11}\Gamma^1_{11} + T^{11}\Gamma^2_{12} = 0, \tag{8.82}$$

$$T^{11}\Gamma^2_{11} = 0, \tag{8.83}$$

$$K_{11}T^{11} + p = 0. \tag{8.84}$$

Remembering (8.28), we see that (8.83) implies that $\dfrac{\partial h_1}{\partial x^2} = 0$, so we can choose x^1 to be arc length measured along the tension rays and set $h_1 = 1$. It follows from (8.62) that $\kappa_g = 0$ along the x^1 coordinate curves, so the tension rays are geodesics on the surface over which the membrane slides. The first equilibrium equation (8.82) now simplifies to

$$\frac{\partial}{\partial x^1}(h_2 T_{11}) = 0$$

and can be easily integrated to give

$$T_{11} = \frac{f(x^2)}{h_2}, \tag{8.85}$$

where $f(x^2)$ is an arbitrary function of x^2. The third equilibrium equation (8.84) predicts the pressure p exerted by the foundation upon the membrane. Knowing the stress components, we can obtain *two* of the strain components from *two* of the constitutive relations (8.77) (the third constitutive relation is not obeyed):

$$\gamma_{11} = \frac{T_{11}}{wY}, \qquad \gamma_{12} = 0.$$

The displacement components u_α are found by integrating, from (8.78) with $u_3 = 0$,

$$u_{1,1} = \frac{T_{11}}{wY}, \qquad u_{1,2} + u_{2,1} = 0. \tag{8.86}$$

The directions of the tension rays and the arbitrary functions in the solution are determined by the boundary conditions. A more accurate model for the skin would include the effects of large deformations and nonlinearities in the stress-strain law.

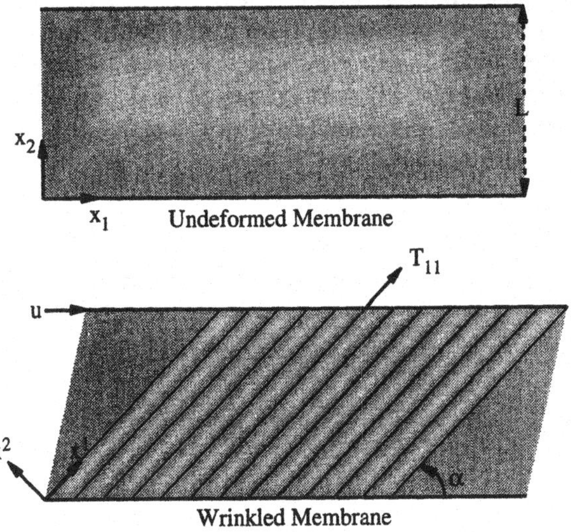

Figure 8.11. Diagonal tension field.

PROBLEMS 8.4

1. Show that the membrane stress tensor is symmetric.

2. Find the pressure difference between the inside and outside of a soda water bubble of diameter 0.1 mm. Assume that the surface tension $T = 0.07\ \frac{\text{N}}{\text{m}}$.

3. A *right helicoid* is the surface determined by $x_1 = \rho\cos\theta$, $x_2 = \rho\sin\theta$, $x_3 = b\theta$, where b is a constant. Show that a right helicoid is a minimal surface.

4. For a closed shell of revolution under uniform internal pressure, write the membrane stresses in terms of components of the surface curvature tensor.

5. Determine the membrane stresses and deflections in a thin spherical shell of radius a, and in a thin cylindrical shell of radius a, when each is subjected to a uniform internal pressure p.

6. An infinite membrane strip occupies the planar region $-\infty < x_1 < \infty$, $0 \le x_2 \le L$, where (x_1, x_2) are Cartesian coordinates (see Fig. 8.11). The bottom edge $x_2 = 0$ is prevented from moving while the top edge $x_2 = L$ is displaced a distance u parallel to the x_1 axis. Find the direction of the tension rays, the membrane stress in the tension rays, and the displacement of points on the membrane.

7. A spherical membrane annulus lies over the region $\ell_1 \le \ell \le \pi a/2, 0 \le \theta < 2\pi$ of a spherical foundation of radius a. The bottom edge $\ell = \pi a/2$ is prevented from moving while points on the top edge $\ell = \ell_1$ are displaced a distance u towards the top pole. Find the direction of the tension rays, the membrane stress in the tension rays, the displacement of points on the membrane, and the pressure exerted by the foundation on the membrane.

8.5 NOTATION OF OTHERS

Table 8

OUR NOTATION	OUR NAME	OTHER NOTATIONS	OTHER NAMES
(x^1, x^2, x^3)	Curvilinear coordinates.	(u_1, u_2, u_3), $(\theta_1, \theta_2, \theta_3)$, (u, v, w)	
Γ^k_{ij}	Christoffel symbol.	$\left\{ {k \atop ij} \right\}$	Christoffel symbol of the second kind, affine connection coefficient.
$u_{i,j}$	Covariant derivative.	$u_{i\|j}$, $u_{i;j}$, $\nabla_j u_i$	
(r, ϕ, θ)	Spherical coordinates.	(ρ, θ, ϕ)	
$g_{\alpha\beta}, K_{\alpha\beta}$	Components of the surface metric and curvature tensors.	$a_{\alpha\beta}$, $b_{\alpha\beta}$ (respectively)	

Chapter 9

FOUR-DIMENSIONAL SPACETIME

9.1 SPECIAL RELATIVITY

According to the relation (5.17) from Newtonian mechanics, the speeds of a particle in two coordinate systems in relative motion are not necessarily the same. However, the speed c of an electromagnetic wave, which may also be regarded as a stream of particles called ***photons***, has been found to always have the same value when measured experimentally in different coordinate systems. Newtonian mechanics is actually erroneous for particles moving at a significant fraction of the speed of light. In this chapter we study the relativistically correct equations for the motion of moving particles.

It turns out that the right mathematics for relativity theory is tensor analysis in four-dimensional spacetime. Tensor analysis in four-dimensional spacetime may be developed by analogy with our previous work in three dimensions. We associate with each ***event*** or point in spacetime four coordinates (x^0, x^1, x^2, x^3), where $x^0 = ct$ is the temporal coordinate and (x^1, x^2, x^3) are spatial coordinates. Here and throughout this chapter, we adopt the convention that *Greek indices can have the values 0, 1, 2, or 3*, whereas as usual Latin indices can have the values 1, 2, or 3. The temporal as well as the spatial coordinates may change when general spacetime coordinates $(\bar{x}^0, \bar{x}^1, \bar{x}^2, \bar{x}^3)$ are introduced by the transformations

$$x^\alpha = x^\alpha(\bar{x}^0, \bar{x}^1, \bar{x}^2, \bar{x}^3), \qquad \alpha = 0, 1, 2, 3.$$

The components of a four-dimensional tensor may be defined as sets of numbers that are related by the appropriate laws under transformations of coordinates. For example, the contravariant components (f^0, f^1, f^2, f^3) and $(\bar{f}^0, \bar{f}^1, \bar{f}^2, \bar{f}^3)$ of a ***four vector*** in the two coordinate systems (x^0, x^1, x^2, x^3) and $(\bar{x}^0, \bar{x}^1, \bar{x}^2, \bar{x}^3)$ are related by the familiar laws

$$\bar{f}^\alpha = \frac{\partial \bar{x}^\alpha}{\partial x^\beta} f^\beta, \qquad f^\alpha = \frac{\partial x^\alpha}{\partial \bar{x}^\beta} \bar{f}^\beta. \tag{9.1}$$

In relativistic mechanics the spacetime coordinates $x^\alpha(\tau)$ of a point mass are regarded as being functions of the ***proper time*** τ. The proper time is a scalar invariant that can be physically interpreted as the time measured by a clock that is carried along by the particle. The proper time interval $\Delta\tau$ between two adjacent events (x^0, x^1, x^2, x^3) and $(x^0 + \Delta x^0, x^1 + \Delta x^1, x^2 + \Delta x^2, x^3 + \Delta x^3)$ is defined by the quadratic form

$$\Delta\tau^2 = \frac{g_{\alpha\beta}}{c^2} \Delta x^\alpha \Delta x^\beta. \tag{9.2}$$

Here $g_{\alpha\beta}$ are components of the ***spacetime metric tensor.***

In *special relativity* theory the components of the Riemann curvature tensor (as defined by (8.6) and (8.20) with Latin indices replaced by Greek) are all zero, and it is possible to choose a coordinate system in which $g_{\alpha\beta}$ equals $\eta_{\alpha\beta}$ at all points in spacetime, where

$$\eta_{\alpha\beta} = \begin{cases} +1 \text{ if } \alpha = \beta = 0, \\ -1 \text{ if } \alpha = \beta = 1, 2, \text{ or } 3, \\ \ \ 0 \text{ if } \alpha \neq \beta. \end{cases}$$

In such a coordinate system, (9.2) reduces to

$$(\Delta\tau)^2 = (\Delta t)^2 - \frac{(\Delta x_1)^2 + (\Delta x_2)^2 + (\Delta x_3)^2}{c^2}$$

and we can interpret $(x^1, x^2, x^3) = (x_1, x_2, x_3)$ as Cartesian spatial coordinates. Since

$$v = \sqrt{\left(\frac{dx_1}{dt}\right)^2 + \left(\frac{dx_2}{dt}\right)^2 + \left(\frac{dx_3}{dt}\right)^2}$$

is the speed of a particle, the proper time interval $\Delta\tau$ becomes

$$\Delta\tau = \sqrt{1 - \frac{v^2}{c^2}}\,\Delta t. \tag{9.3}$$

An interesting physical consequence of (9.3) is *motional time dilation*: when a clock moves with a speed v, it runs slower by the factor $\sqrt{1 - \frac{v^2}{c^2}}$.

For a particle with nonzero mass, $v < c$ so $\Delta\tau > 0$; whereas for a photon, $v = c$ so $\Delta\tau = 0$. We can illustrate a four-dimensional particle path or *worldline* on a

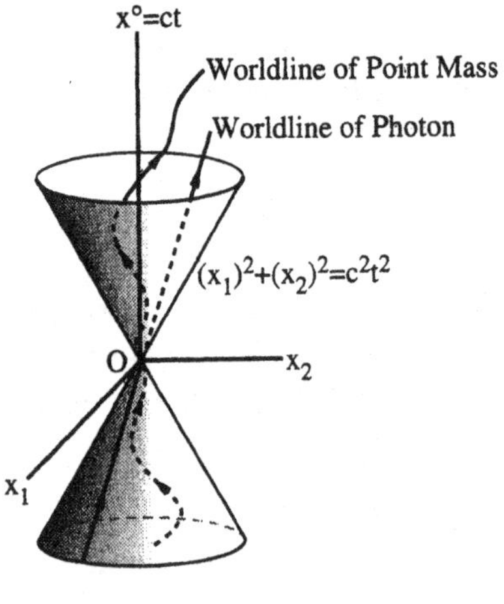

Figure 9.1. Spacetime diagram showing light cone and worldlines of two particles.

spacetime diagram such as that shown in Fig. 9.1, wherein the spatial coordinates x_1 and x_2 are plotted on the horizontal axes, the temporal coordinate x^0 is plotted on the vertical axis, and the x_3 coordinate is omitted. Note that the worldline of a point mass lies inside the conical surface that is the set of all worldlines of light rays leaving from or arriving at the origin.

Those spacetime coordinate transformations which result in the new components of the spacetime metric tensor remaining equal to $\eta_{\alpha\beta}$ are called *Lorentz transformations*. It can be shown that every Lorentz transformation with a single-valued inverse has the form

$$\bar{x}^\alpha = \Lambda_\beta{}^\alpha x^\beta + a^\alpha, \tag{9.4}$$

where a^α and $\Lambda_\beta{}^\alpha$ are constants restricted by the conditions

$$\eta_{\gamma\delta}\Lambda_\alpha{}^\gamma\Lambda_\beta{}^\delta = \eta_{\alpha\beta}. \tag{9.5}$$

An obvious example of a Lorentz transformation is a purely spatial rotation and/or reflection that does not vary with time, that is,

$$\Lambda_0^0 = 1, \qquad \Lambda_0^i = \Lambda_i^0 = 0, \qquad \Lambda_i{}^j = Q_i{}^j,$$

where $Q_i{}^j$ are components of an orthogonal matrix. Another example of a Lorentz transformation gives a *boost* to the old Cartesian coordinate system (x_1, x_2, x_3) so that the new Cartesian coordinate system $(\bar{x}_1, \bar{x}_2, \bar{x}_3)$ is moving with constant velocity components $(v^1, v^2, v^3) = (v_1, v_2, v_3)$ relative to the old:

$$\begin{aligned} &\Lambda_0^0 = \frac{1}{\sqrt{1 - \frac{v^2}{c^2}}}, \qquad \Lambda_0^i = -\frac{v^i}{c\sqrt{1 - \frac{v^2}{c^2}}}, \\ &\Lambda_i^0 = -\frac{v_i}{c\sqrt{1 - \frac{v^2}{c^2}}}, \qquad \Lambda_j^i = \delta_j^i + \left(\frac{1}{\sqrt{1 - \frac{v^2}{c^2}}} - 1\right)\frac{v^i v_j}{v^2}. \end{aligned} \tag{9.6}$$

In the special case of $v_1 = v$, $v_2 = v_3 = 0$, (9.4) and (9.6) reduce to the well-known Lorentz transformation

$$\bar{t} = \frac{t - \frac{vx_1}{c^2}}{\sqrt{1 - \frac{v^2}{c^2}}}, \qquad \bar{x}_1 = \frac{x_1 - vt}{\sqrt{1 - \frac{v^2}{c^2}}}, \qquad \bar{x}_2 = x_2, \qquad \bar{x}_3 = x_3. \tag{9.7}$$

Formulas for the *relativistic addition of velocities* may be obtained by differentiating (9.7) with respect to t and solving for $\frac{dx_i}{dt}$:

$$\frac{dx_1}{dt} = \frac{v + \dfrac{d\bar{x}_1}{d\bar{t}}}{1 + \dfrac{v}{c^2}\dfrac{d\bar{x}_1}{d\bar{t}}}, \qquad \frac{dx_2}{dt} = \frac{\sqrt{1 - \dfrac{v^2}{c^2}}\dfrac{d\bar{x}_2}{d\bar{t}}}{1 + \dfrac{v}{c^2}\dfrac{d\bar{x}_1}{d\bar{t}}},$$
$$\frac{dx_3}{dt} = \frac{\sqrt{1 - \dfrac{v^2}{c_2}}\dfrac{d\bar{x}_3}{d\bar{t}}}{1 + \dfrac{v}{c^2}\dfrac{d\bar{x}_1}{d\bar{t}}}. \tag{9.8}$$

Here $\dfrac{d\bar{x}_i}{d\bar{t}}$ are a particle's velocity components in the new coordinate system, and $\dfrac{dx_i}{dt}$ are its velocity components in the old coordinate system.

When the spacetime metric tensor is $\eta_{\alpha\beta}$, the relativistically correct equations for the motion of a point mass are

$$\frac{d}{d\tau}\left(m\frac{dx^\alpha}{d\tau}\right) = f^\alpha. \tag{9.9}$$

Here m is called the ***rest mass*** of the particle and f^α are components of the ***relativistic force four-vector***. If we give the (x_1, x_2, x_3) system a boost of velocity components $v_i = \dfrac{dx_i}{dt}$ so that the particle is instantaneously at rest in the new system $(\bar{x}_1, \bar{x}_2, \bar{x}_3)$, then $\bar{f}^0 = 0$ and $\bar{f}^i$ can be identified as the Cartesian components of the Newtonian force. Knowing $\bar{f}^i$, we can calculate f^α from (9.1) and the transformation laws (9.4) and (9.6):

$$f^0 = \frac{v_i \bar{f}^i}{c\sqrt{1 - \dfrac{v^2}{c^2}}},$$
$$f^i = \bar{f}^i + \left(\frac{1}{\sqrt{1 - \dfrac{v^2}{c^2}}} - 1\right)\frac{v^i v_j}{v^2}\bar{f}^j. \tag{9.10}$$

The temporal and spatial parts of the relativistic force law (9.9) can then be written as

$$\frac{d}{dt}\left(\frac{mc^2}{\sqrt{1 - \dfrac{v^2}{c^2}}}\right) = v_i \bar{f}^i, \tag{9.11}$$

$$\frac{d}{dt}\left(\frac{m}{\sqrt{1-\frac{v^2}{c^2}}}\frac{dx^i}{dt}\right)=\sqrt{1-\frac{v^2}{c^2}}\,\bar{f}^i+\left(1-\sqrt{1-\frac{v^2}{c^2}}\right)\frac{v^i v_j}{v^2}\bar{f}^j. \tag{9.12}$$

It can easily be seen that the spatial part (9.12) may be approximated by Newton's law (5.12b) in an inertial coordinate system when $\frac{v}{c} \ll 1$ and m is constant. The temporal part (9.11) is actually the relativistic form of the energy-work theorem. Equation (9.11) says that the rate of change of the particle's *total energy* $\frac{mc^2}{\sqrt{1-\frac{v^2}{c^2}}}$ per unit t is equal to the rate at which work is being done by the force. The total energy can be decomposed into

$$\frac{mc^2}{\sqrt{1-\frac{v^2}{c^2}}}=mc^2+mc^2\left(\frac{1}{\sqrt{1-\frac{v^2}{c^2}}}-1\right).$$

The part mc^2 is the *rest energy* or *intrinsic energy*, whereas the remaining part is the *relativistic kinetic energy* and is approximately $\frac{1}{2}mv^2$ for $\frac{v}{c} \ll 1$. In the nuclear fission process the uranium nuclei break up into fragments having a smaller rest mass; the decrease in rest energy is initially equal to the increase in kinetic energy of the fragments.

Unlike Newton's law for a point mass, Maxwell's equations of electromagnetism need no modification when the spacetime metric tensor is $\eta_{\alpha\beta}$. We can write them in four-dimensional notation by introducing the antisymmetrical *electromagnetic field tensor* with components defined by

$$F^{0i}=\frac{\mathcal{E}^i}{c},\qquad F^{12}=\mathcal{B}_3,\qquad F^{31}=\mathcal{B}_2,\qquad F^{23}=\mathcal{B}_1,\qquad F^{\alpha\beta}=-F^{\beta\alpha}$$

and the *charge-current four-vector* with components defined by

$$J^0=c\rho,\qquad J^i=\mathcal{J}^i.$$

You can then easily verify that Maxwell's equations (7.61a)–(7.64a) and (7.67) and (7.68) in Cartesian components are equivalent to

$$\frac{\partial F^{\alpha\beta}}{\partial x^\beta}=\mu J^\alpha,\qquad \frac{\partial F_{\beta\gamma}}{\partial x^\alpha}+\frac{\partial F_{\gamma\alpha}}{\partial x^\beta}+\frac{\partial F_{\alpha\beta}}{\partial x^\gamma}=0. \tag{9.13}$$

Here the indices on $F^{\alpha\beta}$ are lowered in the usual way by means of the metric $\eta_{\alpha\beta}$. The continuity equation (7.70a) now takes the concise form

$$\frac{\partial J^\alpha}{\partial x^\alpha}=0.$$

The components of the relativistic force acting on a single particle of charge q are

$$f^\alpha = qF_\beta{}^\alpha \frac{dx^\beta}{d\tau}. \tag{9.14}$$

That this is correct can be ascertained by noting that, in a Cartesian coordinate system moving with the charge q, $\bar{f}^0 = 0$ and $\bar{f}^i = q\bar{\mathcal{E}}^i$, then by using (9.10) to find f^α. The momentum conservation law (7.73) and energy conservation law (7.76) for a distribution of charges and currents are the spatial and temporal components of a single equation:

$$\iiint_V F_\beta{}^\alpha J^\beta dV = \frac{1}{c}\frac{d}{dt}\iiint_V T^{\alpha 0} dV + \oiint_S T^{\alpha i} n_i dS.$$

Here $T^{\alpha\beta}$ are components of the *energy-momentum tensor* of the electromagnetic field:

$$T^{\alpha\beta} = \frac{1}{\mu}\left(F^{\alpha\gamma}F^\beta_{\ \gamma} - \frac{1}{4}\eta^{\alpha\beta}F^{\gamma\delta}F_{\gamma\delta}\right).$$

The spatial components of the energy-momentum tensor are just the Cartesian components of Maxwell's stress tensor (7.74), while the other components are proportional to the energy density and Cartesian components of the Poynting vector.

You can check that Maxwell's equations (9.13) maintain the same form under a Lorentz transformation (9.4)–(9.5), so in the present framework the speed c of light does not depend upon the coordinate system. We can obtain the correct transformation laws for the electric field, magnetic field, charge, and current from the transformation laws for the electromagnetic field tensor and the charge-current four-vector. Let us consider again the case in which the old Cartesian coordinate system (x_1, x_2, x_3) is given a boost by the Lorentz transformation (9.7) so that the new system $(\bar{x}_1, \bar{x}_2, \bar{x}_3)$ is moving with constant velocity components $v_1 = v$, $v_2 = v_3 = 0$ relative to the old. For this case the fields transform as

$$\bar{\mathcal{E}}_1 = c\bar{F}^{01} = c\frac{\partial \bar{x}^0}{\partial x^\alpha}\frac{\partial \bar{x}^1}{\partial x^\beta}F^{\alpha\beta} = \frac{\partial \bar{t}}{\partial t}\frac{\partial \bar{x}^1}{\partial x^1}\mathcal{E}_1 - \frac{\partial \bar{t}}{\partial x^1}\frac{\partial \bar{x}^1}{\partial t}\mathcal{E}_1 = \mathcal{E}_1, \tag{9.15}$$

$$\bar{\mathcal{E}}_2 = \frac{\mathcal{E}_2 - v\mathcal{B}_3}{\sqrt{1 - \dfrac{v^2}{c^2}}}, \qquad \bar{\mathcal{E}}_3 = \frac{\mathcal{E}_3 + v\mathcal{B}_2}{\sqrt{1 - \dfrac{v^2}{c^2}}}, \tag{9.16}$$

$$\bar{\mathcal{B}}_1 = \mathcal{B}_1, \qquad \bar{\mathcal{B}}_2 = \frac{\mathcal{B}_2 + \dfrac{v}{c^2}\mathcal{E}_3}{\sqrt{1 - \dfrac{v^2}{c^2}}}, \qquad \bar{\mathcal{B}}_3 = \frac{\mathcal{B}_3 - \dfrac{v}{c^2}\mathcal{E}_2}{\sqrt{1 - \dfrac{v^2}{c^2}}}, \tag{9.17}$$

$$\bar{\rho} = \frac{\bar{J}^0}{c} = \frac{1}{c}\frac{\partial \bar{x}^0}{\partial x^\alpha}J^\alpha = \frac{\partial \bar{t}}{\partial t}\rho + \frac{\partial \bar{t}}{\partial x^1}\mathcal{J}_1 = \frac{\rho - \dfrac{v}{c^2}\mathcal{J}_1}{\sqrt{1 - \dfrac{v^2}{c^2}}}, \tag{9.18}$$

$$\bar{\mathcal{J}}_1 = \frac{\mathcal{J}_1 - v\rho}{\sqrt{1 - \dfrac{v^2}{c^2}}}, \qquad \bar{\mathcal{J}}_2 = \mathcal{J}_2, \qquad \bar{\mathcal{J}}_3 = \mathcal{J}_3. \tag{9.19}$$

Electromagnetic field of a uniformly moving charged particle. We can use these transformation laws to find the electric and magnetic fields around a particle of charge q moving with constant velocity v. Suppose the charge is moving along the x_1-axis of a Cartesian coordinate system (x_1, x_2, x_3), as shown in Fig. 9.2. We construct another Cartesian coordinate system $(\bar{x}_1, \bar{x}_2, \bar{x}_3)$ with origin at the charge and moving with velocity components $v_1 = v$, $v_2 = v_3 = 0$ relative to the (x_1, x_2, x_3) system. In the $(\bar{x}_1, \bar{x}_2, \bar{x}_3)$ system, the electric field is given by Coulomb's law and there is no magnetic field:

$$\bar{\mathcal{E}}_i = \frac{q\bar{x}_i}{4\pi\epsilon[(\bar{x}_1)^2 + (\bar{x}_2)^2 + (\bar{x}_3)^2]^{3/2}}, \qquad \bar{\mathcal{B}}_i = 0. \tag{9.20}$$

The fields in the (x_1, x_2, x_3) system may be obtained from (9.7), (9.15)–(9.17), and (9.20):

$$\mathcal{E}_1 = \frac{q(x_1 - vt)}{4\pi\epsilon\sqrt{1 - \dfrac{v^2}{c^2}}\left[\dfrac{(x_1 - vt)^2}{1 - \dfrac{v^2}{c^2}} + (x_2)^2 + (x_3)^2\right]^{3/2}},$$

$$\mathcal{E}_2 = \frac{qx_2}{4\pi\epsilon\sqrt{1 - \dfrac{v^2}{c^2}}\left[\dfrac{(x_1 - vt)^2}{1 - \dfrac{v^2}{c^2}} + (x_2)^2 + (x_3)^2\right]^{3/2}},$$

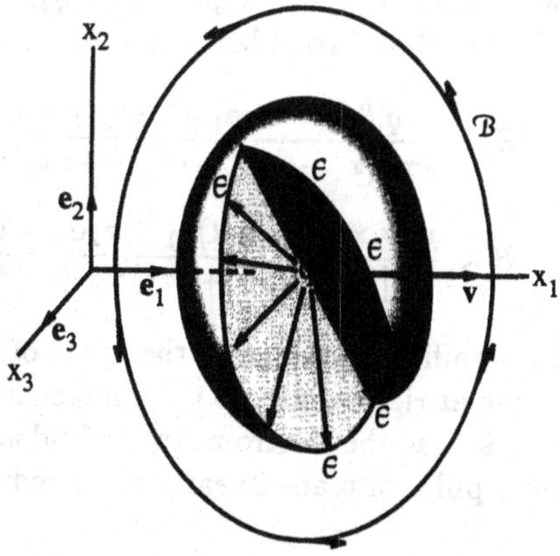

Figure 9.2. Electromagnetic field of a positively charged particle q moving with constant speed v along the x_1-axis.

$$\mathcal{E}_3 = \frac{q x_3}{4\pi\epsilon\sqrt{1-\dfrac{v^2}{c^2}}\left[\dfrac{(x_1-vt)^2}{1-\dfrac{v^2}{c^2}} + (x_2)^2 + (x_3)^2\right]^{3/2}},$$

$$\mathcal{B}_1 = 0, \qquad \mathcal{B}_2 = -\frac{v\mathcal{E}_3}{c^2}, \qquad \mathcal{B}_3 = \frac{v\mathcal{E}_2}{c^2}.$$

We can write these formulas in the vector form

$$\boldsymbol{\mathcal{E}} = \frac{q[(x_1-vt)\mathbf{e}_1 + x_2\mathbf{e}_2 + x_3\mathbf{e}_3]}{4\pi\epsilon\sqrt{1-\dfrac{v^2}{c^2}}\left[\dfrac{(x_1-vt)^2}{1-\dfrac{v^2}{c^2}} + (x_2)^2 + (x_3)^2\right]^{3/2}}, \qquad \boldsymbol{\mathcal{B}} = \frac{\mathbf{v}\times\boldsymbol{\mathcal{E}}}{c^2}.$$

Thus the electric field is still directed radially away from the instantaneous position of the charge, but is stronger than the Coulomb law on the sides of the charge and weaker than the Coulomb law ahead and behind the charge:

$$\boldsymbol{\mathcal{E}}(x_1 = vt, x_2, x_3) = \frac{q(x_2\mathbf{e}_2 + x_3\mathbf{e}_3)}{4\pi\epsilon\sqrt{1-\dfrac{v^2}{c^2}}\,[(x_2)^2 + (x_3)^2]^{3/2}},$$

$$\boldsymbol{\mathcal{E}}(x_1, x_2 = 0, x_3 = 0) = \frac{q\left(1-\dfrac{v^2}{c^2}\right)\mathbf{e}_1}{4\pi\epsilon(x_1-vt)^2}.$$

The field lines of $\boldsymbol{\mathcal{B}}$ are circles around the x_1 axis and the magnitude $\mathcal{B}$ is constant on each field line. For a slowly moving charge, the electric field is approximately given by Coulomb's law (7.80) (with $\rho dV'$ replaced by q), and the magnetic field is approximately given by the Biot-Savart law (7.86) (with $\mathbf{J}dV'$ replaced by $q\mathbf{v}$):

$$\frac{v}{c} \ll 1: \ \boldsymbol{\mathcal{E}} \approx \frac{q[(x_1-vt)\mathbf{e}_1 + x_2\mathbf{e}_2 + x_3\mathbf{e}_3]}{4\pi\epsilon[(x_1-vt)^2 + (x_2)^2 + (x_3)^2]^{3/2}}$$

$$\boldsymbol{\mathcal{B}} \approx \frac{\mu q\mathbf{v}\times[(x_1-vt)\mathbf{e}_1 + x_2\mathbf{e}_2 + x_3\mathbf{e}_3]}{4\pi[(x_1-vt)^2 + (x_2)^2 + (x_3)^2]^{3/2}}.$$

For a charge moving at a significant fraction of the speed of light, the electric field is concentrated in the plane at right angles to the charge and the magnetic field is approximately given by (7.96), so the electromagnetic field at a point fixed in space is indistinguishable from a pulse of plane linearly polarized radiation propagating in the x_1 direction.

PROBLEMS 9.1

1. How many components does a tensor of order n in a space of dimension m have?

2. The length of a four-vector $\mathbf{u}$ may be defined as $|\mathbf{u}| = \sqrt{|\eta_{\alpha\beta}u^\alpha u^\beta|}$. If $\mathbf{u}$ and $\mathbf{v}$ are two four-vectors, show that any one of the three possibilities $|\mathbf{u}+\mathbf{v}| < |\mathbf{u}| + |\mathbf{v}|$, $|\mathbf{u}+\mathbf{v}| = |\mathbf{u}| + |\mathbf{v}|$, $|\mathbf{u}+\mathbf{v}| > |\mathbf{u}| + |\mathbf{v}|$ is feasible.

3. Show that (9.6) represents a Lorentz transformation and gives a boost in velocity to the coordinate system (x_1, x_2, x_3).

4. A nuclear particle that has an average lifetime of 10^{-6} seconds is moving at a speed of $0.99c$. According to a stationary clock, how long will the particle last?

5. Show that the components $\dfrac{dx^\alpha}{d\tau}$ of the *relativistic velocity four-vector* obey the transformation laws (9.1).

6. Use the formulas (9.8) to show that if the speed of a photon is equal to c in the $(\bar{x}_1, \bar{x}_2, \bar{x}_3)$ coordinate system, then its speed is also equal to c in the (x_1, x_2, x_3) system.

7. Show that the temporal part of the relativistic force law (9.9) is actually a consequence of the spatial part.

8. A point mass is acted upon by a constant force $\bar{f}^1 = F$, $\bar{f}^2 = \bar{f}^3 = 0$ in a coordinate system $(\bar{x}_1, \bar{x}_2, \bar{x}_3)$ moving with the particle. Determine the relativistically correct motion in a fixed coordinate system (x_1, x_2, x_3).

9. Show that the relativistically correct equation governing the motion of a particle of charge q in an electric field $\boldsymbol{\mathcal{E}}$ and magnetic field $\boldsymbol{B}$ is

$$\frac{d}{dt}\left(\frac{m\mathbf{v}}{\sqrt{1-\dfrac{v^2}{c^2}}}\right) = q(\boldsymbol{\mathcal{E}} + \mathbf{v}\times\boldsymbol{B}).$$

10. Redo Problem 6 in Section 5.4 using the relativistically correct equation.

11. Show that in regions of space free of charges and currents the electromagnetic field tensor obeys the equation $\Box^2 F^{\alpha\beta} = 0$, where

$$\Box^2 = \eta^{\alpha\beta}\frac{\partial^2}{\partial x^\alpha \partial x^\beta}$$

is the *four-dimensional Laplacian operator.*

12. Defining the four-vector potential function by $A^\alpha = \left(\frac{\phi}{c}, A^i\right)$, where

$$\mathcal{E} = -\nabla\phi - \frac{\partial \mathbf{A}}{\partial t} \quad \text{and} \quad \boldsymbol{B} = \nabla \times \mathbf{A},$$

find the equations satisfied by A^α if the *Lorentz gauge condition* $\frac{\partial A^\alpha}{\partial x^\alpha} = 0$ is imposed.

13. A long straight wire of circular cross section has a constant charge density λ (charge per unit length). Find the electric and magnetic fields outside the wire in a coordinate system moving with constant speed v in a direction parallel to the wire.

9.2 GENERAL RELATIVITY

General relativity theory includes the effects of gravitation within the framework of the preceding section. If there is a gravitational field, then the components of the Riemann curvature tensor are not all zero, and it is not possible to find a coordinate system in which $g_{\alpha\beta}$ equals $\eta_{\alpha\beta}$ everywhere. The situation is very similar to the Riemannian geometry of surfaces developed in Section 8.3.

Even in the presence of a gravitational field, it *is* possible to construct a geodesic coordinate system in which $g_{\alpha\beta}$ equals $\eta_{\alpha\beta}$ and the derivatives of $g_{\alpha\beta}$ are zero at any particular point O in spacetime:

$$g_{\alpha\beta}(O) = \eta_{\alpha\beta}, \qquad \frac{\partial g_{\alpha\beta}}{\partial x^\gamma}(O) = 0.$$

The equations for the motion of a particle in the geodesic coordinate system reduce at O to the special relativistic form (9.9). The geodesic coordinate system can be thought of as a coordinate system that is attached to a particle at O falling freely in the gravitational field. Transforming the special relativistic form (9.9) from the geodesic system to an arbitrary coordinate system, we obtain the general relativistic form of the equations for the motion of a point mass:

$$\frac{d}{d\tau}\left(m\frac{dx^\alpha}{d\tau}\right) + \underline{m\Gamma^\alpha_{\beta\gamma}\frac{dx^\beta}{d\tau}\frac{dx^\gamma}{d\tau}} = f^\alpha. \tag{9.21}$$

Of course, in (9.21) the proper time integral is given by the general quadratic form (9.2), and the Christoffel symbols are related to $g_{\alpha\beta}$ by (8.6) with Latin indices replaced by Greek. The underlined terms are accelerations that will arise in curvilinear spatial coordinates (c.f. 8.27), in a rotating spatial Cartesian coordinate system, in a gravitational field, and so on. When $f^\alpha = 0$ and the mass m of the particle is a constant, (9.21) reduces to

$$\frac{d^2x^\alpha}{d\tau^2} + \Gamma^\alpha_{\beta\gamma}\frac{dx^\beta}{d\tau}\frac{dx^\gamma}{d\tau} = 0. \tag{9.22}$$

Since these equations are the same as (8.64), we can conclude that the trajectory of a freely falling particle is simply a geodesic curve in four-dimensional spacetime. It can also be shown that the trajectory of a freely falling particle joining two events is the spacetime curve along which the proper time elapsed is a maximum.

We can deduce the form of Maxwell's equations in an arbitrary coordinate system by similar means. Maxwell's equations reduce at the origin of a geodesic coordinate system to the special relativistic form (9.13). Transforming the special relativistic form (9.13) from the geodesic system to an arbitrary coordinate system, we obtain the general relativistic form of Maxwell's equations:

$$F^{\alpha\beta}{}_{,\beta} = \mu J^{\alpha}, \qquad F_{\beta\gamma,\alpha} + F_{\gamma\alpha,\beta} + F_{\alpha\beta,\gamma} = 0. \tag{9.23}$$

Here, as always, the commas indicate covariant derivatives.

In general relativity theory the components $g_{\alpha\beta}$ of the spacetime metric tensor are governed by the distribution of matter in the universe. In empty space, the *Einstein field equations* reduce to

$$R_{\alpha\beta} = 0. \tag{9.24}$$

Here $R_{\alpha\beta}$ are components of the symmetric *Ricci tensor*, defined by contracting the first and last indices of the Riemann curvature tensor:

$$R_{\alpha\beta} = R^{\gamma}{}_{\alpha\beta\gamma}. \tag{9.25}$$

With the use of (8.6) and (8.20), we can write the Einstein equations (9.24) as 10 partial differential equations for the 10 unknown functions $g_{\alpha\beta}$. These equations are not all independent since the 10 $R_{\alpha\beta}$ are related by four Bianchi identities, obtained from (8.21):

$$R^{\beta}_{\alpha,\beta} - \frac{1}{2}R^{\gamma}_{\gamma,\alpha} = 0.$$

The four degrees of freedom correspond to the fact that, if $g_{\alpha\beta}$ are a solution of Einstein's equations, then so are the $\bar{g}_{\alpha\beta}$ obtained from $g_{\alpha\beta}$ by a general coordinate transformation $\bar{x}^{\alpha} = x^{\alpha}(x^0, x^1, x^2, x^3)$.

In the remainder of this section we suppose that the gravitational field is produced by a spherically symmetric body fixed in space. For this case it is possible to derive an analytical solution to the Einstein field equations. We choose spherical coordinates $(x^1, x^2, x^3) = (r, \phi, \theta)$ with origin at the center of the spherical body, and guess that the components of the metric tensor have the form

$$g_{00} = g_{00}(r), \quad g_{11} = g_{11}(r), \quad g_{22} = -r^2, \quad g_{33} = -r^2\sin^2\phi, \quad g_{\alpha\beta} = 0 \text{ for } \alpha \neq \beta.$$

The nonvanishing Christoffel symbols are then, from (8.6),

$$\left.\begin{aligned}
&\Gamma^0_{01} = \Gamma^0_{10} = \frac{1}{2g_{00}}\frac{dg_{00}}{dr}, \quad \Gamma^1_{00} = -\frac{1}{2g_{11}}\frac{dg_{00}}{dr}, \quad \Gamma^1_{11} = \frac{1}{2g_{11}}\frac{dg_{11}}{dr}, \\
&\Gamma^1_{22} = \frac{r}{g_{11}}, \quad \Gamma^1_{33} = \frac{r\sin^2\phi}{g_{11}}, \quad \Gamma^2_{12} = \Gamma^2_{21} = \Gamma^3_{13} = \Gamma^3_{31} = \frac{1}{r}, \\
&\Gamma^2_{33} = -\sin\phi\cos\phi, \quad \Gamma^3_{23} = \Gamma^3_{32} = \cot\phi.
\end{aligned}\right\} \tag{9.26}$$

The nonvanishing components of the Ricci tensor are, from (8.20) and (9.25)–(9.26):

$$R_{00} = \frac{1}{2g_{11}} \frac{d^2 g_{00}}{dr^2} + \frac{1}{rg_{11}} \frac{dg_{00}}{dr} - \frac{1}{4g_{11}} \frac{dg_{00}}{dr} \left(\frac{1}{g_{00}} \frac{dg_{00}}{dr} + \frac{1}{g_{11}} \frac{dg_{11}}{dr} \right),$$

$$R_{11} = \frac{1}{2g_{00}} \frac{d^2 g_{00}}{dr^2} - \frac{1}{rg_{11}} \frac{dg_{11}}{dr} - \frac{1}{4g_{00}} \frac{dg_{00}}{dr} \left(\frac{1}{g_{00}} \frac{dg_{00}}{dr} + \frac{1}{g_{11}} \frac{dg_{11}}{dr} \right),$$

$$R_{22} = -1 - \frac{1}{g_{11}} - \frac{r}{2g_{11}} \left(\frac{1}{g_{00}} \frac{dg_{00}}{dr} - \frac{1}{g_{11}} \frac{dg_{11}}{dr} \right),$$

$$R_{33} = \sin^2 \phi R_{22}.$$

Since

$$\frac{R_{00}}{g_{00}} - \frac{R_{11}}{g_{11}} = \frac{1}{rg_{11}} \left(\frac{1}{g_{00}} \frac{dg_{00}}{dr} + \frac{1}{g_{11}} \frac{dg_{11}}{dr} \right),$$

(9.24) requires that

$$\frac{1}{g_{00}} \frac{dg_{00}}{dr} = -\frac{1}{g_{11}} \frac{dg_{11}}{dr}$$

whence

$$g_{11} = \frac{C_1}{g_{00}}.$$

Since then

$$R_{22} = -1 - \frac{g_{00}}{C_1} - \frac{r}{C_1} \frac{dg_{00}}{dr},$$

(9.24) requires that

$$r\frac{dg_{00}}{dr} + g_{00} = -C_1,$$

whence

$$g_{00} = -C_1 + \frac{C_2}{r}.$$

Requiring the metric components to approach those of flat spacetime as $r \to \infty$ sets the constant $C_1 = -1$. As we shall subsequently see, requiring the orbit of a point mass to be approximated by the Newtonian result sets the constant $C_2 = -\frac{2GM}{c^2}$, where G is the gravitational constant and M is the mass of the spherical body. To recapitulate, the metric coefficients corresponding to the proper time interval

$$\Delta\tau^2 = \left(1 - \frac{2GM}{c^2 r}\right) \Delta t^2 - \frac{1}{c^2} \left(\frac{\Delta r^2}{1 - \dfrac{2GM}{c^2 r}} + r^2 \Delta\phi^2 + r^2 \sin^2 \phi \Delta\theta^2 \right) \quad (9.27)$$

are an exact solution to the Einstein equations (9.24) known as the *Schwarzschild solution*. An interesting physical consequence of (9.27) is *gravitational time dilation*: when a clock is at rest a distance r from the center of a spherical mass M, it runs slower by the factor $\sqrt{1 - 2GM/c^2r}$.

We can now determine the general relativistic prediction for the orbit of a point mass moving under the influence of the gravitational field produced by a spherically symmetric body. Inserting the Christoffel symbols (9.26) corresponding to the Schwarzschild solution into the geodesic equations (9.22), we obtain

$$\frac{d^2t}{d\tau^2} + \frac{1}{g_{00}}\frac{dg_{00}}{dr}\frac{dt}{d\tau}\frac{dr}{d\tau} = 0 \tag{9.28}$$

$$\frac{d^2r}{d\tau^2} - \frac{c^2}{2g_{11}}\frac{dg_{00}}{dr}\left(\frac{dt}{d\tau}\right)^2 + \frac{1}{2g_{11}}\frac{dg_{11}}{dr}\left(\frac{dr}{d\tau}\right)^2 + \frac{r}{g_{11}}\left(\frac{d\phi}{d\tau}\right)^2 + \frac{r\sin^2\phi}{g_{11}}\left(\frac{d\theta}{d\tau}\right)^2 = 0 \tag{9.29}$$

$$\frac{d^2\phi}{d\tau^2} + \frac{2}{r}\frac{dr}{d\tau}\frac{d\phi}{d\tau} - \sin\phi\cos\phi\left(\frac{d\theta}{d\tau}\right)^2 = 0 \tag{9.30}$$

$$\frac{d^2\theta}{d\tau^2} + \frac{2}{r}\frac{dr}{d\tau}\frac{d\theta}{d\tau} + 2\cot\phi\frac{d\phi}{d\tau}\frac{d\theta}{d\tau} = 0. \tag{9.31}$$

The first equation (9.28) can immediately be integrated to yield

$$\frac{dt}{d\tau} = \frac{C_3}{g_{00}}. \tag{9.32}$$

The third equation (9.30) is satisfied by

$$\phi = \frac{\pi}{2}$$

so, without loss of generality, we assume that the orbit is confined to the equatorial plane. The fourth equation (9.31) can then be integrated to yield

$$\frac{d\theta}{d\tau} = \frac{C_4}{r^2}. \tag{9.33}$$

Requiring the relativistic orbit to agree with the two-body solution at $\theta = 0$ sets the constants $C_3 = 1 - 2GM/[a(1-e)c^2]$ and $C_4 = h$, where (a, e, h) are elements of the Newtonian orbit. With the use of these relations and the Schwarzschild solution (9.27), the second equation (9.29) reduces to

$$\frac{d^2r}{d\tau^2} - \frac{h^2}{r^3} = -\frac{GM}{r^2} - \frac{3(GM)^2a(1-e^2)}{c^2r^4}.$$

Introducing the variable $u = a(1 - e^2)/r$ and the small parameter $\varepsilon = 3GM/[a(1-e^2)c^2]$, we can cast this equation into the more convenient form

$$\frac{d^2u}{d\theta^2} + u = 1 + \underline{\varepsilon u^2} \tag{9.34}$$

Equations (9.32)–(9.34) together with appropriate initial conditions suffice to determine the orbit. These equations differ from the Newtonian results by the presence of $d\tau$ instead of dt in (9.33) and by the underlined term in (9.34) (cf. Problem 6 in Section 8.2).

Although the relativistic equation (9.34) can be integrated exactly with the aid of elliptic functions, for purposes of comparison with the Newtonian theory a perturbation procedure known as the *method of strained coordinates* is sufficient. It is reasonable to expect that the relativistic solution for $u(\theta)$ will be close to the Newtonian two-body solution $1 + e\cos\theta$; the differences can be accounted for by letting

$$u = 1 + e\cos y(\theta) + \varepsilon u_1(\theta) + \varepsilon^2 u_2(\theta) + \cdots,$$
$$y = \theta(1 + \varepsilon y_1 + \varepsilon^2 y_2 + \cdots). \tag{9.35}$$

An algorithm for the perturbation procedure is:

Let $n = 1$
Substitute expressions (9.35) into equation (9.34)
Equate the coefficients of ε^n
Choose the constant y_n so secular terms will not arise
Solve for the function $u_n(\theta)$
Iterate on n

Beginning by substituting (9.35) into (9.34) and equating the coefficients of ε, we obtain

$$\frac{d^2u_1}{d\theta^2} + u_1 = 1 + \frac{e^2}{2} + \frac{e^2}{2}\cos 2y + 2e(y_1 + 1)\cos y. \tag{9.36}$$

In (9.36) the $\cos y$ term would produce a secular term $\theta \sin y$ in u_1; to eliminate these possibilities we choose

$$y_1 = -1.$$

It suffices for our purposes to terminate the sequence of steps in the perturbation procedure at this stage and to write

$$u = 1 + e\cos[(1-\varepsilon)\theta] + \cdots$$

or

$$r = \frac{a(1-e^2)}{1 + e\cos[(1-\varepsilon)\theta] + \cdots}. \tag{9.37}$$

Equation (9.37) represents an orbit that is a precessing conic section, such as the precessing ellipse shown in Fig. 9.3. The argument of the cosine changes by 2 when θ changes by $\dfrac{2\pi}{1-\varepsilon} \approx 2\pi(1+\varepsilon)$, so the angular distance between one perihelion point and the next is larger than 2π by the factor

$$2\pi\varepsilon = \frac{6\pi GM}{a(1-e^2)c^2}. \tag{9.38}$$

The agreement of the prediction (9.38) with the actually observed angular precession of perihelion in the orbits of certain celestial objects is a stunning validation of the general theory of relativity.

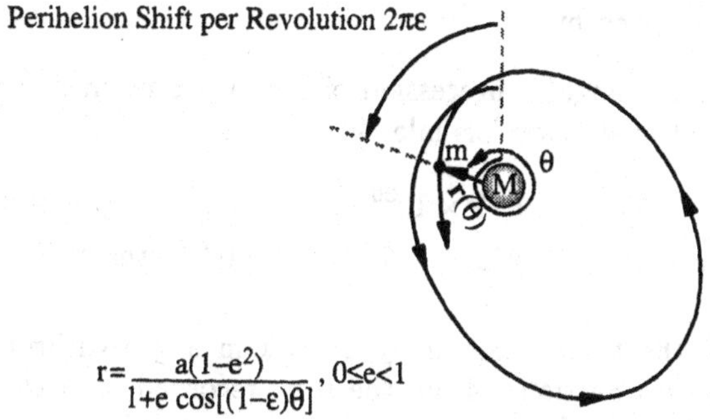

Figure 9.3. General relativity theory predicts that the orbit of a point mass m about a spherical body M is a precessing conic section.

PROBLEMS 9.2

1. Show that the equations (9.9), which are valid at the origin O of a geodesic coordinate system (x^0, x^1, x^2, x^3), transform into the general relativistic form (9.21) at the corresponding point $\bar{O}$ of an arbitrary coordinate system $(\bar{x}^0, \bar{x}^1, \bar{x}^2, \bar{x}^3)$.

2. Show that the equations (9.13), which are valid at the origin O of a geodesic coordinate system (x^0, x^1, x^2, x^3), transform into the general relativistic form (9.23) at the corresponding point $\bar{O}$ of an arbitrary coordinate system $(\bar{x}^0, \bar{x}^1, \bar{x}^2, \bar{x}^3)$.

3. Neglecting gravity, and assuming that the speed of the particle is much less than c, show that the relativistic equations (9.21) for the motion of a particle of constant mass in a coordinate system $(x^0, \bar{x}^1, \bar{x}^2, \bar{x}^3)$ which is spatially accelerating with respect to a geodesic coordinate system (x^0, x^1, x^2, x^3) reduce to the Newtonian equations (5.19).

4. Neglecting gravity, write down the correct form of Maxwell's equations in a coordinate system that is spatially accelerating with respect to a geodesic coordinate system.

5. Assuming that the gravitational field is weak, so that $g_{\alpha\beta} = \eta_{\alpha\beta} + h_{\alpha\beta}$ where $|h_{\alpha\beta}| \ll 1$, find the equations satisfied by $h_{\alpha\beta}$ if the *harmonic coordinate condition* $g^{\alpha\beta}\Gamma^{\gamma}_{\alpha\beta} = 0$ is imposed.

6. Assuming that the grativational field is weak and stationary, so that $g_{\alpha\beta} = \eta_{\alpha\beta} + h_{\alpha\beta}(x^1, x^2, x^3)$ where $|h_{\alpha\beta}| \ll 1$, and that the speed of the particle is much less than c, show that the general relativistic equations for the motion of a freely falling particle reduce to the Newtonian equations.

7. Obtain the complete first order perturbation solution to (9.34), including all terms multiplied by ε.

8. Calculate the angular precession of Mercury's perihelion per century from the general relativistic formula (9.38). Note:

$$M_{\text{Sun}} = 1.99 \times 10^{30}\ \text{kg}, \qquad a_{\text{Mercury}} = 5.79 \times 10^{10}\ \text{m},$$
$$e_{\text{Mercury}} = 0.206, \qquad T_{\text{Mercury}} = 0.241\ \text{years}$$

9. Compare the time recorded by a clock in a geosynchronous satellite (see Problem 4 in Section 7.4) to the time recorded by a clock at rest on the surface of the earth.

9.3 NOTATION OF OTHERS

Since (9.9) contains no extraneous accelerations, a geodesic coordinate system is sometimes said to be *locally inertial* near its origin O. However, since general relativity theory accounts for the effects of gravitation by the metric tensor and not by a force, a geodesic coordinate system has the same acceleration as a coordinate system attached to a freely falling particle at O, unlike the inertial coordinate systems of Newtonian mechanics which are unaffected by the gravitational field. To avoid confusion, we have used the word "inertial" only within the context of Newtonian mechanics.

Table 9

OUR NOTATION	OUR NAME	OTHER NOTATIONS	OTHER NAMES
$x^0 = ct$	Temporal spacetime coordinate.	$x^4 = ct$	
$g_{\alpha\beta}$	Components of spacetime metric tensor.	$-g_{\alpha\beta}$	Same name and symbol may be negative of ours.
$F^{\alpha\beta}$	Components of electro-magnetic field tensor.	$F^{\beta\alpha}$	Same name and symbol may be transpose of ours.

Nature is the realization of the simplest conveivable mathematical ideas. I am convinced we can discover by means of purely mathematical constructions the concepts and laws connecting them with each other, which furnish the key to the understanding of natural phenomena.

– Albert Einstein, 1933

REFERENCES

HISTORICAL

1. Crowe, M. J., *A History of Vector Analysis: The Evolution of the Idea of a Vectorial System*, 1985, Dover.
2. Einstein, A., *The Meaning of Relativity*, 1956, Princeton University Press.
3. Euclid, *The Thirteen Books of Euclid's Elements*, translated by T. L. Heath, 1925, Dover.
4. Gibbs, J. W., and Wilson, E. B., *Vector Analysis*, 1901, Yale University Press.
5. Hamilton, W. R., *Elements of Quaternions*, 1899, Chelsea.
6. Heaviside, O., *Electromagnetic Theory*, 1912, Chelsea.
7. Kline, M., *Mathematical Thought from Ancient to Modern Times*, 1972, Oxford University Press.
8. Maxwell, J. C., *A Treatise on Electricity and Magnetism*, 1891, Dover.
9. Newton, I., *Principia*, Motte's translation revised by F. Cajori, 1934, University of California Press.
10. Ricci and Levi-Civita's, *Tensor Analysis Paper*, translated by R. Hermann, 1975, Math Sci Press.
11. Riemann, *On the Hypotheses which lie at the Foundations of Geometry*, translated in *A Comprehensive Introduction to Differential Geometry*, vol. II, by Michael Spivak, 1979, Publish or Perish, Inc.
12. Timoshenko, S. P., *History of Stength of Materials*, 1953, Dover.

MATHEMATICAL

1. Abraham, R., Marsden, J. E., and Ratiu, T., *Manifolds, Tensor Analysis, and Applications*, 1993, Springer-Verlag.
2. Akivis, M. A., and Goldberg, V. V., *An Introduction to Linear Algebra and Tensors*, 1977, Dover.
3. Billings, A. R., *Tensor Properties of Materials*, 1969, Wiley.
4. Block, H. D., *Introduction to Tensor Analysis*, 1962, Merrill.

5. Borisenko, A. I., and Tarapov, I. E., *Vector and Tensor Analysis with Applications*, 1979, Dover.

6. Brand, L., *Vector and Tensor Analysis*, 1947, Wiley.

7. Brillouin, L., *Tensors in Mechanics and Elasticity*, 1964, Academic Press.

8. Budiansky, B., *Tensor Analysis*, Ch. 4 of *Handbook of Applied Mathematics*, Pearson, C. E., ed., 1990, Chapman and Hall.

9. Coburn, *Vector and Tensor Analysis*, 1955, Macmillan.

10. Davis, H. F., and Snider, A. D., *Introduction to Vector Analysis*, 1995, Brown.

11. Dodson, T., and Poston, C. T., *Tensor Geometry: The Geometric Viewpoint and Its Uses*, 1991, Springer-Verlag.

12. Eisele, J. A., and Mason, R. M., *Applied Matrix and Tensor Analysis*, 1970, Wiley.

13. Greenspan, H. P., and Benney, D. J., *Calculus, An Introduction to Applied Mathematics*, 1986, McGraw-Hill.

14. Hildebrand, F. B., *Advanced Calculus for Applications*, 1976, Prentice-Hall

15. Kovach, L. D., *Advanced Engineering Mathematics*, 1991, Addison-Wesley.

16. Kron, G., *Tensors for Circuits*, 1959, Dover.

17. Lass, H., *Vector and Tensor Analysis*, 1950, McGraw-Hill.

18. Lawden, D. F., *An Introduction to Tensor Calculus, Relativity, and Cosmology*, 1982, Wiley.

19. Lewis, P. E., and Ward, J. P., *Vector Analysis for Engineers and Scientists* 1989, Addison-Wesley.

20. Lichnerowicz, A., *Elements of Tensor Calculus*, 1962, Methuen.

21. Lingren, B. W., *Vector Calculus*, 1964, Macmillan.

22. Lovelock, D., and Rund, H., *Tensors, Differential Forms, and Variational Principles*, 1975, Wiley.

23. Marsden, J. E., and Tromba, A. J., *Vector Calculus*, 1995, Freeman.

24. McConnell, A. J., *Applications of the Absolute Differential Calculus*, 1931 Blackie.

25. McQuistan, R. B., *Scalar and Vector Fields: A Physical Interpretation*, 1965 Wiley.

26. Nayfeh, A., *Perturbation Methods*, 1973, Wiley.

27. Newell, H. E., Jr., *Vector Analysis*, 1955, McGraw-Hill.

28. Schouten, J. A., *Tensor Analysis for Physicists*, 1954, Dover.

29. Segal, L. A., *Mathematics Applied to Continuum Mechanics*, 1977, Macmillan.

30. Simmonds, J. G., *A Brief on Tensor Analysis*, 1993, Springer-Verlag.

31. Simmonds, J. G., and Mann, J. E., Jr., *A First Look at Perturbation Theory*, 1986, Krieger.

32. Smith, M., *Principles and Applications of Tensor Analysis*, 1963, H. W. Sams.

33. Sokolnikoff, I. S., *Tensor Analysis Theory and Applications to Geometry and Mechanics of Continua*, 1964, Books on Demand.

34. Spiegel, M. R., *Schaum's Vector Analysis*, 1959, McGraw-Hill.

35. Struik, D. J., *Lectures on Classical Differential Geometry*, 1988, Dover.

36. Synge, J. L., and Schild, A., *Tensor Calculus*, 1949, Dover.

37. Teichmann, H., *Physical Applications of Vectors and Tensors*, 1969, Harrap.

38. Temple, G., *Catesian Tensors*, 1960, Methuen.

39. Thomas, G. B., Jr., and Finney, R. L., *Calculus and Analytic Geometry*, 1988, Addison-Wesley.

40. Tyldesley, J. R., *An Introduction to Tensor Analysis for Engineers and Applied Scientists*, 1975, Books on Demand.

41. Wilde, C., *Linear Algebra*, 1988, Addison-Wesley.

42. Willis, A. P., *Vector Analysis with an Introduction to Tensor Analysis*, 1940, Prentice-Hall.

43. Wrede, R. C., *Introduction to Vector and Tensor Analysis*, 1972, Dover.

44. Yaglom, I. M., *A Simple Non-Euclidean Geometry and Its Physical Basis*, 1979, Springer-Verlag.

PHYSICAL

1. Agrawal, B. N., *Design of Geosynchronous Spacecraft*, 1986, Prentice-Hall.

2. Anderson, J. D., *Fundamentals of Aerodynamics*, 1991, McGraw-Hill.

3. Barger, V., and Olsson, M., *Classical Mechanics, A Modern Perspective*, 1973, McGraw-Hill.

4. Batchelor, G. K., *An Introduction to Fluid Dynamics*, 1967, Cambridge University Press.

5. Bate, R. R., Mueller, D. D., and White, J. E., *Fundamentals of Astrodynamics*, 1971, Dover.

6. Beer, F. P., and Johnston, E. R., Jr., *Mechanics of Materials*, 1981, McGraw-Hill.

7. Cesarone, R. J., "A Gravity Assist Primer," *AIAA Student Journal*, Spring 1989, pp. 16–22.

8. Danielson, D. A., "Tension Field Theories for Soft Tissues," *Bulletin of Mathematical Biology*, vol. 40, 1978, pp. 161–182.

9. Feynman, R. P., *"Surely You're Joking, Mr. Feynman,"* 1986, Bantam.

10. Feynman, R. P., Leighton, R. B., and Sands, M., *The Feynman Lectures on Physics*, vol. I and II, 1964, Addison-Wesley.

11. Fung, Y. C., *A First Course in Continuum Mechanics*, 1994, Prentice-Hall.

12. Goldstein, H., *Classical Mechanics*, 1980, Addison-Wesley.

13. Halliday, D., and Resnick, R., *Fundamentals of Physics*, 1989, Wiley.

14. Haltiner, G. J., and Williams, R. T., *Numerical Prediction and Dynamic Meteorology*, 1980, Wiley.

15. Hughes, P. C., *Spacecraft Attitude Dynamics*, 1986, Wiley.

16. Incropera, F. P., and DeWitt, D. P., *Introduction to Heat Transfer*, 1990, Wiley.

17. Jackson, J. D., *Classical Electrodynamics*, 1975, Wiley.

18. Kaplan, M. H., *Modern Spacecraft Dynamics and Control*, 1976, Wiley.

19. Kittel, C., *Introduction to Solid State Physics*, 1986, Wiley.

20. Lugt, H. J., *Vortex Flow in Nature and Technology*, 1983, Wiley.

21. Magid, L. M., *Electromagnetic Fields, Energy, and Waves*, 1972, Wiley.

22. Malvern, L. E., *Introduction to the Mechanics of a Continuous Medium*, 1969, Prentice-Hall.

23. Malvern, L. E., *Engineering Mechanics*, vol. I and II, 1976, Prentice-Hall.

24. Mason, W. P., *Crystal Physics of Interaction Processes*, 1966, Academic Press.

25. Misner, C. W., Thorne, K. S., and Wheeler, J. A., *Gravitation*, 1973, Freeman.

26. Morse, P. M., and Feshback, H., *Methods of Theoretical Physics*, Part I, 1953, McGraw-Hill.

27. Niordson, F. I., *Shell Theory*, 1985, North Holland.

28. Nye, J. F., *Physical Properties of Crystals*, 1987, Oxford.

29. Ogden, R. W., *Non-linear Elastic Deformations*, 1984, Wiley.

30. Ohanian, H. C., *Gravitation and Spacetime*, 1976, Norton.

31. Reitz, J. R., Milford, F. J., and Christy, R. W., *Foundations of Electromagnetic Theory*, 1980, Addison-Wesley.

32. Resnick, D., and Halliday, D., *Basic Concepts in Relativity and Early Quantum Theory*, 1985, Wiley.

33. Robertson, H. P., and Noonan, T. W., *Relativity and Cosmology*, 1968, Saunders.

34. Roy, A. E., *Orbital Motion*, 1988, Adam Hilger.

35. Sears, F. W., Zemansky, M. W., and Young, H. D., *University Physics*, 1987, Addision-Wesley.

36. Symon, K. R., *Mechanics*, 1971, Addison-Wesley.

37. Timoshenko, S. P., and Goodier, J. N., *Theory of Elasticity*, 1970, McGraw-Hill.

38. Timoshenko, S., Young, D. H., and Weaver, W., Jr., *Vibration Problems in Engineering*, 1974, Wiley.

39. von Schwind, J. J., *Geophysical Fluid Dynamics for Oceanographers*, 1980, Prentice-Hall.

40. Weinberg, S., *Gravitation and Cosmology Principles and Appplications of the General Theory of Relativity*, 1972, Wiley.

41. Wetty, J. R., Wicks, C. E., and Wilson, R. E., *Fundamentals of Momentum, Heat, and Mass Transfer*, 1969, Wiley.

42. White, F. M., *Fluid Mechanics*, 1994, McGraw-Hill.

43. Wilson, O. B., *An Introduction to the Theory and Design of Sonar Transducers*, 1988, Peninsula Publishing.

ANSWERS TO PROBLEMS

Problems 1.3

1. A unit vector in direction of $\mathbf{v}$.
2. Direction northeast, magnitude $2\sqrt{2}$.
3. $|\mathbf{e}_1 + \mathbf{e}_2 + \mathbf{e}_3| = \sqrt{3}$.
4. The result follows from the geometrical interpretation of the equation

$$\mathbf{a} + \mathbf{b} + \mathbf{c} + \mathbf{d} = ((\mathbf{a} + \mathbf{b}) + \mathbf{c}) + \mathbf{d}.$$

5. Since $\mathbf{a}$, $\mathbf{b}$, $\mathbf{c}$, $\mathbf{d}$ are sides of a quadrilateral, $\mathbf{a} + \mathbf{b} + \mathbf{c} + \mathbf{d} = 0$. If the quadrilateral is a parallelogram, $\mathbf{a}$ must be parallel to $\mathbf{c}$ and in the opposite direction so $\mathbf{a} = -\mathbf{c}$, and conversely.
6. If $\mathbf{a} + \mathbf{b} = 0$, $\mathbf{a}$ and $\mathbf{b}$ must lie along the same line when added in the head-to-tail method. If $\mathbf{a} + \mathbf{b} + \mathbf{c} = 0$, $\mathbf{a}$, $\mathbf{b}$, $\mathbf{c}$ must lie in the same plane when added in the head-to-tail method.
7. Label vectors eminating from the point of intersection of the diagonals to the vertices by $\mathbf{a}$, $\mathbf{a}'$, $\mathbf{b}$, $\mathbf{b}'$. Since vectors along the opposite sides of the parallelogram must be equal, we must have $\mathbf{a} + \mathbf{a}' = \mathbf{b} + \mathbf{b}'$. But $\mathbf{a} + \mathbf{a}'$ is not parallel to $\mathbf{b} + \mathbf{b}'$ and hence each must be equal to zero: $\mathbf{a} = -\mathbf{a}'$, $\mathbf{b} = -\mathbf{b}'$.
8. Look at a triangle with sides $\mathbf{u}$, $\mathbf{v}$, $\mathbf{u} + \mathbf{v}$. The side of length $|\mathbf{u} + \mathbf{v}|$ must be greater than or equal to the difference and less than or equal to the sum of the other two sides.
9. Let P denote the point $\frac{3}{4}$ of the distance along a line joining a vertex to the point of intersection of the medians on the opposite side. Label vectors emanating from P to the vertices by $\mathbf{a}$, $\mathbf{b}$, $\mathbf{c}$ and to the intersections of the medians by $\mathbf{a}'$, $\mathbf{b}'$, $\mathbf{c}'$. By vector addition,

$$\mathbf{a}' = \frac{\mathbf{b} + \mathbf{c} + \mathbf{d}}{3} \quad \text{and} \quad \mathbf{b}' = \frac{\mathbf{a} + \mathbf{c} + \mathbf{d}}{3.}$$

Eliminating $\mathbf{c} + \mathbf{d}$ yields $\mathbf{a}' + \frac{\mathbf{a}}{3} = \mathbf{b}' + \frac{\mathbf{b}}{3}$. Since $\mathbf{a}' = -\frac{\mathbf{a}}{3}$, $\mathbf{b}' = -\frac{\mathbf{b}}{3}$. Thus the straight line emanating from the $\mathbf{b}$-vertex passes through the point P at $\frac{3}{4}$ of the distance to the point of intersection of the medians on the opposite side.

Problems 1.4

1. Yes.

2. Construct vectors **a**, **b**, **c**, **d** from the point of intersection to the vertices. By symmetry, the lengths of these vectors and angle θ between them is the same and $\mathbf{a}+\mathbf{b}+\mathbf{c}+\mathbf{d}=0$. Dot with **a** to obtain

$$a^2+3a^2\cos\theta=0, \text{ so } \theta=\cos^{-1}\left(-\frac{1}{3}\right)=109.5^\circ.$$

3. Let **a**, **b**, and $\mathbf{c}=\mathbf{a}+\mathbf{b}$ be vectors along the sides of the triangle, where $\mathbf{a}\cdot\mathbf{b}=0$. Then

$$c^2=\mathbf{c}\cdot\mathbf{c}=(\mathbf{a}+\mathbf{b})\cdot(\mathbf{a}+\mathbf{b})=\mathbf{a}\cdot\mathbf{a}+2\mathbf{a}\cdot\mathbf{b}+\mathbf{b}\cdot\mathbf{b}=a^2+b^2$$

4. $|\mathbf{u}\cdot\mathbf{v}|=uv|\cos\theta|\le uv$, $|\mathbf{u}\times\mathbf{v}|=uv|\sin\theta|\le uv$.

5. $(\mathbf{u}\times\mathbf{v})\cdot(\mathbf{u}\times\mathbf{v})>0$, $(\mathbf{u}\times\mathbf{v})\cdot(\mathbf{v}\times\mathbf{u})<0$.

6. If the vectors are as shown in Fig. 1.5(a),

$$(\mathbf{u}\times\mathbf{v})\cdot\mathbf{w}=|\mathbf{u}\times\mathbf{v}|w\cos\beta=(\text{area of base})(\text{height})=\text{volume}.$$

7. $\mathbf{v}=\dfrac{\mathbf{v}\times\mathbf{g}_2\cdot\mathbf{g}_3}{\mathbf{g}_1\times\mathbf{g}_2\cdot\mathbf{g}_3}\mathbf{g}_1+\dfrac{\mathbf{g}_1\times\mathbf{v}\cdot\mathbf{g}_3}{\mathbf{g}_1\times\mathbf{g}_2\cdot\mathbf{g}_3}\mathbf{g}_2+\dfrac{\mathbf{g}_1\times\mathbf{g}_2\cdot\mathbf{v}}{\mathbf{g}_1\times\mathbf{g}_2\cdot\mathbf{g}_3}\mathbf{g}_3.$

8. $\mathbf{u}\times(\mathbf{v}\times\mathbf{w})-(\mathbf{u}\times\mathbf{v})\times\mathbf{w}=\mathbf{v}(\mathbf{u}\cdot\mathbf{w})-\mathbf{w}(\mathbf{u}\cdot\mathbf{v})+\mathbf{u}(\mathbf{v}\cdot\mathbf{w})-\mathbf{v}(\mathbf{u}\cdot\mathbf{w})=\mathbf{u}(\mathbf{v}\cdot\mathbf{w})-\mathbf{w}(\mathbf{u}\cdot\mathbf{v})=\mathbf{v}\times(\mathbf{u}\times\mathbf{w})$.

9. The first result follows from (1.4) with **a** replaced by $\mathbf{a}+\mathbf{b}$, **b** replaced by **c**, and **c** replaced by **d**. The second result follows from (1.3) and (1.4):

$$\begin{aligned}(\mathbf{a}\times\mathbf{b})\cdot(\mathbf{c}\times\mathbf{d})&=\mathbf{a}\cdot[\mathbf{b}\times(\mathbf{c}\times\mathbf{d}]\\&=\mathbf{a}\cdot[\mathbf{c}(\mathbf{b}\cdot\mathbf{d})-\mathbf{d}(\mathbf{b}\cdot\mathbf{c})]\\&=(\mathbf{a}\cdot\mathbf{c})(\mathbf{b}\cdot\mathbf{d})-(\mathbf{a}\cdot\mathbf{d})(\mathbf{b}\cdot\mathbf{c})\end{aligned}$$

10. In the triangle of Fig. 1.6, $\mathbf{a}\times\mathbf{c}=\mathbf{a}\times(\mathbf{a}+\mathbf{b})=\mathbf{a}\times\mathbf{b}$, so $ac\sin B=ab\sin C$.

11. From Fig. 1.7, $\sin A=\dfrac{|(\mathbf{a}\times\mathbf{b})\times(\mathbf{a}\times\mathbf{c})|}{|\mathbf{a}\times\mathbf{b}|\,|\mathbf{a}\times\mathbf{c}|}=\dfrac{|\mathbf{a}\times\mathbf{b}\cdot\mathbf{c}|}{\sin\gamma\sin\beta}$ and

$$\sin B=\frac{|(\mathbf{b}\times\mathbf{c})\times(\mathbf{b}\times\mathbf{a})|}{|\mathbf{b}\times\mathbf{c}|\,|\mathbf{b}\times\mathbf{a}|}=\frac{|\mathbf{a}\times\mathbf{b}\cdot\mathbf{c}|}{\sin\gamma\sin\alpha},\quad\text{so}\quad\frac{\sin A}{\sin\alpha}=\frac{\sin B}{\sin\beta}$$

Problems 1.5

1. See Fig. 1.10. Choose $\delta_2=180^\circ$ so increase is

$$v+\bar{v}-\sqrt{v^2+\bar{v}^2-2v\bar{v}\cos\delta_1}.$$

2. Distance 2 meters, velocity 0.

3. $\tan\theta=\dfrac{v}{c}$, so $\theta\approx\dfrac{v}{c}$ radians.

4. Force is $q\mathbf{v} \times \boldsymbol{B}$, q is negative for an electron, so particle is deflected upwards from the surface of the earth.

5. $\mathbf{r} \times (k\mathbf{r}) = k\mathbf{r} \times \mathbf{r} = 0$.

6. Let $\mathbf{r}_1$ and $\mathbf{r}_2 = \mathbf{r}_1 + K\mathbf{f}$ denote position vectors from O to two different points on the line of action of $\mathbf{f}$. Then

$$\mathbf{r}_2 \times \mathbf{f} = \mathbf{r}_1 \times \mathbf{f} + K\mathbf{f} \times \mathbf{f} = \mathbf{r}_1 \times \mathbf{f}.$$

However, by moving the origin O of $\mathbf{r}$ we can give any value to $\mathbf{r} \times \mathbf{f}$.

7. Let $\mathbf{a}$ be the vector from another point O' to the point O, so that $\mathbf{r}_1' = \mathbf{r}_1 + \mathbf{a}$, $\mathbf{r}_2' = \mathbf{r}_2 + \mathbf{a}, \cdots$, are the position vectors from O' to the points of application of the forces $\mathbf{f}_1, \mathbf{f}_2, \cdots$. Then, using (1.7) and (1.8), we have

$$\mathbf{r}_1' \times \mathbf{f}_1 + \mathbf{r}_2' \times \mathbf{f}_2 + \cdots = \mathbf{r}_1 \times \mathbf{f}_1 + \mathbf{r}_2 \times \mathbf{f}_2 + \cdots + \mathbf{a} \times (\mathbf{f}_1 + \mathbf{f}_2 + \cdots) = 0.$$

8. To keep the center of gravity over the wire.

$$f_1 = \frac{wd_1}{s\left(1 + \sqrt{\dfrac{d_1^2 - s^2}{d_2^2 - s^2}}\right)}, \qquad f_2 = \frac{wd_2}{s\left(1 + \sqrt{\dfrac{d_2^2 - s^2}{d_1^2 - s^2}}\right)},$$

where $d_1 + d_2 = \ell$.

9. Isolate the lower block including the weight to obtain the rope tension $f = \dfrac{w}{2n}$.

10. Isolate the upper half of the body including the weight to obtain $f = \dfrac{\ell_1}{\ell_2}w$. Actual back tension f is lessened by abdominal pressure.

Problems 2.1

1. $\mathbf{f}(c\mathbf{u}+\mathbf{v}) = \mathbf{a}\cdot(c\mathbf{u}+\mathbf{v})+d = c\mathbf{a}\cdot\mathbf{u}+\mathbf{a}\cdot\mathbf{v}+d$, $c\mathbf{f}(\mathbf{u})+\mathbf{f}(\mathbf{v}) = c\mathbf{a}\cdot\mathbf{u}+cd+\mathbf{a}\cdot\mathbf{v}+d$, so $\mathbf{f}$ is linear only when $d = 0$.

2. All vectors $\mathbf{a} \times \mathbf{v}$ lie in a plane orthogonal to $\mathbf{a}$. All vectors $\mathbf{ab} \cdot \mathbf{v}$ are parallel to $\mathbf{a}$.

3. $(\mathbf{e}\cdot\mathbf{u})(\mathbf{e}\cdot\mathbf{v})$ is the product of the components of $\mathbf{u}$ and $\mathbf{v}$ in the direction of $\mathbf{e}$.

4. $\mathbf{P}' \cdot \mathbf{v} = (\mathbf{e}_1\mathbf{e}_1 + \mathbf{e}_2\mathbf{e}_2) \cdot \mathbf{v} = v_1\mathbf{e}_1 + v_2\mathbf{e}_2 =$ orthogonal projection of $\mathbf{v}$ onto plane formed by $\mathbf{e}_1$ and $\mathbf{e}_2$.

5. $K(\mathbf{e}_1\mathbf{e}_1 + \mathbf{e}_2\mathbf{e}_2)$, where $\mathbf{e}_1$ and $\mathbf{e}_2$ are orthogonal vectors in the plane of the slide.

6. $\mathbf{T}(\mathbf{a} \cdot \mathbf{v})$ is a tensor of order two and $\mathbf{Ta} \cdot (c\mathbf{u} + \mathbf{v}) = c\mathbf{Ta} \cdot \mathbf{u} + \mathbf{Ta} \cdot \mathbf{v}$.

7. $\mathbf{ab}\times$, where $\mathbf{ab} \times \mathbf{v} = \mathbf{a}(\mathbf{b} \times \mathbf{v})$.

8. A tensor of order four is a linear function that maps every vector into a triadic. A polyad of order n is n adjacent vectors (a polyad of order two is a dyad, and a polyad of order three is a triad).
9. $[(\mathbf{B}\cdot\mathbf{u})\cdot\mathbf{v}]\cdot\mathbf{w}$.

Problems 2.2

1. -1.
2. $\mathbf{1}^2\cdot\mathbf{T}=(\mathbf{1}\cdot\mathbf{1})\cdot\mathbf{T}=\mathbf{1}\cdot(\mathbf{1}\cdot\mathbf{T})=\mathbf{1}\cdot\mathbf{T}=\mathbf{T}$, $\mathbf{P}^2=\mathbf{ee}\cdot\mathbf{ee}=(\mathbf{e}\cdot\mathbf{e})\mathbf{ee}=\mathbf{ee}=\mathbf{P}$. Further projection $\mathbf{P}\cdot(\mathbf{P}\cdot\mathbf{v})$ causes no change in $\mathbf{P}\cdot\mathbf{v}$.
3. $\overline{\mathbf{1}}\cdot\mathbf{v}=\overline{\mathbf{1}}\cdot(\mathbf{1}\cdot\mathbf{v})=(\overline{\mathbf{1}}\cdot\mathbf{1})\cdot\mathbf{v}=\mathbf{1}\cdot\mathbf{v}$ for all vectors $\mathbf{v}$.
4. $\mathbf{T}=\mathbf{a}\times$.
5. Both sides are $(\mathbf{a}\cdot\mathbf{b})(\mathbf{c}\cdot\mathbf{d})$.
6. $(\mathbf{b}\times\mathbf{a})\times\mathbf{v}=\mathbf{a}(\mathbf{b}\cdot\mathbf{v})-\mathbf{b}(\mathbf{a}\cdot\mathbf{v})=(\mathbf{ab}-\mathbf{ba})\cdot\mathbf{v}$ for all vectors $\mathbf{v}$.
7. Only when $\mathbf{a}=k\mathbf{b}$ or $\mathbf{b}=0$.
8. $(\mathbf{e}\times)^2\mathbf{v}=\mathbf{e}\times(\mathbf{e}\times\mathbf{v})=\mathbf{e}(\mathbf{e}\cdot\mathbf{v})-\mathbf{v}(\mathbf{e}\cdot\mathbf{e})=\mathbf{ee}\cdot\mathbf{v}-\mathbf{v}=(\mathbf{ee}-\mathbf{1})\cdot\mathbf{v}$ for all vectors $\mathbf{v}$. $(\mathbf{e}\times)^3\mathbf{v}=\mathbf{e}\times[\mathbf{e}\times(\mathbf{e}\times\mathbf{v})]=\mathbf{e}\times[\mathbf{e}(\mathbf{e}\cdot\mathbf{v})-\mathbf{v}]=-\mathbf{e}\times\mathbf{v}$.

Problems 2.3

1. $\mathbf{v}\cdot\mathbf{T}^t\cdot\mathbf{u}=(\mathbf{v}\cdot\mathbf{T}^t)\cdot\mathbf{u}=(\mathbf{T}\cdot\mathbf{v})\cdot\mathbf{u}=\mathbf{u}\cdot\mathbf{T}\cdot\mathbf{v}$.
2. Take the transpose of $\mathbf{T}\cdot\mathbf{T}^{-1}=\mathbf{1}$.
3. $\mathbf{T}^{-n}=\mathbf{T}^{-n+1}\cdot\mathbf{T}^{-1}$.
4. Eigenvectors are parallel to $\mathbf{a}$ with eigenvalue $\lambda=\mathbf{a}\cdot\mathbf{b}$, or perpendicular to $\mathbf{b}$ with eigenvalue $\lambda=0$.
5. $\mathbf{u}\cdot\mathbf{P}'\cdot\mathbf{v}=(\mathbf{e}_1\cdot\mathbf{u})(\mathbf{e}_1\cdot\mathbf{v})+(\mathbf{e}_2\cdot\mathbf{u})(\mathbf{e}_2\cdot\mathbf{v})=\mathbf{v}\cdot\mathbf{P}'\cdot\mathbf{u}$, $\mathbf{P}'\cdot(\mathbf{e}_1\times\mathbf{e}_2)=0$.
6. By Problem 1, if $\mathbf{A}$ is any tensor $\mathbf{v}\cdot\mathbf{A}\cdot\mathbf{v}=\mathbf{v}\cdot\mathbf{A}^t\cdot\mathbf{v}$. If $\mathbf{A}$ is antisymmetric, $\mathbf{v}\cdot\mathbf{A}^t\cdot\mathbf{v}=-\mathbf{v}\cdot\mathbf{A}\cdot\mathbf{v}$. Hence $\mathbf{v}\cdot\mathbf{A}\cdot\mathbf{v}=-\mathbf{v}\cdot\mathbf{A}\cdot\mathbf{v}$.
7. $(\mathbf{Q}\cdot\mathbf{u})\cdot(\mathbf{Q}\cdot\mathbf{v})=(\mathbf{u}\cdot\mathbf{Q}^t)\cdot(\mathbf{Q}\cdot\mathbf{v})=\mathbf{u}\cdot(\mathbf{Q}^t\cdot\mathbf{Q})\cdot\mathbf{v}=\mathbf{u}\cdot(\mathbf{Q}^{-1}\cdot\mathbf{Q})\cdot\mathbf{v}=\mathbf{u}\cdot\mathbf{1}\cdot\mathbf{v}=\mathbf{u}\cdot\mathbf{v}$.
8. If $\mathbf{T}$ is symmetric, $\mathbf{T}^n$ is symmetric. If $\mathbf{T}$ is antisymmetric, $\mathbf{T}^n$ is symmetric for n even, and antisymmetric for n odd. If $\mathbf{T}$ is orthogonal, $\mathbf{T}^n$ is orthogonal.

Problems 2.4

1. $m\mathbf{r}\times\mathbf{v}=m\mathbf{r}\times(\boldsymbol{\omega}\times\mathbf{r})=m[r^2\boldsymbol{\omega}-\mathbf{r}(\mathbf{r}\cdot\boldsymbol{\omega})]=m(r^2\mathbf{1}-\mathbf{rr})\cdot\boldsymbol{\omega}=\mathbf{I}\cdot\boldsymbol{\omega}$.
2. $-m(\mathbf{r}\times)^2\cdot\boldsymbol{\omega}=-m\mathbf{r}\times(\mathbf{r}\times\boldsymbol{\omega})=m\mathbf{r}\times\mathbf{v}=\mathbf{I}\cdot\boldsymbol{\omega}$.

3. From (2.7), $I = m[r^2 - (\mathbf{e}\cdot\mathbf{r})^2] = md^2$, $\frac{1}{2}mv^2 = \frac{1}{2}md^2\omega^2 = \frac{1}{2}I\omega^2 = \frac{\omega^2}{2}\mathbf{e}\cdot\mathbf{I}\cdot\mathbf{e}$ $= \frac{1}{2}\boldsymbol{\omega}\cdot\mathbf{I}\cdot\boldsymbol{\omega}$. If O is at the center of the circle traced by the particle (Fig. 2.4), $\mathbf{h} = m\mathbf{r}\times\mathbf{v} = mdv\mathbf{e} = md^2\boldsymbol{\omega} = I\boldsymbol{\omega}$.

4. $\mathbf{t}\cdot\mathbf{e} = \mathbf{e}\cdot\mathbf{S}\cdot\mathbf{e}$.

5. The only nonzero tractions acting on a small cube within the wire are perpendicular to the faces with normals $\pm\mathbf{e}$ where $\mathbf{e}$ is directed along the wire, and have magnitude $\frac{f}{A}$ where A is the wire cross-sectional area, so (see Figure 3.9) $\mathbf{S} = \frac{f}{A}\mathbf{ee}$. If $\mathbf{e}'$ denotes a surface normal making an angle ϕ with $\mathbf{e}$, $\mathbf{t} = \mathbf{S}\cdot\mathbf{e}' = \frac{f}{A}\cos\phi\,\mathbf{e}$.

6. The $\mathbf{e}_1$ component of momentum transferred through surface with normal $\mathbf{e}_2$ equals the $\mathbf{e}_2$ component of momentum transferred through surface with normal $\mathbf{e}_1$.

7. The mass per unit time ρvA passing through any cross-sectional area A of the stream is a constant, so as the speed v increases the area A must decrease.

8. The normal component $(\rho\mathbf{vv}\cdot\mathbf{e})\cdot\mathbf{e}$ of the momentum per unit time and area of the gas particles impacting the wall undergoes a sign change, whereas the other components remain unchanged. The particles exert a pressure equal to the magnitude of the change of their momentum per unit time and area: $2(\rho\mathbf{vv}\cdot\mathbf{e})\cdot\mathbf{e} = 2\rho v^2\cos^2\phi$.

9. It follows from the symmetry of the stress tensor:

$$\mathbf{e}_1\cdot\mathbf{S}\cdot\mathbf{e}_2 = \mathbf{e}_2\cdot(\mathbf{e}_1\cdot\mathbf{B})\cdot\boldsymbol{\mathcal{E}} = \mathbf{e}_2\cdot\mathbf{S}\cdot\mathbf{e}_1 = \mathbf{e}_1\cdot(\mathbf{e}_2\cdot\mathbf{B})\cdot\boldsymbol{\mathcal{E}}.$$

Problems 3.1

1. $v_i = \mathbf{v}\cdot\mathbf{e}_i = v\cos\theta_i$, so $v = v_i\mathbf{e}_i = v\cos\theta_i\,\mathbf{e}_i$.

2. Suppose the cube has sides of length 1 parallel to $(\mathbf{e}_1, \mathbf{e}_2, \mathbf{e}_3)$. Then $\mathbf{v} = \mathbf{e}_1 + \mathbf{e}_2 + \mathbf{e}_3$ is a diagonal vector and $\mathbf{v}\cdot\mathbf{e}_1 = v\cos\theta = 1$ so

$$\theta = \cos^{-1}\left(\frac{1}{\sqrt{3}}\right) = 54.7^\circ.$$

3. $\delta_{ii} = \delta_{11} + \delta_{22} + \delta_{33} = 3$, $\epsilon_{ijk}\epsilon_{ijk} = \delta_{ii}\delta_{jj} - \delta_{ij}\delta_{ij} = (3)(3) - 3 = 6$.

4. $\epsilon_{ijk}a_ib_j\epsilon_{mnk}c_md_n = (\delta_{im}\delta_{jn} - \delta_{in}\delta_{jm})a_ib_jc_md_n = a_ic_ib_jd_j - a_id_ib_jc_j$.

5. $\mathbf{u}\times\mathbf{v}\cdot\mathbf{w} = \begin{vmatrix} 1 & 1 & 1 \\ 0 & 1 & 2 \\ -1 & 0 & 2 \end{vmatrix} = 1.$

6. $(\mathbf{a}\times\mathbf{b}\cdot\mathbf{c})(\mathbf{u}\times\mathbf{v}\cdot\mathbf{w}) =$

$$\begin{vmatrix} a_1 & a_2 & a_3 \\ b_1 & b_2 & b_3 \\ c_1 & c_2 & c_3 \end{vmatrix}\begin{vmatrix} u_1 & u_2 & u_3 \\ v_1 & v_2 & v_3 \\ w_1 & w_2 & w_3 \end{vmatrix} = \begin{vmatrix} a_1 & a_2 & a_3 \\ b_1 & b_2 & b_3 \\ c_1 & c_2 & c_3 \end{vmatrix}\begin{vmatrix} u_1 & v_1 & w_1 \\ u_2 & v_2 & w_2 \\ u_3 & v_3 & w_3 \end{vmatrix}$$

$$= \left|\begin{bmatrix} a_1 & a_2 & a_3 \\ b_1 & b_2 & b_3 \\ c_1 & c_2 & c_3 \end{bmatrix}\begin{bmatrix} u_1 & v_1 & w_1 \\ u_2 & v_2 & w_2 \\ u_3 & v_3 & w_3 \end{bmatrix}\right| = \begin{vmatrix} \mathbf{a}\cdot\mathbf{u} & \mathbf{a}\cdot\mathbf{v} & \mathbf{a}\cdot\mathbf{w} \\ \mathbf{b}\cdot\mathbf{u} & \mathbf{b}\cdot\mathbf{v} & \mathbf{b}\cdot\mathbf{w} \\ \mathbf{c}\cdot\mathbf{u} & \mathbf{c}\cdot\mathbf{v} & \mathbf{c}\cdot\mathbf{w} \end{vmatrix}$$

7. $3^4 = 81$.

8. $\dfrac{x_1 - 1}{3} = \dfrac{x_2 - 2}{2} = x_3 - 3,\ 3x_1 + 2x_2 + x_3 = 10.$

9. Since the line and plane must pass through the origin, their equations are of the form
$$\frac{x_1}{v_1} = \frac{x_2}{v_2} = \frac{x_3}{v_3} \quad \text{and} \quad n_1x_1 + n_2x_2 + n_3x_3 = 0\,.$$
Substitute the coordinates of the other points to obtain $6x_1 = 3x_2 = 2x_3$ and $x_1 - 2x_2 + x_3 = 0$.

10. $\dfrac{x_1 - a_1}{n_1} = \dfrac{x_2 - a_2}{n_2} = \dfrac{x_3 - a_3}{n_3}$, where $n_ia_i = c$.

11. $\dfrac{|\mathbf{a}\times\mathbf{v}|}{|\mathbf{v}|}$, $|c|$.

Problems 3.2

1. $P_{ij} = \mathbf{e}_i\cdot\mathbf{P}\cdot\mathbf{e}_j = \cos\theta_i\cos\theta_j$, where $\cos\theta_i = \mathbf{e}\cdot\mathbf{e}_i$ are the direction cosines of $\mathbf{e}$.

2. 9, 6, 3, 3.

3. 3^n.

4. $\det T = \epsilon_{pqr}T_{1p}T_{2q}T_{3r} = \epsilon_{123}\epsilon_{pqr}T_{1p}T_{2q}T_{3r} = \frac{1}{6}\epsilon_{ijk}\epsilon_{pqr}T_{ip}T_{jq}T_{kr}$.

5. $\mathbf{T}\cdot\mathbf{v} = (\mathbf{1} + \mathbf{e}_1\mathbf{e}_2 + \mathbf{e}_2\mathbf{e}_1)\cdot(v_1\mathbf{e}_1 + v_2\mathbf{e}_2 + v_3\mathbf{e}_3) = (v_1 + v_2)\mathbf{e}_1 + (v_1 + v_2)\mathbf{e}_2 + v_3\mathbf{e}_3 = \mathbf{e}_1 + \mathbf{e}_2 + \mathbf{e}_3$. Equate components to obtain $v_1 + v_2 = 1$, $v_3 = 1$. Thus $\mathbf{v} = v_1\mathbf{e}_1 + (1 - v_1)\mathbf{e}_2 + \mathbf{e}_3$, where v_1 is an arbitrary constant.

6. The columns of T are the components of orthonormal vectors, so from (4.4) T is an orthogonal matrix. Hence $\mathbf{T}^{-1} = \mathbf{T}^t = \mathbf{e}_1\mathbf{e}_2 + \mathbf{e}_2\mathbf{e}_3 + \mathbf{e}_3\mathbf{e}_1$.

7. $B_{ijk} = B_{123}\epsilon_{ijk}$.

Problems 3.3

1. Expand (3.21) to get $\lambda^3 - T_{ii}\lambda^2 + \frac{1}{2}[T_{ii}T_{jj} - T_{ij}T_{ji}]\lambda - \det T = 0$. Expand $(\lambda - \lambda_1)(\lambda - \lambda_2)(\lambda - \lambda_3) = 0$ to get $\lambda^3 - (\lambda_1 + \lambda_2 + \lambda_3)\lambda^2 + (\lambda_1\lambda_2 + \lambda_1\lambda_3 + \lambda_2\lambda_3)\lambda - \lambda_1\lambda_2\lambda_3 = 0$.

2. $\lambda = 0$ with eigenvector $\dfrac{\mathbf{e}_1 - \mathbf{e}_2}{\sqrt{2}}$, $\lambda = 1$ with eigenvector $\mathbf{e}_3$, $\lambda = 2$ with eigenvector $\dfrac{\mathbf{e}_1 + \mathbf{e}_2}{\sqrt{2}}$.

3. $J_1 = 2,\ J_2 = -1,\ J_3 = 0.$

4. $\text{tr}\,(\mathbf{S}\cdot\mathbf{T}) = \text{tr}(S_{ik}T_{kj}\mathbf{e}_i\mathbf{e}_j) = S_{ik}T_{ki} = S_{ij}T_{ji}.$

5. $\det(T - \lambda I) = \begin{vmatrix} -\lambda & 0 & -c \\ 1 & -\lambda & -b \\ 0 & 1 & -a-\lambda \end{vmatrix} = -(\lambda^3 + a\lambda^2 + b\lambda + c).$

6. $\omega_1 = -\frac{1}{2}\epsilon_{123}A_{23} - \frac{1}{2}\epsilon_{132}A_{32} = -A_{23}$, etc., $\det(\mathbf{1}+\mathbf{A}) = 1 + \omega_1^2 + \omega_2^2 + \omega_3^2 > 0.$

7. $\cos\theta = \dfrac{1}{2}(\text{tr}\,\mathbf{R} - 1)$, $\mathbf{e}\times = \dfrac{\mathbf{R}-\mathbf{R}^t}{2\sin\theta}(\theta \neq 0 \text{ or } \pi).$

8. Use Cartesian components with $\mathbf{e} = \mathbf{e}_1$ in (3.25) and (3.26) to obtain

$$\det\ R = \begin{vmatrix} 1 & 0 & 0 \\ 0 & \cos\theta & -\sin\theta \\ 0 & \sin\theta & \cos\theta \end{vmatrix} = 1, \quad \det H = \begin{vmatrix} -1 & 0 & 0 \\ 0 & 1 & 0 \\ 0 & 0 & 1 \end{vmatrix} = -1$$

9. $\mathbf{H}\cdot\mathbf{v} = \mathbf{v} - 2(\mathbf{v}\cdot\mathbf{e})\mathbf{e}$ = reflected vector (see Fig. 3.6). $\mathbf{H}\cdot\mathbf{e} = -\mathbf{e}$, so $\mathbf{e}$ is an eigenvector with corresponding eigenvalue -1. $\mathbf{H}^2 = \mathbf{1}$, so $\mathbf{H}^{-1} = \mathbf{H}$. $\mathbf{H}^t = \mathbf{H} = \mathbf{H}^{-1}$, so $\mathbf{H}$ is orthogonal.

10. $(\mathbf{T}\cdot\mathbf{T}^t)^t = (\mathbf{T}^t)^t\cdot\mathbf{T}^t = \mathbf{T}\cdot\mathbf{T}^t, \mathbf{r}\cdot(\mathbf{T}\cdot\mathbf{T}^t)\cdot\mathbf{r} = \mathbf{r}\cdot(\mathbf{T}^t\cdot\mathbf{T})\cdot\mathbf{r} = (\mathbf{r}\cdot\mathbf{T}^t)\cdot(\mathbf{T}\cdot\mathbf{r}) = |\mathbf{T}\cdot\mathbf{r}|^2 > 0$ for $\mathbf{r} \neq 0$.

11. Linear form is a_ix_i, linear surface is plane $a_ix_i = 1$, cubic form is $B_{ijk}x_ix_jx_k$, cubic surface is $B_{ijk}x_ix_jx_k = 1$.

12. Use Problem 8 in Section 2.2 together with the series expansions of $\cos\theta$ and $\sin\theta$:

$$e^{\theta\mathbf{e}\times} = \mathbf{1} + \theta\mathbf{e}\times + \frac{\theta^2}{2}(\mathbf{e}\times)^2 + \frac{\theta^3}{6}(\mathbf{e}\times)^3 + \cdots$$

$$= \left(1 - \frac{\theta^2}{2} + \cdots\right)\mathbf{1} + \left(\frac{\theta^2}{2} - \cdots\right)\mathbf{ee} + \left(\theta - \frac{\theta^3}{6} + \cdots\right)\mathbf{e}\times$$

$$= \cos\theta\mathbf{1} + (1-\cos\theta)\mathbf{ee} + \sin\theta\mathbf{e}\times.$$

Problems 3.4

1. Let $\mathbf{y}$ denote the vector from O to a second fixed point $\overline{O}$, and let $\overline{\mathbf{r}} = \mathbf{r} - \mathbf{y}$ denote the vectors from $\overline{O}$ to the given region (see Fig. 4.2). Then $\mathbf{T}\cdot\overline{\mathbf{r}} = \mathbf{T}\cdot\mathbf{r} - \mathbf{T}\cdot\mathbf{y}$, so the image regions differ only by a translation.

2. $\mathbf{T} = K\mathbf{1}$, $K > 1$.

3. If $\mathbf{T}$ is nonsingular, then every vector $\mathbf{x}$ from a fixed point O is the image of another vector $\mathbf{r} = \mathbf{T}^{-1} \cdot \mathbf{x}$. If $\mathbf{T}$ is singular, then the volume of the image region must be zero, and any linear combination of the image vectors is also an image vector; the three possibilities are a plane, a line, and a point.

4. Apply $\mathbf{T}$ to equations (3.15a) and (3.16a).

5. See the first three equations in the Proof of (3.33).
$\mathbf{T}^{1/2} = \mathbf{e}_1\mathbf{e}_1 + 2\mathbf{e}_2\mathbf{e}_2 + 3\mathbf{e}_3\mathbf{e}_3$. $\mathbf{T}^{-1} = \mathbf{e}_1\mathbf{e}_1 + \frac{1}{4}\mathbf{e}_2\mathbf{e}_2 + \frac{1}{9}\mathbf{e}_3\mathbf{e}_3$.

6. Line $\mathbf{r} - \mathbf{a} = t\mathbf{v}$ gets mapped into line $\mathbf{x} - \mathbf{T} \cdot \mathbf{a} = t\mathbf{T} \cdot \mathbf{v}$ and plane $\mathbf{n} \cdot \mathbf{r} = c$ gets mapped into plane $(\mathbf{n} \cdot \mathbf{T}^{-1}) \cdot \mathbf{x} = c$, so parallel edges of cube get mapped into parallel lines and parallel faces of cube get mapped into parallel planes.

7. $\lambda_1^n x_1^2 + \lambda_2^n x_2^2 + \lambda_3^n x_3^2 = 1$ is an ellipsoid for any n.

8. Surface $\mathbf{x} \cdot \mathbf{x} = 1$ is image of surface $\mathbf{r} \cdot (\mathbf{T}^t \cdot \mathbf{T}) \cdot \mathbf{r} = \mathbf{r} \cdot (\mathbf{S}')^2 \cdot \mathbf{r}$, which is ellipsoid with semi-axes coinciding with principal axes of $\mathbf{S}' = (\mathbf{T}^t \cdot \mathbf{T})^{\frac{1}{2}}$ and having magnitude $\left(\dfrac{1}{\lambda_1'}, \dfrac{1}{\lambda_2'}, \dfrac{1}{\lambda_3'}\right)$, where $(\lambda_1', \lambda_2', \lambda_3')$ are eigenvalues of $\mathbf{S}'$.

9. $\mathbf{T} = \frac{1}{3}(\mathrm{tr}\mathbf{T})\mathbf{1} + \frac{1}{2}(\mathbf{T} - \mathbf{T}^t) + \left[\frac{1}{2}(\mathbf{T} + \mathbf{T}^t) - \frac{1}{3}(\mathrm{tr}\mathbf{T})\mathbf{1}\right]$.

Problems 3.5

1. $I_{11} = \rho \displaystyle\iiint_{\text{sphere}} (x_2^2 + x_3^2)dx_1dx_2dx_3 = \rho \int_0^{2\pi}\int_0^{\pi}\int_0^{a} r^4 \sin^3\phi \, dr d\phi d\theta =$ $\dfrac{2ma^2}{5}$ (do integral in spherical coordinates, see Section 8.2), by symmetry $I_{22} = I_{33} = I_{11}$ and $I_{12} = I_{13} = I_{23} = 0$, so $\mathbf{I} = \frac{2}{5}ma^2\mathbf{1}$.

2. See Problem 3 in Section 2.4. If $\boldsymbol{\omega} \cdot \mathbf{I} \cdot \boldsymbol{\omega} = 1$, the rotational kinetic energy is $\frac{1}{2}$.

3. $S_{ij} = -p\delta_{ij}$.

4. Unit vector $\mathbf{e}$ gets mapped into $\mathbf{t} = \mathbf{S} \cdot \mathbf{e}$, so sphere $\mathbf{r} \cdot \mathbf{r} = 1$ gets mapped into the ellipsoid (3.35), with $\mathbf{t} = x_i\mathbf{e}_i$ and $(\lambda_1, \lambda_2, \lambda_3)$ the absolute values of the normal stresses in the principal axes.

5. One component is ρvv, all others are zero.

6. $\mathbf{R} = \mathbf{e}_1\mathbf{e}_2 + \mathbf{e}_2\mathbf{e}_3 + \mathbf{e}_3\mathbf{e}_1$ turns the body through an angle $\theta = \dfrac{2\pi}{3}$ about $\mathbf{e} = -\dfrac{1}{\sqrt{3}}(\mathbf{e}_1 + \mathbf{e}_2 + \mathbf{e}_3)$.

7. $\det \mathbf{T} = \det(\mathbf{S} \cdot \mathbf{R}) = \det \mathbf{S} \det \mathbf{R} = \det \mathbf{S} = \det(\mathbf{1} + \boldsymbol{\Gamma}) \approx 1 + \mathrm{tr}\,\boldsymbol{\Gamma}$.

8. $\mathbf{T} \cdot \mathbf{r} = \mathbf{r} + x_1\gamma_{12}\mathbf{e}_2 + x_2\gamma_{12}\mathbf{e}_1$, so deformed solid is a parallelepiped with angles $\dfrac{\pi}{2} \pm 2\tan^{-1}\gamma_{12}$ between adjacent oblique faces.

9. $\gamma_{11} = \dfrac{R\Omega}{L}$, where R is the radius of the joint, Ω is the angle of rotation, and L is the undeformed length of the skin. Actual strain is less due to wrinkles in the skin.

Problems 4.1

1. $Q_{ij} = \mathbf{e}_i \cdot \mathbf{Q} \cdot \mathbf{e}_j = \mathbf{e}_i \cdot \overline{\mathbf{e}}_j = \overline{\mathbf{e}}_i \cdot \mathbf{Q} \cdot \overline{\mathbf{e}}_j = \overline{Q}_{ij}$.

2. Let $\overline{\mathbf{v}} = \overline{v}_i \mathbf{e}_i = \mathbf{Q} \cdot \mathbf{v} = Q_{ji} v_j \mathbf{e}_i$. Then $\overline{v}_i = Q_{ji} v_j$, which is the same as (4.5). The vector $\mathbf{v}$ is rotated and/or reflected by $\mathbf{Q}^t$ to the vector $\overline{\mathbf{v}}$.

3.
$$Q_{ki} Q_{\ell j} \delta_{k\ell} = Q^t_{ik} Q_{kj} = \delta_{ij},$$
$$Q_{\ell i} Q_{mj} Q_{nk} \epsilon_{\ell mn} = (\det Q)\epsilon_{ijk} \quad \text{(see Problem 4 in Section 3.2)}$$
$$= \begin{cases} \epsilon_{ijk} & \text{if } \mathbf{Q} \text{ is a rotation tensor,} \\ -\epsilon_{ijk} & \text{if } \mathbf{Q} \text{ is a reflection tensor.} \end{cases}$$

4. From (3.37), $\mathbf{I} = \dfrac{m\ell^2}{6}(\mathbf{e}_1\mathbf{e}_1 + \mathbf{e}_2\mathbf{e}_2 + \mathbf{e}_3\mathbf{e}_3) = \dfrac{m\ell^2}{6}\mathbf{1}$.

5. The unit sphere is deformed into an ellipsoid (see Fig. 3.7).
$$S' = R^t S R = \begin{bmatrix} 2 & -1 & 0 \\ -1 & 3 & -1 \\ 0 & -1 & 2 \end{bmatrix}.$$

6. The eigenvalues of $\mathbf{S}$ are $\lambda_1 = 2$, $\lambda_2 = 2$, $\lambda_3 = 8$ corresponding to normalized eigenvectors
$$\overline{\mathbf{e}}_1 = \frac{-\mathbf{e}_1 + \mathbf{e}_2}{\sqrt{2}}, \qquad \overline{\mathbf{e}}_2 = \frac{-\mathbf{e}_1 - \mathbf{e}_2 + 2\mathbf{e}_3}{\sqrt{6}}, \qquad \overline{\mathbf{e}}_3 = \frac{\mathbf{e}_1 + \mathbf{e}_2 + \mathbf{e}_3}{\sqrt{3}}.$$
Thus $\mathbf{S} = 2\overline{\mathbf{e}}_1\overline{\mathbf{e}}_1 + 2\overline{\mathbf{e}}_2\overline{\mathbf{e}}_2 + 8\overline{\mathbf{e}}_3\overline{\mathbf{e}}_3$.

Problems 4.2

1. Degenerates to a point (all axes are principal).

2. $$\det \mathbf{S} = S_{11}S_{22} - S_{12}^2 = \lambda_1\lambda_2 \begin{cases} > 0 & \text{ellipse} \\ = 0 & \text{parabola or straight line.} \\ < 0 & \text{hyperbola} \end{cases}$$

3. $\mathbf{I} = \dfrac{\ell_2\ell_3^3}{12}\mathbf{e}_2\mathbf{e}_2 + \dfrac{\ell_3\ell_2^3}{12}\mathbf{e}_3\mathbf{e}_3$, where (ℓ_2, ℓ_3) are the lengths of the sides of the cross section.

4. Principal axes $(\overline{x}_2, \overline{x}_3)$ are rotated through angle $\alpha = \dfrac{\pi}{4}$ relative to axes (x_2, x_3) along the sides and
$$\mathbf{I} = \frac{\ell^4}{3}\mathbf{e}_2\mathbf{e}_2 - \frac{\ell^4}{4}\mathbf{e}_2\mathbf{e}_3 - \frac{\ell^4}{4}\mathbf{e}_3\mathbf{e}_2 + \frac{\ell^4}{3}\mathbf{e}_3\mathbf{e}_3 = \frac{7\ell^4}{12}\overline{\mathbf{e}}_2\overline{\mathbf{e}}_2 + \frac{\ell^4}{12}\overline{\mathbf{e}}_3\overline{\mathbf{e}}_3.$$

5. The maximum value of S_{12} occurs at the top point of the Mohr circle (Fig. 4.4), which has coordinates

$$(S_{11}, S_{12}) = \left(\frac{\lambda_1 + \lambda_2}{2}, \frac{|\lambda_1 - \lambda_2|}{2}\right)$$

and makes an angle $2\alpha = \dfrac{\pi}{2}$ with the horizontal axis.

6. From the preceding problem, $\dfrac{|S_{11} - S_{22}|}{2} = K$.

7. The locus of all points is three circles, each passing through two of the eigenvalues of $\mathbf{S}$ on the horizontal axis. The coordinates of terminal points on the diameters of each of the three circles are (S_{11}, S_{12}) and $(S_{22}, -S_{12})$, (S_{22}, S_{23}) and $(S_{33}, -S_{23})$, (S_{11}, S_{13}) and $(S_{33}, -S_{13})$.

Problems 4.3

1. Several possibilities are

$$R = R_3\left(\frac{\pi}{2}\right) R_2\left(\frac{\pi}{2}\right) = R_2\left(\frac{\pi}{2}\right) R_1\left(-\frac{\pi}{2}\right) = R_1\left(-\frac{\pi}{2}\right) R_3\left(\frac{\pi}{2}\right).$$

Nonzero components are $R_{12} = -1$, $R_{23} = 1$, $R_{31} = -1$ (see 3.38).

2. Same as Problem 1.

3. Set $\overline{x}_1 = a, \overline{x}_2 = \overline{x}_3 = 0, \alpha = 90°, \beta = 30°$ in (4.15b) to get $x_1 = 0$, $x_2 = \sqrt{3}a$ $x_3 = a$.

4. Set $x_1 = 2a, x_2 = x_3 = 0, \overline{x}_1 = \overline{x}_2 = 0, \overline{x}_3 = \pm a\sqrt{3}, \alpha = 0$ in (4.15a) to get $\sin\beta = \pm\sqrt{3}/2$, so the maximum latitudes are 60°N and 60°S.

5. See Problem 7 in Section 3.3.

$$\cos\theta = \frac{1}{2}(\cos\theta_1\cos\theta_2 + \cos\theta_1\cos\theta_3 + \cos\theta_2\cos\theta_3 + \sin\theta_1\sin\theta_2\sin\theta_3 - 1).$$

6. Any one of the three body-attached axes can be chosen for the first rotation followed by either of the other two body-attached axes. Similarly for the space-fixed axes. $3\cdot2\cdot2 + 3\cdot2\cdot2 = 24$.

Problems 4.4

1. $v_i v_i = Q_{ij}Q_{ik}\overline{v}_j\overline{v}_k = \delta_{jk}\overline{v}_j\overline{v}_k = \overline{v}_i\overline{v}_i$, etc.

2. Units of A_{ij} are $\dfrac{\text{Pa}}{\text{K}}$, B_{ijk} are $\dfrac{\text{C}}{\text{m}^2}$, C_{ijkl} are Pa, where Pa = pascal = $\dfrac{\text{N}}{\text{m}^2}$ N = newton = $\dfrac{\text{kg}\cdot\text{m}}{\text{s}^2}$, C = coulomb = A · s, K = kelvin, kg = kilogram m = meter, s = second, A = ampere.

3. Since $S_{ij} = S_{ji}$, the ABC's must be symmetric in the first two indices. Since $\gamma_{k\ell} = \gamma_{\ell k}$, we can write $S_{ij} = \frac{1}{2}(C_{ijk\ell} + C_{ij\ell k})\gamma_{k\ell}$, so the C's can always be chosen to be symmetric in the last two indices.

4. There are 8 possible sign combinations $(\pm\mathbf{e}_1, \pm\mathbf{e}_2, \pm\mathbf{e}_3)$ for each of the 6 permutations of $(\mathbf{e}_1, \mathbf{e}_2, \mathbf{e}_3)$, and $8 \cdot 6 = 48$.

5. Solve $C_{ijk\ell}\gamma_{k\ell} - A_{ij}T = 0$ for γ_{ij} using (4.18)–(4.20) to obtain

$$\mathbf{\Gamma} = \frac{A_{11}T\mathbf{1}}{C_{1111} + 2C_{1122}}.$$

6. $\gamma_{11} \approx \dfrac{2s^2}{\ell^2} = 0.0002$, $A = \dfrac{f}{S_{11}} = \dfrac{f}{Y\gamma_{11}} = 3.125 \text{ cm}^2$.

7. See Problem 7 in Section 3.5. $\text{tr}\ \mathbf{\Gamma} = \dfrac{1-2\nu}{Y}(\text{tr}\ \mathbf{S} + 3A_{11}T) = 0 \Longrightarrow \nu = \dfrac{1}{2}$.

8. Setting $Q_{ij} = -\delta_{ij}$ in (4.5) yields $\overline{v}_i = -v_i$, so no vectors are centrosymmetric (other than $\mathbf{0}$). Setting $Q_{ij} = -\delta_{ij}$ in (4.5a) yields $\overline{T}_{ij} = T_{ij}$, so all second-order tensors are centrosymmetric, etc.

9. Require that the components remain unchanged under a rotation characterized by $R_3(\theta)$ (see (4.13)). Nonzero components are $A_{11} = A_{22}$, A_{33}.

10. Use the fact that if any two base vectors are interchanged and any base vector is reversed in direction, the ABCs are unchanged. $\mathbf{S} = 2\mu\mathbf{\Gamma} + \lambda(\text{tr}\ \mathbf{\Gamma})\mathbf{1} + B_{123}(\boldsymbol{\mathcal{E}}\times) - A_{11}T\mathbf{1}$.

11. $\overline{S}_{ij} = Q_{ki}Q_{\ell j}(f_1\delta_{k\ell} + f_2\gamma_{k\ell} + f_3\gamma_{km}\gamma_{\ell m}) = f_1\delta_{ij} + f_2\overline{\gamma}_{ij} + f_3\overline{\gamma}_{in}\overline{\gamma}_{jn}$. When $\mathbf{\Gamma}$ is small, expand f_1, f_2, f_3 in a Taylor series expansion about $(0,0,0)$ to obtain $\mathbf{S} \approx \dfrac{\partial f_1}{\partial J_1}(0,0,0)(\text{tr}\ \mathbf{\Gamma})\mathbf{1} + f_2(0,0,0)\mathbf{\Gamma}$.

Problems 4.5

1. No, we must require that the vectors do not lie in a plane: $\mathbf{g}_1 \times \mathbf{g}_2 \cdot \mathbf{g}_3 \neq 0$.

2. Yes, $\mathbf{g}^i = \frac{1}{2}e^{ijk}\mathbf{g}_j \times \mathbf{g}_k$.

3. Using (3.12), taking the determinant of both sides of (4.26), and using (4.28) yields

$$(\mathbf{g}^2 \times \mathbf{g}^2 \cdot \mathbf{g}^3)^2 = \begin{vmatrix} g^{11} & g^{12} & g^{13} \\ g^{12} & g^{22} & g^{23} \\ g^{13} & g^{23} & g^{33} \end{vmatrix} = \frac{1}{g},$$

so

$$e^{ijk} = \mathbf{g}^i \cdot (\mathbf{E} \cdot \mathbf{g}^k) \cdot \mathbf{g}^j = \mathbf{g}^i \times \mathbf{g}^j \cdot \mathbf{g}^k = \epsilon^{ijk}\mathbf{g}^1 \times \mathbf{g}^2 \cdot \mathbf{g}^3 = \begin{cases} \dfrac{\epsilon^{ijk}}{\sqrt{g}} & \text{if } \mathbf{g}_1 \times \mathbf{g}_2 \cdot \mathbf{g}_3 > 0, \\ -\dfrac{\epsilon^{ijk}}{\sqrt{g}} & \text{if } \mathbf{g}_1 \times \mathbf{g}_2 \cdot \mathbf{g}_3 < 0. \end{cases}$$

4. $T_i^{\ j} - T^j_{\ i} = g^{jk}(T_{ik} - T_{ki})$.

5. 6^n.

6. From $g_{ij} = \mathbf{g}_i \cdot \mathbf{g}_j$: $g_{11} = 1,\ g_{22} = 1,\ g_{33} = 2,\ g_{12} = 0,\ g_{13} = 1,\ g_{23} = 0,$ $g = 1, e_{ijk} = \epsilon_{ijk}$.
From Problems 2-3: $\mathbf{g}^1 = \mathbf{e}_1 - \mathbf{e}_3,\ \mathbf{g}^2 = \mathbf{e}_2,\ \mathbf{g}^3 = \mathbf{e}_3$.
From $g^{ij} = \mathbf{g}^i \cdot \mathbf{g}^j$: $g^{11} = 2,\ g^{22} = 1,\ g^{33} = 1,\ g^{12} = 0,\ g^{13} = -1,\ g^{23} = 0$.
From $v_i = \mathbf{v} \cdot \mathbf{g}_i, v^i = \mathbf{v} \cdot \mathbf{g}^i$: $v_1 = 1,\ v_2 = 1, v_3 = 2,\ v^1 = 0,\ v^2 = 1,\ v^3 = 1$.
From $T_{ij} = \mathbf{g}_i \cdot \mathbf{T} \cdot \mathbf{g}_j$: $T_{11} = 1,\ T_{12} = 0,\ T_{13} = 1,\ T_{21} = 1,\ T_{22} = 1,$ $T_{23} = 1,\ T_{31} = 1,\ T_{32} = 0,\ T_{33} = 2$.
From $T^{ij} = \mathbf{g}^i \cdot \mathbf{T} \cdot \mathbf{g}^j$: $T^{11} = 2,\ T^{12} = 0,\ T^{13} = -1,\ T^{21} = 1,\ T^{22} = 1,$ $T^{23} = 0,\ T^{31} = -1,\ T^{32} = 0,\ T^{33} = 1$.
From $T^i_{\ j} = \mathbf{g}^i \cdot \mathbf{T} \cdot \mathbf{g}_j$: $T^1_{\ 1} = 1,\ T^1_{\ 2} = 0,\ T^1_{\ 3} = 1,\ T^2_{\ 1} = 1,\ T^2_{\ 2} = 1,$ $T^2_{\ 3} = 1,\ T^3_{\ 1} = 0,\ T^3_{\ 2} = 0,\ T^3_{\ 3} = 1$.
From $T_j^{\ i} = \mathbf{g}_i \cdot \mathbf{T} \cdot \mathbf{g}^j$: $T_1^{\ 1} = 1,\ T_1^{\ 2} = 0,\ T_1^{\ 3} = 0,\ T_2^{\ 1} = 1,\ T_2^{\ 2} = 1,$ $T_2^{\ 3} = 0,\ T_3^{\ 1} = 0,\ T_3^{\ 2} = 0,\ T_3^{\ 3} = 1$.

7. $\mathbf{r} = x_i\mathbf{e}_i = x_1\mathbf{g}_1 + (x_2 - x_1)\mathbf{g}_2 + x_3\mathbf{g}_3$, so $x^1 = x_1,\ x^2 = x_2 - x_1,\ x^3 = x_3$, and the coordinate curves are the straight lines $x_1 =$ constant, $x_2 - x_1 =$ constant, $x_3 =$ constant. $\mathbf{g}^1 = \mathbf{e}_1,\ \mathbf{g}^2 = -\mathbf{e}_1 + \mathbf{e}_2,\ \mathbf{g}^3 = \mathbf{e}_3,\ g_{11} = 2,\ g_{22} = g_{33} = 1,$ $g_{12} = 1,\ g_{13} = g_{23} = 0$.

8. $$\mathbf{Q} = \overline{\mathbf{g}}_i\mathbf{g}^i = \overline{\mathbf{g}}^i\mathbf{g}_i,$$
$$Q_{ij} = \mathbf{g}_i \cdot \overline{\mathbf{g}}_j,\ Q^i_{\cdot j} = \mathbf{g}^i \cdot \overline{\mathbf{g}}_j,\ Q^j_{i\cdot} = \mathbf{g}_i \cdot \overline{\mathbf{g}}^j,\ Q^{ij} = \mathbf{g}^i \cdot \overline{\mathbf{g}}^j,$$
$$\overline{v}_i = Q_{ji}v^j = Q^j_{\cdot i}v_j.$$

9. $$\begin{bmatrix} g^{11} & g^{12} \\ g^{12} & g^{22} \end{bmatrix} = \begin{bmatrix} g_{11} & g_{12} \\ g_{12} & g_{22} \end{bmatrix}^{-1} = \frac{1}{g}\begin{bmatrix} g_{22} & -g_{12} \\ -g_{12} & g_{11} \end{bmatrix}$$

10. $\mathbf{e}_1 = \dfrac{\mathbf{g}_1}{|\mathbf{g}_1|},\ \mathbf{e}_2 = \dfrac{\mathbf{g}_2 - (\mathbf{g}_2 \cdot \mathbf{e}_1)\mathbf{e}_1}{|\mathbf{g}_2 - (\mathbf{g}_2 \cdot \mathbf{e}_1)\mathbf{e}_1|}$ (since $\mathbf{g}_2$ minus the vector projection of $\mathbf{g}_2$ onto $\mathbf{e}_1$ is orthogonal to $\mathbf{e}_1$), $\mathbf{e}_3 = \dfrac{\mathbf{g}_3 - (\mathbf{g}_3 \cdot \mathbf{e}_1)\mathbf{e}_1 - (\mathbf{g}_3 \cdot \mathbf{e}_2)\mathbf{e}_2}{|\mathbf{g}_3 - (\mathbf{g}_3 \cdot \mathbf{e}_1)\mathbf{e}_1 - (\mathbf{g}_3 \cdot \mathbf{e}_2)\mathbf{e}_2|}$. Any vector $\mathbf{v} = v_i\mathbf{e}_i$.

Problems 5.1

1. $\dfrac{d\mathbf{u}}{dt} \times \mathbf{v} \cdot \mathbf{w} + \mathbf{u} \times \dfrac{d\mathbf{v}}{dt} \cdot \mathbf{w} + \mathbf{u} \times \mathbf{v} \cdot \dfrac{d\mathbf{w}}{dt}$.

2. $\mathbf{e} \cdot \dfrac{d\mathbf{e}}{dt} = \dfrac{1}{2}\dfrac{d}{dt}(\mathbf{e} \cdot \mathbf{e}) = 0$.

3. $\dfrac{d}{dt}[\mathbf{u} \times \mathbf{v}] = \epsilon_{ijk}\left(\dfrac{du_i}{dt}v_j + u_i\dfrac{dv_j}{dt}\right)\mathbf{e}_k$.

4. $$\frac{d\mathbf{u}}{dt} = \cos t\mathbf{e}_1 + e^t\mathbf{e}_2, \qquad \frac{d^2\mathbf{u}}{dt^2} = -\sin t\mathbf{e}_1 + e^t\mathbf{e}_2,$$

$$\int \mathbf{u}(t)dt = -\cos t\mathbf{e}_1 + e^t\mathbf{e}_2 + t\mathbf{e}_3 + \mathbf{c}.$$

5. $\mathbf{u} = \mathbf{e}_1 \cos t + \mathbf{e}_2 \sin t$.

6. Similar to the definitions and rules for vector fields $\mathbf{u}(t)$ in Section 5.1.

$$\lim_{\Delta t \to 0} \frac{\mathbf{T}(t+\Delta t) - \mathbf{T}(t)}{\Delta t} = \frac{d\mathbf{T}}{dt}(t),$$

$$\frac{d(\mathbf{S}\cdot\mathbf{T})}{dt} = \frac{d\mathbf{S}}{dt}\cdot\mathbf{T} + \mathbf{S}\cdot\frac{d\mathbf{T}}{dt}, \text{ etc.}$$

Problems 5.2

1. From Fig. 4.6, $\vec{\omega} = \overline{\omega}_i\overline{\mathbf{e}}_i = \frac{d\theta_1}{dt}\overline{\mathbf{e}}_1 + \frac{d\theta_2}{dt}(\cos\theta_3\mathbf{e}_2 - \sin\theta_3\mathbf{e}_1) + \frac{d\theta_3}{dt}\mathbf{e}_3$. Use (4.4) and (4.14) to obtain $\overline{\omega}_1 = \frac{d\theta_1}{dt} + \frac{d\theta_2}{dt}(\cos\theta_3 R_{21} - \sin\theta_3 R_{11}) + \frac{d\theta_3}{dt}R_{31}$ $= \frac{d\theta_1}{dt} - \frac{d\theta_3}{dt}\sin\theta_2$, etc.

2. Find the inverse of the matrix in (5.7):

$$\begin{bmatrix} \frac{d\theta_1}{dt} \\ \frac{d\theta_2}{dt} \\ \frac{d\theta_3}{dt} \end{bmatrix} = \frac{1}{\cos\theta_2}\begin{bmatrix} \cos\theta_2 & \sin\theta_1\sin\theta_2 & \cos\theta_1\sin\theta_2 \\ 0 & \cos\theta_1\cos\theta_2 & -\sin\theta_1\cos\theta_2 \\ 0 & \sin\theta_1 & \cos\theta_1 \end{bmatrix}\begin{bmatrix} \overline{\omega}_1 \\ \overline{\omega}_2 \\ \overline{\omega}_3 \end{bmatrix},$$

$$\theta_2 \neq \frac{\pi}{2} \text{ or } \frac{3\pi}{2}.$$

3. $\boldsymbol{\omega}$ is sum of angular velocity about $\mathbf{e}$ plus angular velocities associated with $\frac{d\mathbf{e}}{dt}$. Resolve $\boldsymbol{\omega}$ into orthogonal components: $\boldsymbol{\omega} = \frac{d\theta}{dt}\mathbf{e} + a\frac{d\mathbf{e}}{dt} + b\mathbf{e}\times\frac{d\mathbf{e}}{dt}$. Then use (3.25) and (5.6a): $\boldsymbol{\omega}\times\mathbf{e} = a\frac{d\mathbf{e}}{dt}\times\mathbf{e} + b\frac{d\mathbf{e}}{dt} = \frac{d\mathbf{R}}{dt}\cdot(\mathbf{R}^t\cdot\mathbf{e}) = (1-\cos\theta)\frac{d\mathbf{e}}{dt} + \sin\theta\frac{d\mathbf{e}}{dt}\times\mathbf{e}$. Solve for a and b to obtain $\boldsymbol{\omega} = \frac{d\theta}{dt}\mathbf{e} + \sin\theta\frac{d\mathbf{e}}{dt} + (1-\cos\theta)\mathbf{e}\times\frac{d\mathbf{e}}{dt}$.

4. Let $\overline{\mathbf{e}}'_i = R'_{ji}\overline{\mathbf{e}}_j$, where R'_{ji} is independent of t. Then

$$\frac{d'\mathbf{u}}{dt} = \frac{d'\overline{u}'_i}{dt}\overline{\mathbf{e}}'_i = \frac{d'\overline{u}'_i}{dt}R'_{ji}\overline{\mathbf{e}}_j = \frac{d}{dt}(R'_{ji}\overline{u}'_i)\overline{\mathbf{e}}_j = \frac{d\overline{u}_j}{dt}\overline{\mathbf{e}}_j = \frac{\overline{d}\mathbf{u}}{dt}.$$

5. $$\frac{d\boldsymbol{\omega}}{dt} = \frac{d\overline{\omega}_i}{dt}\overline{\mathbf{e}}_i + \overline{\omega}_i\frac{d\overline{\mathbf{e}}_i}{dt} = \frac{d\overline{\omega}_i}{dt}\overline{\mathbf{e}}_i + \boldsymbol{\omega}\times\boldsymbol{\omega} = \frac{d\overline{\omega}_i}{dt}\overline{\mathbf{e}}_i = \frac{\overline{d}\boldsymbol{\omega}}{dt}.$$

6. The composition of two sequential finite rotations $\mathbf{R}_1$ and $\mathbf{R}_2$ is the product $\mathbf{R}_1\cdot\mathbf{R}_2$, which is in general not commutative: $\mathbf{R}_1\cdot\mathbf{R}_2 \neq \mathbf{R}_2\cdot\mathbf{R}_1$. But if

the rotation is infinitesimal, from (3.25), $\mathbf{R}_1 \cdot \mathbf{R}_2 \approx \mathbf{1} + \Delta\theta_1 \mathbf{e}_1 \times + \Delta\theta_2 \mathbf{e}_2 \times \approx \mathbf{R}_2 \cdot \mathbf{R}_1$.

7. In a basis rotating with the ants, each ant moves directly towards the center of the square with speed $\frac{v}{\sqrt{2}}$. Thus each ant reaches the center at time $\frac{\ell}{v}$. See also Problem 2 in Section 8.2.

Problems 5.3

1. Either $0 < \alpha = \frac{1}{2}\sin^{-1}\left[\frac{g}{v_0^2}(x_2)_{\max}\right] \leq \frac{\pi}{4}$

or $\frac{\pi}{4} \leq \alpha = \frac{\pi}{2} - \frac{1}{2}\sin^{-1}\left[\frac{g}{v_0^2}(x_2)_{\max}\right] < \frac{\pi}{2}$. Smaller muzzle angle results in shorter flight time. Maximum range is $\frac{v_0^2}{g}$ for $\alpha = 45°$.

2. Solve $\frac{d^2x_1}{dt^2} = -g - \frac{k}{m}\frac{dx_1}{dt}$ and $\frac{d^2x_2}{dt^2} = -\frac{k}{m}\frac{dx_2}{dt}$ to obtain

$$x_1 = \left(\frac{m^2g}{k^2} + \frac{mv_0 \sin\alpha}{k}\right)\left(1 - e^{-kt/m}\right) - \frac{mgt}{k},$$

$$x_2 = \frac{mv_0 \cos\alpha}{k}\left(1 - e^{-kt/m}\right).$$

3. Equations are

$$\frac{d^2\overline{x}_1}{dt^2} = -g, \quad \frac{d^2\overline{x}_2}{dt^2} = 2\omega\frac{d\overline{x}_3}{dt}, \quad \frac{d^2\overline{x}_3}{dt^2} = -2\omega\frac{d\overline{x}_2}{dt}.$$

Integrating once gives

$$\frac{d\overline{x}_1}{dt} = v_0 \sin\alpha - gt, \quad \frac{d\overline{x}_2}{dt} = 2\omega\overline{x}_3 + v_0\cos\alpha, \quad \frac{d\overline{x}_3}{dt} = -2\omega\overline{x}_2.$$

Integrate again to obtain

$$\overline{x}_1 = (v_0 \sin\alpha)t - \frac{1}{2}gt^2, \qquad \overline{x}_2 = \frac{v_0\cos\alpha}{2\omega}\sin 2\omega t,$$

$$\overline{x}_3 = \frac{v_0\cos\alpha}{2\omega}(\cos 2\omega t - 1).$$

Neglect terms of order ω^2t^2 compared to 1, since centrifugal force is of thi[s] order and is neglected:

$$\overline{x}_1 = (v_0\sin\alpha)t - \frac{1}{2}gt^2, \ \overline{x}_2 = (v_0\cos\alpha)t, \ \overline{x}_3 = -(\omega v_0 \cos\alpha)t^2.$$

4. The work done in one second is $mgx_1 = 9.8 \times 10^8$ J (joules). Power output i[s] 980 MW (megawatts).

5. The equation of motion of the mass m is $m\dfrac{d^2\mathbf{u}}{dt^2} = -k\mathbf{u}$, where $\mathbf{u}$ is displacement from its rest position. The general solution is

$$\mathbf{u} = \mathbf{c}_1 \cos\sqrt{\frac{k}{m}}t + \mathbf{c}_2 \sin\sqrt{\frac{k}{m}}t,$$

where $\mathbf{c}_1$ and $\mathbf{c}_2$ are constant vectors. The mass traces out an ellipse ($\mathbf{c}_1 \neq 0, \mathbf{c}_2 \neq 0$), circle ($\mathbf{c}_1 = \mathbf{c}_2$), or straight line ($\mathbf{c}_1 = 0$ or $\mathbf{c}_2 = 0$).

6. $$m\frac{d^2\overline{x}_i}{dt^2} = \overline{f}_i + m\overline{\omega}_j\overline{\omega}_j(\overline{x}_i + \overline{y}_i) - m\overline{\omega}_i\overline{\omega}_j(\overline{x}_j + \overline{y}_j)$$
$$-2m\epsilon_{ijk}\overline{\omega}_j\left(\frac{d\overline{x}_k}{dt} + \frac{d\overline{y}_k}{dt}\right) - m\epsilon_{ijk}\frac{d\overline{\omega}_j}{dt}(\overline{x}_k + \overline{y}_k) - m\frac{d^2\overline{y}_i}{dt^2}.$$

7. After one day, the earth has rotated through $1 + \dfrac{1}{365\frac{1}{4}}$ revolutions. Thus $\omega = 366\frac{1}{4}$ revolutions per year.

8. $\omega^2 a < 0.0035g, 2\omega v < 0.0015g$.

9. Only if the weight is dropped at one of the poles, so the Coriolis force is zero.

10. Differentiate (4.15a) to obtain, with $\omega = \dfrac{d\alpha}{dt}$,

$$\begin{bmatrix} \dfrac{d\overline{x}_1}{dt} \\ \dfrac{d\overline{x}_2}{dt} \\ \dfrac{d\overline{x}_3}{dt} \end{bmatrix} = \begin{bmatrix} \cos\alpha\cos\beta & \sin\alpha\cos\beta & \sin\beta \\ -\sin\alpha & \cos\alpha & 0 \\ -\cos\alpha\sin\beta & -\sin\alpha\sin\beta & \cos\beta \end{bmatrix} \begin{bmatrix} \dfrac{dx_1}{dt} \\ \dfrac{dx_2}{dt} \\ \dfrac{dx_3}{dt} \end{bmatrix}$$
$$+\omega \begin{bmatrix} -\sin\alpha\cos\beta & \cos\alpha\cos\beta & 0 \\ -\cos\alpha & -\sin\alpha & 0 \\ \sin\alpha\sin\beta & -\cos\alpha\sin\beta & 0 \end{bmatrix} \begin{bmatrix} x_1 \\ x_2 \\ x_3 \end{bmatrix},$$

$$\begin{bmatrix} \dfrac{d^2\overline{x}_1}{dt^2} \\ \dfrac{d^2\overline{x}_2}{dt^2} \\ \dfrac{d^2\overline{x}_3}{dt^2} \end{bmatrix} = \begin{bmatrix} \cos\alpha\cos\beta & \sin\alpha\cos\beta & \sin\beta \\ -\sin\alpha & \cos\alpha & 0 \\ -\cos\alpha\sin\beta & -\sin\alpha\sin\beta & \cos\alpha \end{bmatrix} \begin{bmatrix} \dfrac{d^2x_1}{dt^2} \\ \dfrac{d^2x_2}{dt^2} \\ \dfrac{d^2x_3}{dt^2} \end{bmatrix}$$
$$+2\omega \begin{bmatrix} -\sin\alpha\cos\beta & \cos\alpha\cos\beta & 0 \\ -\cos\alpha & -\sin\alpha & 0 \\ \sin\alpha\sin\beta & -\cos\alpha\sin\beta & 0 \end{bmatrix} \begin{bmatrix} \dfrac{dx_1}{dt} \\ \dfrac{dx_2}{dt} \\ \dfrac{dx_3}{dt} \end{bmatrix}$$

$$+\omega^2\begin{bmatrix} -\cos\alpha\cos\beta & -\sin\alpha\cos\beta & 0 \\ \sin\alpha & -\cos\alpha & 0 \\ \cos\alpha\sin\beta & \sin\alpha\sin\beta & 0 \end{bmatrix}\begin{bmatrix} x_1 \\ x_2 \\ x_3 \end{bmatrix}.$$

11. Stars rise in the east, travel circular arcs across the sky, set in the west. The center points of the arcs lie on an axis that makes an angle β with a line to the observer's north or south horizon.

12. The component of Coriolis force per unit mass directed towards the center of the circular flow is $2\omega v\sin\beta$, where β is latitude of the flow. This must equal $\Omega^2 r$, where Ω is the angular velocity and r is the radius of the flow. Since $v = \Omega r$, $\Omega = 2\omega\sin\beta$. Direction of flow is clockwise in northern hemisphere and counterclockwise in southern hemisphere.

13. Centrifugal force per unit mass has magnitude $\omega^2 r\cos\theta$ and is directed outward from vertical axis. Coriolis force per unit mass has magnitude $2\omega\dfrac{d\theta}{dt}r\sin\theta$, where r is distance from the blade hinge, and is in the direction of motion of the blade tip.

14. In a coordinate system accelerating with the bucket, the fluid has no weight and exerts no buoyant force on the object. Hence the object will be pulled down by the spring.

Problems 5.4

1. $\dfrac{d\bar{\mathbf{e}}_1}{ds} = \kappa\bar{\mathbf{e}}_2$ by the definition (5.24). Since $\bar{\mathbf{e}}_2\cdot\bar{\mathbf{e}}_2 = 1$, $\dfrac{d\bar{\mathbf{e}}_2}{ds}$ is perpendicular to $\bar{\mathbf{e}}_2$ (by Problem 2 in Section 5.1), and

$$\bar{\mathbf{e}}_1\cdot\frac{d\bar{\mathbf{e}}_2}{ds} = \frac{d}{ds}(\bar{\mathbf{e}}_1\cdot\bar{\mathbf{e}}_2) - \bar{\mathbf{e}}_2\cdot\frac{d\bar{\mathbf{e}}_1}{ds} = -\kappa, \qquad \text{so} \qquad \frac{d\bar{\mathbf{e}}_2}{ds} = -\kappa\bar{\mathbf{e}}_1 + \tau\bar{\mathbf{e}}_3.$$

Then

$$\bar{\mathbf{e}}_1\cdot\frac{d\bar{\mathbf{e}}_3}{ds} = -\bar{\mathbf{e}}_3\cdot\frac{d\bar{\mathbf{e}}_1}{ds} = 0 \quad\text{and}\quad \bar{\mathbf{e}}_2\cdot\frac{d\bar{\mathbf{e}}_3}{ds} = -\bar{\mathbf{e}}_3\cdot\frac{d\bar{\mathbf{e}}_2}{ds} = -\tau, \qquad \text{so} \qquad \frac{d\bar{\mathbf{e}}_3}{ds} = -\tau\bar{\mathbf{e}}_2$$

2. $\boldsymbol{\omega} = \omega\mathbf{e}_3$.

3. $$\bar{\mathbf{e}}_1 = \frac{(\cos t - \sin t)\mathbf{e}_1 + (\cos t + \sin t)\mathbf{e}_2 + \mathbf{e}_3}{\sqrt{3}},$$
$$\bar{\mathbf{e}}_2 = \frac{-(\cos t + \sin t)\mathbf{e}_1 + (\cos t - \sin t)\mathbf{e}_2}{\sqrt{2}},$$
$$\bar{\mathbf{e}}_3 = \frac{-(\cos t - \sin t)\mathbf{e}_1 - (\cos t + \sin t)\mathbf{e}_2 + 2\mathbf{e}_3}{\sqrt{6}},$$
$$\kappa = \frac{\sqrt{2}e^{-t}}{3}, \quad \tau = \frac{e^{-t}}{3}.$$

4. $\dfrac{d^2\mathbf{r}}{dt^2} = \sqrt{3}e^t\bar{\mathbf{e}}_1 + \sqrt{2}e^t\bar{\mathbf{e}}_2.$

5. From (5.27)–(5.28), $\mathbf{v}\times\mathbf{a} = \kappa\left(\frac{ds}{dt}\right)^3 \bar{\mathbf{e}}_3$ and $\mathbf{v}\times\mathbf{a}\cdot\frac{d\mathbf{a}}{dt} = \kappa^2\left(\frac{ds}{dt}\right)^6 \frac{d\bar{\mathbf{e}}_2}{ds}\cdot\bar{\mathbf{e}}_3$, so

$$\kappa = \frac{\left|\frac{d\mathbf{r}}{dt}\times\frac{d^2\mathbf{r}}{dt^2}\right|}{\left|\frac{d\mathbf{r}}{dt}\right|^3} \quad\text{and}\quad \tau = \frac{\frac{d\mathbf{r}}{dt}\times\frac{d^2\mathbf{r}}{dt^2}\cdot\frac{d^3\mathbf{r}}{dt^3}}{\left|\frac{d\mathbf{r}}{dt}\times\frac{d^2\mathbf{r}}{dt^2}\right|^2}.$$

6. $m\kappa v^2 = qvB$ with $v = \frac{\omega}{\kappa}$ leads to $\frac{\omega}{2\pi} = \frac{qB}{2\pi m}$.

7. The component $g\sin\phi$ of gravitational force per unit mass tangent to the road must be equal to the component $\kappa v^2\cos\phi$ of centripetal acceleration (from (5.28)), leading to $v = \sqrt{\frac{g\tan\phi}{\kappa}}$.

8. Force per unit mass felt at top of loop is $-g + \kappa v_1^2 = \frac{g}{2}$, so speed at top is $v_1 = \sqrt{\frac{3g}{2\kappa}} = 478\frac{\text{km}}{\text{hr}}$. Conservation of energy yields corresponding speed at bottom $v_2 = \sqrt{v_1^2 + 2g(x_1 - x_2)} = 915\frac{\text{km}}{\text{hr}}$, so force per unit mass felt is $g + \kappa v_2^2 = 6.5g$.

9. Horizontal force equilibrium is $\frac{w\kappa_1 v^2}{g} = L\sin\phi$, vertical force equilibrium is $\frac{w\kappa_2 v^2}{g} = L\cos\phi - w$, so total curvature is

$$\kappa = \sqrt{\kappa_1^2 + \kappa_2^2} = \frac{g}{v^2}\sqrt{\frac{L^2}{w^2} - \frac{2L\cos\phi}{w} + 1}. \qquad L = \frac{w}{\cos\phi} \text{ for level flight.}$$

Problems 5.5

1. $\mathbf{f} = 0 \Rightarrow m\frac{d\mathbf{y}}{dt} =$ constant, $\mathbf{q} = 0 \Rightarrow \mathbf{h} =$ constant.

2. From the previous problem, the distribution of mass must be symmetric about the axis of rotation.

3. Let $\mathbf{r}_i(t)$ denote the position vector from the point fixed in the body to the ith particle. Then inserting (5.32) with $\mathbf{y} = 0$ into (5.35) and using (5.30)–(531), we obtain (5.36). The remaining derivation of (5.40) is identical to that in the text.

4. The rewinding motion is exactly the reverse of the unwinding motion, except that the string wraps around the axle in the opposite sense.

5. Use the facts that the angular momentum and energy of the satellite plus mass are conserved: $ma^2\omega + I\omega = m(\ell + a)^2\omega_f$, $\frac{1}{2}ma^2\omega^2 + \frac{1}{2}I\omega^2 = \frac{1}{2}m(\ell + a)^2\omega_f^2$, where ω_f is the angular velocity of the cord at release.

$$\Longrightarrow \omega_f = \omega,\ \ell = \sqrt{a^2 + \frac{I}{m}} - a.$$

6. Sliding: Letting $r(t)$ denote the displacement of the center of the sphere, we obtain from (5.33) $m\dfrac{d^2r}{dt^2} = mg\sin\alpha$. Solution for time to slide a length ℓ is $\sqrt{\dfrac{2\ell}{g\sin\alpha}}$.

   Rolling: Letting f denote the frictional force and ω denote the angular velocity of the sphere of radius a, we have $m\dfrac{d^2r}{dt^2} = mg\sin\alpha - f$, $I\dfrac{d\omega}{dt} = af$, $\omega = \dfrac{d\theta}{dt}$, $r = a\theta$. From Problem 1 in Section 3.5, $I = \dfrac{2}{5}ma^2$. Solution for time to roll a length ℓ is $\sqrt{\dfrac{14\ell}{5g\sin\alpha}}$.

7. Apply equations (5.43)–(5.44) about the fixed point with the gravitational torque $mgr\sin\theta$, where θ is the angle the axis makes with the vertical, to obtain $\Omega = \dfrac{mgr}{I\omega}$.

8. Leaning produces a torque about the center of the front wheel which causes it to turn. When going around a curve, enough torque must be applied to the bike to maintain the wheels' precession.

9. Since $\mathbf{I}$ is proportional to $\mathbf{1}$ for the sphere, (5.36) and (5.38) imply $\boldsymbol{\omega} =$ constant. Thus the sphere spins at a constant rate about an axis passing through its center of mass and parallel to a fixed direction in an inertial coordinate system.

10. $\overline{I}_{11} = \dfrac{\pi}{2}\rho t a^4$ and $\overline{I}_{22} \approx \dfrac{\pi}{4}\rho t a^4$, where ρ is density, t is thickness, and a is radius of the plate. From (5.45), $\dfrac{d\alpha}{dt} = \dfrac{\overline{I}_{11}\overline{\omega}_1}{\overline{I}_{22}\sin\beta} \approx \dfrac{\overline{I}_{11}\overline{\omega}_1}{\overline{I}_{22}} \approx 2\overline{\omega}_1$, $\dfrac{d\gamma}{dt} \approx -\overline{\omega}_1$.

    Thus the precession is in the opposite direction to the spin, and the precession rate is twice the magnitude of the spin rate.

Problems 6.1

1. Circles $x_1^2 + x_2^2 =$ constant.
2. Must be maximum, minimum, or saddle point (unless $f =$ constant).
3. $\dfrac{df}{ds} = \mathbf{e}\cdot\nabla f(0,0,0) = \mathbf{e}\cdot(\mathbf{e}_1 + \mathbf{e}_3)$. $\left(\dfrac{df}{ds}\right)_{\max} = \sqrt{2}$ when $\mathbf{e} = \dfrac{\mathbf{e}_1 + \mathbf{e}_3}{\sqrt{2}}$.
4. $d\mathbf{r} \times \nabla f = 0$ implies $\dfrac{dx_1}{x_1} = \dfrac{dx_2}{2x_2}$, so $x_2 = \dfrac{7}{2}x_1^2$, $\sqrt{2} \geq x_1 \geq 0$.
5. $\nabla(fh) = \mathbf{e}_i\dfrac{\partial(fh)}{\partial x_i} = f\mathbf{e}_i\dfrac{\partial h}{\partial x_i} + h\mathbf{e}_i\dfrac{\partial f}{\partial x_i} = f\nabla h + h\nabla f$.
6. Normal line : $x_2 = x_1$, $x_3 = 3 - x_1$.
   Tangent plane : $x_1 + x_2 - x_3 = 0$.

7. Normal line : $x_1 - a_1 = t\dfrac{\partial f}{\partial x_1}(a_1, a_2), \quad x_2 - a_2 = t\dfrac{\partial f}{\partial x_2}(a_1, a_2).$

Tangent line : $(x_1 - a_1)\dfrac{\partial f}{\partial x_1}(a_1, a_2) + (x_2 - a_2)\dfrac{\partial f}{\partial x_2}(a_1, a_2) = 0.$

8. $\cos\theta = \mathbf{n}_1 \cdot \mathbf{n}_2 = \left(\dfrac{x_1\mathbf{e}_1 + x_2\mathbf{e}_2 + x_3\mathbf{e}_3}{2}\right) \cdot (x_1\mathbf{e}_1 + x_2\mathbf{e}_2) = \dfrac{1}{2}$, so $\theta = \dfrac{\pi}{3}$.

9. A normal vector to the surface is $\nabla(\mathbf{r} \cdot \mathbf{S} \cdot \mathbf{r}) = \nabla(S_{ij}x_ix_j) = 2\mathbf{S} \cdot \mathbf{r}$.

10. See the first example in Section 4.2. Rotate to the principal axes $(\overline{x}_1, \overline{x}_2)$ of $\mathbf{S}$.

11. At the maximum or minimum point, a contour line $f(x_1, x_2) =$ constant is tangent to the line $h(x_1, x_2) = 0$, so the normal ∇f to the contour line is parallel to the normal ∇h, and thus $\nabla f = \lambda\, \nabla h$. Equations $\dfrac{\partial f}{\partial x_1} = \lambda\dfrac{\partial h}{\partial x_1}$, $\dfrac{\partial f}{\partial x_2} = \lambda\dfrac{\partial h}{\partial x_2}$, with $f = x_1x_2$ and $h = x_1^2 + x_2^2 - 1$, imply $x_2 = 2\lambda x_1$, $x_1 = 2\lambda x_2$.

$$\text{Solving yields } \lambda = \pm\frac{1}{2}, \quad x_1 = \pm\frac{1}{\sqrt{2}}, \quad x_2 = \pm\frac{1}{\sqrt{2}}.$$

$$\text{Maximum value is } f\left(\frac{1}{\sqrt{2}}, \frac{1}{\sqrt{2}}\right) = f\left(-\frac{1}{\sqrt{2}}, -\frac{1}{\sqrt{2}}\right) = \frac{1}{2}.$$

Problems 6.2

1. $\nabla \cdot \mathbf{u} = x_3, \quad \nabla \times \mathbf{u} = 3x_1\mathbf{e}_1 + (x_1 - 3x_2)\mathbf{e}_2 + 2x_1\mathbf{e}_3,$
$\nabla\mathbf{u} = x_3\mathbf{e}_1\mathbf{e}_1 + 2x_1\mathbf{e}_1\mathbf{e}_2 + 3x_2\mathbf{e}_2\mathbf{e}_3 + 3x_1\mathbf{e}_2\mathbf{e}_3 + x_1\mathbf{e}_3\mathbf{e}_1, \nabla^2\mathbf{u} = 2\mathbf{e}_2.$

2. $\mathrm{tr}\nabla\mathbf{u} = \dfrac{\partial u_i}{\partial x_i} = \nabla \cdot \mathbf{u}.$

3. Use (6.11) with $\mathbf{v} = \mathbf{u}$.

4. Use (6.12) and (6.16) to obtain $\nabla\cdot(\nabla f \times \nabla g) = \nabla g \cdot (\nabla \times \nabla f) - \nabla f \cdot (\nabla \times \nabla g) = 0$. Use (6.10) and (6.16) to obtain $\nabla \times (f\nabla g + g\nabla f) = f\nabla \times \nabla g + \nabla f \times \nabla g + g\nabla \times \nabla f + \nabla g \times \nabla f = 0.$

5. From (6.15), $\nabla^2(fg) = \dfrac{\partial^2(fg)}{\partial x_i \partial x_i} = \dfrac{\partial}{\partial x_i}\left(f\dfrac{\partial g}{\partial x_i} + g\dfrac{\partial f}{\partial x_i}\right)$
$= f\dfrac{\partial^2 g}{\partial x_i\partial x_i} + g\dfrac{\partial^2 f}{\partial x_i \partial x_i} + 2\dfrac{\partial f}{\partial x_i}\dfrac{\partial g}{\partial x_i} = f\nabla^2 g + g\nabla^2 f + 2\nabla f \cdot \nabla g.$

6. $\dfrac{d\mathbf{u}}{ds} = \dfrac{\partial \mathbf{u}}{\partial x_1}\dfrac{dx_1}{ds} + \dfrac{\partial \mathbf{u}}{\partial x_2}\dfrac{dx_2}{ds} + \dfrac{\partial \mathbf{u}}{\partial x_3}\dfrac{dx_3}{ds} = \mathbf{e} \cdot \nabla\mathbf{u}.$

7. $\nabla \cdot \mathbf{r} = 3, \quad \nabla \times \mathbf{r} = 0, \quad \nabla r^n = nr^{n-2}\mathbf{r}, \quad \nabla^2 r^n = n(n+1)r^{n-2}.$

8. $x_1x_2^2 = c_1$, $x_1x_3^2 = c_2$, $c_3x_2 + c_4x_3 = 0$. Three-dimensional stagnation point flow. Flow in x_1-x_2 or x_1-x_3 planes is similar to that shown in Figure 6.15.

9. $x_2 = c_1 e^{x_1}$, $x_3 = c_2 e^{x_1}$. Flow is axisymetric about x_1-axis and diverges from axis as x_1 increases.

10. Use (6.4) and (6.19) to obtain

$$f(x_1, x_2, x_3) = f(a_1, a_2, a_3) + (x_i - a_i)\frac{\partial f}{\partial x_i}(a_1, a_2, a_3)$$
$$+\frac{1}{2}(x_i - a_i)(x_j - a_j)\frac{\partial^2 f}{\partial x_i \partial x_j}(a_1, a_2, a_3) + \cdots.$$

Problems 6.3

1.
$$-\int_1^0 (1 + x_1)dx_1 = \frac{3}{2}, \quad \int_0^{\pi/2} \left(1 + \frac{2}{\pi}\cos t\right) dt = \frac{\pi}{2} + \frac{2}{\pi},$$
$$\int_0^{3\pi/2} \left(-1 + \frac{2}{3\pi}\cos t\right) dt = -\frac{3\pi}{2} - \frac{2}{3\pi}.$$

2. $\displaystyle\int_0^1 x_1 dx_1 = \frac{1}{2}, \quad \int_0^1 (t^2 + 3t^3)dt = \frac{13}{12}, \quad \int_0^{2\pi} (\sin^2 s + \cos^2 s)ds = 2\pi.$

3. $\displaystyle 3\int_0^{2\pi} (\sin^2 t + \cos^2 t)dt = 6\pi, \quad 8\int_0^{2\pi} \sin t \cos t \, dt = 0.$

4. $\displaystyle\iint_{\substack{\text{unit}\\\text{disk}}} (1 + x_1^2 + x_2^2)dx_1 dx_2 = \int_0^{2\pi}\int_0^1 (1 + r^2)r dr d\theta = \frac{3\pi}{2},$

$$\iint_{\substack{\text{upper}\\\text{hemisphere}}} dS = 2\pi, \quad -\iint_{\substack{\text{lower}\\\text{hemisphere}}} dS = -2\pi.$$

5. If we choose the unit normal $\mathbf{n} = \dfrac{2\mathbf{e}_1 + 2\mathbf{e}_2 + \mathbf{e}_3}{3}$ for the triangular surface, $\mathbf{n} = x_1\mathbf{e}_1 + x_2\mathbf{e}_2$ for the cylindrical surface, and $\mathbf{n}$ outward from the cubical volume:

$$2\int_0^1\int_0^{1-x_2} dx_1 dx_2 = 1, \quad \int_0^2\int_0^1 \frac{dx_1 dx_3}{\sqrt{1 - x_1^2}} = \pi,$$
$$\int_0^1\int_0^1 dx_2 dx_3 + \int_0^1\int_0^1 dx_1 dx_3 + \int_0^1\int_0^1 dx_1 dx_2 = 3.$$

6. $\displaystyle\oiint (-x_2\mathbf{e}_1 + x_1\mathbf{e}_2 + x_1\mathbf{e}_3)\cdot d\mathbf{S} = \frac{1}{a}\oiint x_1 x_3 dS = 0, \oiint x_i\mathbf{e}_i \cdot d\mathbf{S} = a\oiint dS = 4\pi a^3.$

7. $\displaystyle\int_{-1}^1\int_{-\sqrt{1-x_1^2}}^{\sqrt{1-x_1^2}}\int_0^{1-x_1^2-x_2^2} dx_3 dx_2 dx_1 = \int_0^{2\pi}\int_0^1 (1 - r^2)r dr d\theta = \frac{\pi}{2}.$

Problems 6.4

1. $$\oint_C (-x_2 dx_1 + x_1 dx_2) = \int_0^{2\pi} (\sin^2 t + \cos^2 t)dt = 2\pi,$$
$$2\int\int_{\substack{\text{unit}\\ \text{disk}}} dx_1 dx_2 = 2\pi.$$

2. $\oint_C ds = 2\pi, \quad 2\int\int_{\substack{\text{unit}\\ \text{disk}}} dx_1 dx_2 = 2\pi.$

3. For each example, the stream function ψ is a single-valued function, so from (6.34), $\oint_C \mathbf{u}\cdot\mathbf{n}ds = 0.$

4. $\phi = x_1x_2$, $\psi = \frac{1}{2}(-x_1^2 + x_2^2)$, $\int_C \mathbf{u}\cdot d\mathbf{r} = \phi\left(\frac{\pi}{4}\right) - \phi(0) = \frac{1}{8}$,
$\int_C \mathbf{u}\cdot\mathbf{n}ds = \psi\left(\frac{\pi}{4}\right) - \psi(0) = 0.$

5. $$\psi = \begin{cases} -\dfrac{(x_1^2+x_2^2}{2} + \dfrac{1}{2}, & x_1^2 + x_2^2 \le 1, \\ -\dfrac{\ln(x_1^2+x_2^2)}{2}, & x_1^2 + x_2^2 \ge 1. \end{cases}$$

6. $\oint \mathbf{u}\cdot\mathbf{n}ds.$

7. $$\phi = \tfrac{1}{2}\ln(x_1^2+x_2^2), \quad \psi = \tan^{-1}\left(\frac{x_2}{x_1}\right), \quad \oint_C \mathbf{u}\cdot d\mathbf{r} = 0$$
$$\oint_C \mathbf{u}\cdot\mathbf{n}ds = \begin{cases} 0 & \text{if } C \text{ does not encircle the } x_3 \text{ axis,} \\ 2\pi & \text{if } C \text{ encircles the } x_3 \text{ axis.} \end{cases}$$

8. Choose $u_1 = -x_2$ and $u_2 = x_1$ in (6.29) to obtain formula for A. Use parameterization $x_1 = a\cos t$, $x_2 = b\sin t$ of ellipse to get
$$A = \frac{ab}{2}\int_0^{2\pi} (\cos^2 t + \sin^2 t)dt = \pi ab.$$

9. $f = x_1 + ix_2 \Longrightarrow \phi = x_1 \Longrightarrow \mathbf{u} = \nabla\phi = \mathbf{e}_1$. $f = \frac{1}{2}(x_1+ix_2)^2 \Longrightarrow$ $\phi = \frac{1}{2}(x_1^2 - x_2^2) \Longrightarrow \mathbf{u} = x_1\mathbf{e}_1 - x_2\mathbf{e}_2$. $f = -i\ln\left(\sqrt{x_1^2+x_2^2}\,e^{i\theta}\right)$ (see problem 12 in Section 3.3) $\Longrightarrow \phi = \theta = \tan^{-1}\dfrac{x_2}{x_1} \Longrightarrow \mathbf{u} = \dfrac{-x_2\mathbf{e}_1 + x_1\mathbf{e}_2}{x_1^2+x_2^2}.$

10. $$\frac{\partial T_{11}}{\partial x_1} + \frac{\partial T_{12}}{\partial x_2} = \frac{\partial^3\Phi}{\partial x_1\partial x_2^2} - \frac{\partial^3\Phi}{\partial x_1\partial x_2^2} = 0, \quad \frac{\partial T_{22}}{\partial x_2} + \frac{\partial T_{12}}{\partial x_1} = \frac{\partial^3\Phi}{\partial x_1^2\partial x_2} - \frac{\partial^3\Phi}{\partial x_1^2\partial x_2} = 0.$$

Problems 6.5

1. $$\int\int \nabla\times\mathbf{u}\cdot d\mathbf{S} = 2\int\int_S \frac{(x_2+1)dS}{\sqrt{1+4x_1^2+4x_2^2}} = 2\int\int_{\substack{\text{unit}\\ \text{disk}}}(x_2+1)dx_1dx_2$$
$$= 2\int_0^{2\pi}\int_0^1 (r\sin\theta+1)rdrd\theta = 2\pi.$$

2. $$\int\int\int \nabla\cdot\mathbf{r}dV = 3\int_0^1\int_0^1\int_0^1 dx_1dx_2dx_3 = 3.$$

3. Let $\mathbf{u} = u_1(x_1,x_2)\mathbf{e}_1 + u_2(x_1,x_2)\mathbf{e}_2$, apply Stokes' theorem to an area in the x_1-x_2 plane, apply the divergence theorem to a cylinder of unit height.

4. The quantity $\frac{\mathbf{r}}{r}\cdot\mathbf{n}\Delta S$ is the projection of ΔS on a sphere of radius r.

5. From (6.41) $\phi = c_ix_i$. From (6.43) $\psi = c_3x_1\mathbf{e}_2 - (c_1x_2+c_2x_1)\mathbf{e}_3 + \nabla f$.

6. A scalar potential is $\phi = x_1^2e^{x_2}\sin\frac{\pi x_3}{2}$, so $\int \mathbf{u}\cdot d\mathbf{r} = \phi(1,1,1)-\phi(0,0,0) = e$.

7. If $\boldsymbol{\psi}$ exists, then
$$\int\int_S \mathbf{u}\cdot d\mathbf{S} = \oint_C \boldsymbol{\psi}\cdot d\mathbf{r}$$
must be the same for all simply connected surfaces S having C as a perimeter. This violates the result (6.47).

8. Replace $\mathbf{u}$ in the divergence theorem (6.39) by $f\mathbf{e}_i$, $\epsilon_{ijk}u_j\mathbf{e}_k$, $f\nabla g - g\nabla f$, or $T_{ij}\mathbf{e}_j$, then use the vector identity (6.9).

9. $\nabla^2\phi = \nabla\cdot\mathbf{u}$, $\nabla\nabla\cdot\boldsymbol{\psi} - \nabla^2\boldsymbol{\psi} = \nabla\times\mathbf{u}$.

10. Differentiation of the expression for a component of $\mathbf{w}$ yields
$$\nabla w_1(x_1,x_2) = \frac{1}{2\pi}\int\int_A \frac{u_1(x_1',x_2')(\mathbf{r}-\mathbf{r}')}{|\mathbf{r}-\mathbf{r}'|^2}dx_1'dx_2'.$$
The flux of ∇w_1 through any curve C is the net algebraic strength of all the two-dimensional sources enclosed by C,
$$\oint_C \nabla w_1\cdot\mathbf{n}ds = \int\int_{A_1} u_1(x_1',x_2')dx_1'dx_2'$$
where A_1 is the area enclosed by an arbitrary curve C, so by (6.30b) $\nabla^2 w_1 = u_1$. By an analogous argument $\nabla^2 w_2 = u_2$, so $\nabla^2\mathbf{w} = \mathbf{u}$. Then (6.18) implies $\mathbf{u} = \nabla(\nabla\cdot\mathbf{w}) - \nabla\times(\nabla\times\mathbf{w}) = \nabla\phi + \nabla\times\boldsymbol{\psi}$.

Problems 7.1

1. $S = \int \frac{dQ}{T} = mc \int_{T_0}^{T_1} \frac{dT}{T} = mc \ln \frac{T_1}{T_0}.$

2. $d\mathbf{r} \times \nabla T = 0$ implies

$$\frac{dx_1}{x_1} = \frac{dx_2}{2x_2} = \frac{dx_3}{3x_3},$$

so $x_1 = 1-t$, $x_2 = (1-t)^2$, $x_3 = (1-t)^3$, $0 \le t \le 1$.

3. $\mathrm{Bi} = \frac{\ell/k}{1/h}.$

4. $T^* = T_1^* - x_1^*$ where $T^* = \frac{T}{T_1 - T_2}$, $T^* = e^{-t^*\mathrm{Bi}}$ where $\ell = \frac{V}{S}$.

5. Since the equations for each turkey in the nondimensionalized variable $t^* = \frac{\alpha t}{\ell^2}$ are the same, the cooking time t is proportional to ℓ^2. Since the volume V of each turkey is proportional to ℓ^3, the weight $w = \rho V$ is also proportional to ℓ^3. Thus $\frac{t}{w}$ is proportional to $w^{-1/3}$.

6. $\mathrm{Bi} = \frac{hV}{kS} = 0.0008 << 1$, so we can assume Newtonian cooling. Decay time $\tau = \frac{\rho c V}{hS} = 400$ seconds.

7. Consider a wall made of two layers of thicknesses ℓ_1 and ℓ_2 and resistances R_1 and R_2. Let T_3 be the temperature at the interface between the two layers. The rate of heat transfer through the two layers must be equal:

$$q_1 = \frac{T_1 - T_3}{R_1} = \frac{T_3 - T_2}{R_2}.$$

Eliminating T_3 yields

$$q_1 = \frac{T_1 - T_2}{R_1 + R_2}.$$

8. $q_1 = \frac{kA}{\ell}(T_1 - T_2) = hA(T_2 - T_\infty)$ leads to $\frac{T_1 - T_2}{T_2 - T_\infty} = \frac{h\ell}{k}.$

9. $q_1 = \frac{k}{\ell}(T_1 - T_2) = h(T_{1\infty} - T_1) = h(T_2 - T_{2\infty})$ leads to $q_1 = \frac{h(T_{1\infty} - T_{2\infty})}{2 + \mathrm{Bi}}.$

10. Substitute $T(x_1, t) = f(t) \sin \frac{\pi x_1}{\ell}$ into the heat equation to obtain

$$\frac{df}{dt} + \frac{\alpha\pi^2}{\ell^2} f = 0.$$

Solve for $f(t)$ with initial condition $f(0) = T_0$ to obtain

$$T(x_1, t) = T_0 e^{-\alpha\pi^2 t/\ell^2} \sin \frac{\pi x_1}{\ell}.$$

11. The boundary conditions are

$$\frac{\partial T}{\partial x}(0,t) = \frac{\partial T}{\partial x}(\ell,t) = 0,$$

$$T(x_1,t) = T_0 e^{-9\pi^2 \alpha t/\ell^2} \cos \frac{3\pi x}{\ell}.$$

12. Energy conservation for a spherical surface S surrounding the point source is

$$\oiint_S \mathbf{q} \cdot d\mathbf{S} = Q.$$

By symmetry, $\mathbf{q} = -k\nabla T = \dfrac{k|\nabla T(r)|\mathbf{r}}{r}$. This leads to

$$\nabla T = -\frac{Q\mathbf{r}}{4\pi k r^3} \qquad \text{and} \qquad T = \frac{Q}{4\pi k r}$$

($\mathbf{q}$ field is sketched in Fig. 6.6).

Problems 7.2

1. Bring time derivative inside integrand on left, use divergence theorem on right:

$$\int\int\int_V \left(\rho \frac{\partial u_i^2}{\partial t^2}\frac{\partial u_i}{\partial t} + S_{ij}\frac{\partial^2 u_i}{\partial x_j \partial t}\right) dV = \int\int\int_V \frac{\partial}{\partial x_j}\left(S_{ij}\frac{\partial u_i}{\partial t}\right) dV.$$

These are equal by (7.17b).

2. Since at $(x_1, x_2, x_3) = (0, 0, -a)$ the deformation gradient $\nabla \mathbf{u} = 2\gamma_{12}\mathbf{e}_1\mathbf{e}_2$ where $\gamma_{12} = \dfrac{1}{2}\dfrac{\partial u_2}{\partial x_1} = \dfrac{a}{2}\dfrac{d\beta}{dx_1} = \dfrac{Ma}{2\mu J} = \dfrac{M}{\mu \pi a^3}$, $\mathbf{T} = \mathbf{1} + 2\gamma_{12}\mathbf{e}_2\mathbf{e}_1$ so the cube shears through an angle $2\gamma_{12}$. Since $\mathbf{T} = \mathbf{S}\cdot\mathbf{R}$ where $\mathbf{R} = \mathbf{1} + \gamma_{12}\mathbf{e}_3\times$ and $\mathbf{S} = \mathbf{1} + \gamma_{12}\mathbf{e}_1\mathbf{e}_2 + \gamma_{12}\mathbf{e}_2\mathbf{e}_1$, the deformation may be may be accomplished by a rigid rotation through an angle γ_{12} (see (3.25)) plus a symmetrical deformation with the eigenvalues $(\gamma_{12}, -\gamma_{12}, 0)$ of the strain tensor along principal axes making a 45° angle with the x_1-x_2 axes (see Fig. 4.5c).

3. Substitute into (7.19b) to obtain

$$\nabla^2\phi = \frac{\partial^2\phi}{\partial x_2^2} + \frac{\partial^2\phi}{\partial x_3^2} = 0.$$

Use (7.22) to obtain

$$J = \iint_A \left(x_2^2 + x_3^2 + x_2\frac{\partial\phi}{\partial x_3} - x_3\frac{\partial\phi}{dx_2}\right) dx_2 dx_3.$$

4. From Problem 3 in Section 4.2, $I = \dfrac{\ell_2^3 \ell_3}{12}$ so $S_{11} = \dfrac{-12M}{\ell_2^3 \ell_3} x_2$.

$$\text{Maximum} \quad S_{11} = \frac{6M}{\ell_2^2 \ell_3} \quad \text{at} \quad x_2 = \frac{-\ell_2}{2}.$$

$$\text{Minimum} \quad S_{11} = \frac{-6M}{\ell_2^2 \ell_3} \quad \text{at} \quad x_2 = \frac{\ell_2}{2}.$$

5. The particles in a line $x_2 = 0$ within the cross section $x_1 = c =$ constant move into the arc $x_1 = c$,

$$x_2 = \frac{c^2 - \nu x_3^2}{2R}.$$

The particles in a line $x_3 = d =$ constant within the cross section move into the line

$$x_1 = c\left(1 - \frac{x_2}{R}\right), \qquad x_3 = d\left(1 + \frac{\nu x_2}{R}\right).$$

6. The only nonzero stress is $S_{11} = \dfrac{f}{A}$, where A is the cross sectional area. From (7.18b)

$$\gamma_{11} = \frac{f}{YA}, \qquad \gamma_{22} = \gamma_{33} = \frac{-\nu f}{YA},$$

so from (7.16b)

$$u_1 = \frac{f x_1}{YA}, \qquad u_2 = \frac{-\nu f x_2}{YA}, \qquad u_3 = \frac{-\nu f x_3}{YA}.$$

7. Superimpose the solutions in the text and the result of the preceding example.

8. $\dfrac{\partial u_1}{\partial x_1} = \dfrac{df}{dy} + \dfrac{dg}{dz}$,

$$\frac{\partial u_1}{\partial t} = -c\frac{df}{dy} + c\frac{dg}{dz}, \ \frac{\partial^2 u_1}{\partial x_1^2} = \frac{d^2 f}{dy^2} + \frac{d^2 g}{dz^2}, \ \frac{\partial^2 u_1}{\partial t^2} = c^2\frac{d^2 f}{dy^2} + c^2\frac{d^2 g}{dz^2},$$

where $y = x_1 - ct$, $z = x_1 + ct$.

9. Longitudinal: $S_{11} = (\lambda + 2\mu)\dfrac{du_1(y)}{dy}$, $S_{22} = S_{33} = \lambda\dfrac{du_1(y)}{dy}$, $y = x_1 - c_\ell t$.

$$\text{Transverse}: \quad S_{12} = \mu\frac{du_2(y)}{dy}, \quad S_{13} = \mu\frac{du_3(y)}{dy}, \quad y = x_1 - c_t t.$$

10. Substitute $u_1 = 0$, $u_2 = -\beta(x_1, t)x_3$, $u_3 = \beta(x_1, t)x_2$ into (7.19b) to obtain

$$\frac{\partial^2 \beta}{\partial t^2} = c_t^2 \frac{\partial^2 \beta}{\partial x_1^2}.$$

11. From (7.17b) and the results of Problem 13 in Section 5.3 (with $\theta = 0$ and $r = x_1$), $\frac{dS_{11}}{dx_1} + \rho\omega^2 x_1 = 0$. Integrating and applying boundary condition $S_{11}(\ell) = 0$ yields $S_{11} = \frac{\rho\omega^2}{2}(\ell^2 - x_1^2)$.

12. The bending moment $M(x_1)$, which counteracts forces acting on the end of a beam, increases with distance from the free end. The maximum bending stress S_{11} in the outer fibers of a cross section is proportional to M so is lowered by increasing the diameter of the cross section.

13. Moment equilibrium of the beam segment in the interval $[\ell - x, \ell]$ yields $M = -W(L-x)$. Integrate (7.26) and apply boundary conditions to obtain

$$u_2(x_1, 0, 0) = \frac{Wx_1^3}{6YI} - \frac{W\ell x_1^2}{2YI}.$$

Problems 7.3

1. Bring time derivative inside integrand and use divergence theorem:

$$\iiint_V \left[\frac{\partial}{\partial t}(\rho\mathbf{v}) + \nabla\cdot(\rho\mathbf{v}\mathbf{v}) - \rho\mathbf{g} - \nabla\cdot\mathbf{S}\right] dV = 0.$$

Use (6.14) and (7.35b) to obtain

$$\iiint_V \left[\rho\frac{\partial\mathbf{v}}{\partial t} + \rho\mathbf{v}\cdot\nabla\mathbf{v} - \rho\mathbf{g} - \nabla\cdot\mathbf{S}\right] dV = 0.$$

The integrand is zero by (7.33) and (7.34).

2. (a) $\dfrac{30 \times \frac{1000}{3600} \times 100}{10^{-6}} = 8.3 \times 10^8$, (b) $\dfrac{800 \times \frac{1000}{3600} \times 3}{1.5 \times 10^{-5}} = 4.4 \times 10^7$,
(c) $\dfrac{\frac{2}{\pi(.005)^2} \times \frac{1}{60} \times \frac{1}{1000} \times .005}{10^{-6}} = 2.1 \times 10^3$ for the 1-cm hose, 1.5×10^3 for the 9-mm hose.

3. The inertial force of a fluid upon your body is proportional to the density ρ of the fluid. You must push aside a mass of fluid that is roughly 10^3 times greater in water than in air.

4. By flapping its wings, a bird adds a component of air velocity normal to its wings. Since dynamic lift is perpendicular to the air velocity and viscous drag is relatively small, a bird in forward flight is thus provided with a component of force in the direction of flight.

5. Center of gravity must be a distance $\ell > 0$ below axis of cylinder. Moment equilibrium leads to $\sin\phi = \frac{M}{mg\ell}$.

6. From (7.43) $\frac{dp}{dx_1} = -\frac{Mg}{RT}p$, so $p = p_0 e^{-Mgx_1/RT}$.

7. Nonzero components are

$$\frac{d\gamma_{12}}{dt} = -\frac{x_2(p_1-p_2)}{4\mu\ell}, \qquad \frac{d\gamma_{13}}{dt} = -\frac{x_3(p_1-p_2)}{4\mu\ell},$$

$$S_{12} = -\frac{x_2(p_1-p_2)}{2\ell}, \qquad S_{13} = -\frac{x_3(p_1-p_2)}{2\ell},$$

$$S_{11} = S_{22} = S_{33} = -p.$$

$$\text{Net force } = \mathbf{e}_1 \int\int_{\substack{\text{pipe}\\\text{surface}}} (S_{12}n_2 + S_{13}n_3)dS = -\pi a^2(p_1-p_2)\mathbf{e}_1,$$

$$4\mu \int\int\int_{\substack{\text{fluid}\\\text{volume}}} \left[\left(\frac{d\gamma_{12}}{dt}\right)^2 + \left(\frac{d\gamma_{13}}{dt}\right)^2\right] dV = \frac{(p_1-p_2)^2}{4\mu\ell}\int_0^{2\pi}\int_0^a r^3 dr d\theta$$

$$= I^2R,$$

$$\int\int_{\substack{\text{pipe}\\\text{ends}}} S_{11}v_1n_1 dS = \frac{p_1}{4\mu\ell}(p_1-p_2)\int_0^{2\pi}\int_0^a (a^2-r^2) r dr d\theta$$

$$-\frac{p_2}{4\mu\ell}(p_1-p_2)\int_0^{2\pi}\int_0^a (a^2-r^2) r dr d\theta = I^2R.$$

8. $v_1 = \frac{1}{4\mu}\left(\frac{p_1-p_2}{\ell} - \rho g\right)(a^2 - x_2^2 - x_3^2)$, $p_1 - p_2 = IR + \rho g\ell$.

9. From (7.46) $\frac{I'}{I} = \frac{R}{R'} = \left(\frac{a'}{a}\right)^4 = \left(\frac{9}{10}\right)^4$, so $I' = .656 \ \frac{\text{liters}}{\text{minute}}$.

10. $p = c - \frac{\rho K^2}{2}(x_1^2 + x_2^2)$. Solve

$$\frac{dx_1}{dt} = Kx_1, \qquad \frac{dx_2}{dt} = -Kx_2$$

to obtain

$$x_1(t) = x_1(0)e^{Kt}, \qquad x_2(t) = x_2(0)e^{-Kt} \qquad \text{(see Fig. 6.15).}$$

11. $$\mathbf{v} = \frac{Vx_2}{\ell}\mathbf{e}_1, \qquad p = \text{constant},$$

$$x_1(t) = \frac{Vtx_2}{\ell} + x_1(0), \qquad x_2 = x_2(0), \qquad x_3 = x_3(0).$$

Nonzero components are

$$\gamma_{12} = \frac{Vt}{2\ell}, \qquad S_{12} = \frac{\mu V}{\ell}, \qquad S_{11} = S_{22} = S_{33} = -p.$$

12. For $r = \sqrt{x_1^2 + x_2^2} \le 1$: $\begin{cases} \mathbf{v} = -x_2\mathbf{e}_1 + x_1\mathbf{e}_2, \\ p = c - \rho + \dfrac{\rho r^2}{2}, \\ r(t) = r(0), \quad \theta(t) = \theta(0) + t. \end{cases}$

For $r \ge 1$: $\begin{cases} \mathbf{v} = \dfrac{-x_2\mathbf{e}_1 + x_1\mathbf{e}_2}{x_1^2 + x_2^2}, \quad p = c - \dfrac{\rho}{2r^2}, \\ r(t) = r(0), \quad \theta(t) = \theta(0) + \dfrac{t}{r^2(0)}. \end{cases}$

13. $p = p_0 - \rho g x_3 + \dfrac{\rho\omega^2}{2}(x_1^2 + x_2^2)$. Requiring pressure to be constant at free surface and total volume of fluid to be unchanged leads to

$$x_3 = \frac{\omega^2}{4g}(2x_1^2 + 2x_2^2 - a^2),$$

where x_3 is height measured from the location of the free surface when the glass and fluid are at rest.

14. $2\rho\boldsymbol{\omega} \times \mathbf{v} + \nabla p = 0$ leads to $2\rho\omega\cos\theta\bar{\mathbf{e}}_1 \times \mathbf{v} + \nabla p = 0$, where θ is the angle between the earth's angular velocity $\boldsymbol{\omega}$ and its outward normal $\bar{\mathbf{e}}_1$. Cross the latter equation with $\bar{\mathbf{e}}_1$ to obtain

$$\mathbf{v} = \frac{\bar{\mathbf{e}}_1 \times \nabla p}{2\rho\omega\cos\theta}.$$

15. Linearize (7.35b) and let $\dfrac{\partial\rho}{\partial t} = \dfrac{\partial p}{\partial t}\Big/\dfrac{dp}{d\rho}$ to obtain: $\dfrac{1}{c^2}\dfrac{\partial p}{\partial t} + \rho\nabla\cdot\mathbf{v} = 0$. Linearize (7.34) and set $\mu = \lambda = 0$ in (7.37) to obtain from (7.33): $\rho\dfrac{\partial\mathbf{v}}{\partial t} = -\nabla p$. Then take the time derivative of the first equation, the divergence of the second equation, and eliminate the term $\rho\nabla\cdot\dfrac{\partial\mathbf{v}}{\partial t}$.

Problems 7.4

1. Let $\mathbf{r}_m$ denote the position vector from the origin of an inertial coordinate system to m, and let $\mathbf{r}_M$ denote the position vector from the same origin to M. Letting $\mathbf{r} = \mathbf{r}_m - \mathbf{r}_M$, we have

$$\frac{d^2\mathbf{r}_m}{dt^2} = -\frac{GM\mathbf{r}}{r^3} \quad \text{and} \quad \frac{d^2\mathbf{r}_M}{dt^2} = \frac{Gm\mathbf{r}}{r^3}, \quad \text{so} \quad \frac{d^2\mathbf{r}}{dt^2} = \frac{-G(M+m)\mathbf{r}}{r^3}.$$

This is the same as (7.55) with M replaced by $M + m$.

2. The equation of the orbit in terms of coordinates (x_1, x_2) along the major and minor axes of the ellipse is $x_1 = a\cos E$, $x_2 = a\sqrt{1-e^2}\sin E$. Then $r = \sqrt{(x_1 - ae)^2 + x_2^2} = a(1 - e\cos E)$. Substitute

$$d\theta = \frac{\sqrt{1-e^2}}{1 - e\cos E}dE \quad \text{in (7.58) to obtain Kepler's equation.}$$

3. Newton's law is $\dfrac{d^2\mathbf{r}}{dt^2} = f\mathbf{r}$. Cross multiplying by $\mathbf{r}$, we obtain

$$\mathbf{r} \times \frac{d^2\mathbf{r}}{dt^2} = \frac{d}{dt}\left(\mathbf{r} \times \frac{d\mathbf{r}}{dt}\right) = 0.$$

Thus $\mathbf{r} \times \dfrac{d\mathbf{r}}{dt} =$ constant $\mathbf{h}$.

4. Period $T = \frac{365.25}{366.25}$ days $=$ 23 h 56 min (see Problem 7 in Section 5.3).

$$a = (GM)^{1/3}\left(\frac{T}{2\pi}\right)^{2/3} \approx 42,200 \text{ km},$$

$$v = \sqrt{\frac{GM}{a}} = 3.07\frac{\text{km}}{s}.$$

5. Escape trajectory is parabola, so from (7.54) and (7.60) with $a = \infty$

$$v_{\text{escape}} = \sqrt{\frac{2GM}{R_{\text{earth}}}} = (4GMg)^{1/4} = 11.2\ \frac{\text{km}}{s}.$$

6. $a = \dfrac{-GM}{v^2} = -1.08 \times 10^6$ km, $e = \csc\left(\dfrac{\delta_2 - \delta_1}{2}\right) = 1.32,$

$r_{\text{perigee}} = a(1-e) = 3.45 \times 10^5$ km,

$$v_{\text{perigee}} = \frac{h}{r_{\text{perigee}}} = \sqrt{\frac{GM(1+e)}{a(1-e)}} = 29.2\ \frac{\text{km}}{s},$$

$$\sqrt{v^2 + \bar{v}^2 - 2v\bar{v}\cos\left(\frac{\delta_1 + \delta_2}{2}\right)} = 36.2\ \frac{\text{km}}{s}.$$

7. Introduce an orthonormal set of base vectors $(\bar{\mathbf{e}}_1, \bar{\mathbf{e}}_2, \bar{\mathbf{e}}_3)$ which move with the satellite so that $\bar{\mathbf{e}}_1$ is the direction of $\mathbf{r}$, $\bar{\mathbf{e}}_2$ is in the orbital plane, and $\bar{\mathbf{e}}_3 = \bar{\mathbf{e}}_1 \times \bar{\mathbf{e}}_2$. The inertial basis $(\mathbf{e}_1, \mathbf{e}_2, \mathbf{e}_3)$ may be transformed into the moving basis $(\bar{\mathbf{e}}_2, \bar{\mathbf{e}}_2, \bar{\mathbf{e}}_3)$ by a succession of three rotations (see (4.2) and (4.13)):

$$\begin{bmatrix} \bar{\mathbf{e}}_1 \\ \bar{\mathbf{e}}_2 \\ \bar{\mathbf{e}}_3 \end{bmatrix} = R_3^t(\theta + \omega)R_1^t(i)R_3^t(\Omega)\begin{bmatrix} \mathbf{e}_1 \\ \mathbf{e}_2 \\ \mathbf{e}_3 \end{bmatrix}.$$

Then use

$$\mathbf{r} = r\bar{\mathbf{e}}_1, \quad \mathbf{v} = \left(\frac{dr}{d\theta}\bar{\mathbf{e}}_1 + r\bar{\mathbf{e}}_2\right)\frac{d\theta}{dt}, \qquad \text{and (7.57)–(7.58).}$$

8. Satellite is in a polar orbit and initial point is either perigee or apogee. Set $i = \dfrac{\pi}{2}$, $\Omega = 0$, $w = 0$ or π in previous problem. Use

$$a = \frac{1}{\dfrac{2}{r_0} - \dfrac{v_0^2}{GM}}, \qquad e = \sqrt{1 - \frac{r_0^2 v_0^2}{GMa}}.$$

9. f causes a change in v equal to $\Delta v = -f\Delta t$. The solution will still be given by the two-body formulas if we choose $\Delta v = \dfrac{GM}{2va^2}\Delta a$ (take the differential of (7.60) with r held constant).

10. In (7.50), if $\mathbf{r} >> \mathbf{r}'$ we can replace $\mathbf{r} - \mathbf{r}'$ by $\mathbf{r}$.

11. See (6.52). Set $M = \frac{4}{3}\pi r^3 \rho$ in (7.54) to obtain $\mathbf{g} = \dfrac{-4\pi G\rho \mathbf{r}}{3}$. $\nabla \mathbf{p} = \rho \mathbf{g}$ leads to $p = c - \dfrac{2\pi G\rho^2 r^2}{3}$.

Problems 7.5

1. Let $\mathbf{r} = \overline{\mathbf{r}} + \mathbf{y}$, where $\mathbf{y}$ is a constant vector. Then

$$\mathbf{p} = \iiint_V \mathbf{r}\rho dV = \iiint_V \overline{\mathbf{r}}\rho dV + \mathbf{y}Q, \text{ from (7.65).}$$

2. Substitute the expansion

$$\frac{1}{|\mathbf{r} - \mathbf{r}'|} = \frac{1}{r}\left(1 + \frac{\mathbf{r}\cdot\mathbf{r}'}{r^2} + \frac{3(\mathbf{r}\cdot\mathbf{r}')^2}{2r^4} - \frac{r'^2}{2r^2} + \cdots\right)$$

into (7.79).

3. $C = \dfrac{q_S A}{\phi_1 - \phi_2} = \dfrac{\epsilon A \mathcal{E}}{\phi_1 - \phi_2} = \dfrac{\epsilon A}{d}$, where A is the surface area of one plate.

4. Equations (7.61a), (7.67), (7.69), and (7.70a) yield $\dfrac{\partial \rho}{\partial t} + \dfrac{\sigma}{\epsilon}\rho = 0$. Solution is $\rho = \rho_0 e^{-\sigma t/\epsilon}$. Decay time $\tau = \dfrac{\epsilon}{\sigma}$.

5. Replace $\mathcal{J} dV'$ by $I d\mathbf{r}'$ in (7.86) to obtain $\mathcal{B}$. Replace $\mathcal{J} dV$ by $I_1 d\mathbf{r}_1$ in (7.71) to obtain $\mathbf{f}$.

6. $\mathbf{m} = \dfrac{I}{2}\oint_C \mathbf{r} \times d\mathbf{r} = \dfrac{I\mathbf{e}_3}{2}\oint_C (x_1 dx_2 - x_2 dx_1) = IA\mathbf{e}_3$,

where A is the area enclosed by the loop, by Problem 8 in Section 6.4.

7. $\sigma \iiint_{\substack{\text{current}\\ \text{volume}}} \mathcal{E}^2 dV = \sigma \mathcal{E}^2 A\ell = I^2 R$,

$-\iiint_{\substack{\text{wire}\\ \text{surface}}} \mathcal{E} \times \mathcal{H} \cdot d\mathbf{S} = \dfrac{1}{\mu}\mathcal{E}B(a)2\pi a\ell = I^2 R.$

8. Faraday's law (7.63b) yields

$$\oint_C \mathcal{E} \cdot d\mathbf{r} = -\frac{d}{dt}(BA\cos\omega t) = \omega BA \sin\omega t,$$

where A is the area enclosed by C.

9. Area ΔA swept out in time Δt is

$$\begin{aligned}\Delta A &= \frac{1}{2}|\mathcal{E} \times \Delta\mathcal{E}| \\ &= \frac{1}{2}|[a\sin(kx_1 - \omega t + \alpha)\mathbf{e}_2 + b\sin(kx_1 - \omega t + \beta)\mathbf{e}_3] \\ &\quad \times[-a\omega\cos(kx_1 - \omega t + \alpha)\mathbf{e}_2 - b\omega\cos(kx_1 - \omega t + \beta)\mathbf{e}_3]\Delta t| \\ &= \frac{1}{2}ab\omega\sin(\beta - \alpha)\Delta t.\end{aligned}$$

The statement also follows from Problem 5 in Section 5.3 and Problem 3 in Section 7.4.

10. Choose $\mathbf{v} = ae^{i\alpha}\mathbf{e}_2 + be^{i\beta}\mathbf{e}_3$. Then (see Problem 12 in Section 3.3) $ae^{i(kx-\omega t+\alpha)}\mathbf{e}_2 + be^{i(kx-\omega t+\beta)}\mathbf{e}_3 = [a\cos(kx - \omega t + \alpha)\mathbf{e}_2$ $+ b\cos(kx - \omega t + \beta)\mathbf{e}_3] + i\,[a\sin(kx - \omega t + \alpha)\mathbf{e}_2 + b\sin(kx - \omega t + \beta)\mathbf{e}_3]$

11. Rate of momentum transferred per unit area $\Delta x_2 \Delta x_3$ for a wave propagating in the x_1 direction is, from (7.74) and (7.96),

$$-\mathbf{e}_1 \cdot \boldsymbol{T} \cdot \mathbf{e}_1 = \frac{1}{2}(\mathcal{E}\mathcal{D} + \mathcal{B}\mathcal{H}) = \frac{\mathcal{E}\mathcal{H}}{c}.$$

Rate of energy transferred per unit area is, from (7.76) and (7.96), $\boldsymbol{\mathcal{E}} \times \boldsymbol{\mathcal{H}} \cdot \mathbf{e}_1 = \mathcal{E}\mathcal{H}$.

12. If the waves are totally absorbed, the pressure is, from the preceding problem,

$$\frac{\mathcal{E}\mathcal{H}}{c} = \frac{1400}{3 \times 10^8} = 4.7 \times 10^{-6}\ \text{Pa} = 4.6 \times 10^{-11}\ \text{atm}.$$

If a portion of the waves is reflected, the pressure will be greater. The greatest possible pressure is, for the case of total reflection, $\frac{2\mathcal{E}\mathcal{H}}{c}$.

13. (7.62a) implies $\mathcal{B} = \boldsymbol{\nabla} \times \mathbf{A}$. Then from (7.63a) $\boldsymbol{\nabla} \times \left(\mathcal{E} + \frac{\partial \mathbf{A}}{\partial t}\right) = 0$, so $\mathcal{E} + \frac{\partial \mathbf{A}}{\partial t} = -\boldsymbol{\nabla}\phi$. Then (7.61a), (7.67), and the Lorentz gauge condition imply $\frac{1}{c^2}\frac{\partial^2\phi}{\partial t^2} - \nabla^2\phi = \frac{\rho}{\epsilon}$. Similarly (7.64a), (7.68), the Lorentz gauge condition, and (6.15) imply $\frac{1}{c^2}\frac{\partial^2 \mathbf{A}}{\partial t^2} - \nabla^2\mathbf{A} = \mu\mathbf{J}$.

Problems 8.1

1. Use (3.9), (3.11), and (4.28).

2. We can interpret $g = gg^{1i}g_{1i}$ as the expansion of the determinant g by the first row, so gg^{1i} is the cofactor of g_{1i}. Since the cofactors of the first row do not contain g_{1i}, $\dfrac{\partial g}{\partial g_{1i}} = gg^{1i}$. Then

$$\frac{\partial g}{\partial x^i} = \frac{\partial g}{\partial g_{kj}} \frac{\partial g_{kj}}{\partial x^i} = gg^{kj} \frac{\partial g_{kj}}{\partial x^i}.$$

Thus from (8.6)

$$\frac{1}{2g} \frac{\partial g}{\partial x^i} = \frac{1}{2} g^{kj} \frac{\partial g_{kj}}{\partial x^i} = \Gamma^j_{ij}.$$

3. Formula for $\nabla \cdot \mathbf{u}$ follows from (8.8), (8.14), and identity in last problem. Formula for $\nabla^2 f$ is obtained by letting

$$\mathbf{u} = \nabla f = g^{ij} \frac{\partial f}{\partial x^j} \mathbf{g}_i.$$

4. R_{1212}, R_{1313}, R_{2323}, R_{1213}, R_{1223}, R_{1323}.

5. $\oint_C u_i dx^i = \int\int_S e^{ijk} u_{j,i} n_k dS$, $\oiint_S u_i n^i dS = \int\int\int_V u^i{}_{,i} dV$.

6. Differentiate (8.25) to obtain

$$\frac{\partial \bar{u}_i}{\partial \bar{x}^j} = \frac{\partial x^k}{\partial \bar{x}^i} \frac{\partial x^\ell}{\partial \bar{x}^j} \frac{\partial u_k}{\partial x^\ell} + \underline{\frac{\partial^2 x^k}{\partial \bar{x}^i \partial \bar{x}^j} u_k}.$$

From (8.4),

$$\bar{\Gamma}^k_{ij} = \bar{\mathbf{g}}^k \cdot \frac{\partial \bar{\mathbf{g}}_i}{\partial \bar{x}^j} = \bar{\mathbf{g}}^k \cdot \frac{\partial}{\partial \bar{x}^j} \left(\frac{\partial x^r}{\partial \bar{x}^i} \mathbf{g}_r \right) = \frac{\partial \bar{x}^k}{\partial x^\ell} \mathbf{g}^\ell \cdot \left(\frac{\partial x^m}{\partial \bar{x}^i} \frac{\partial x^n}{\partial \bar{x}^j} \Gamma^r_{mn} \mathbf{g}_r + \frac{\partial^2 x^r}{\partial \bar{x}^i \partial \bar{x}^j} \mathbf{g}_r \right)$$

$$= \frac{\partial \bar{x}^k}{\partial x^\ell} \frac{\partial x^m}{\partial \bar{x}^i} \frac{\partial x^n}{\partial \bar{x}^j} \Gamma^\ell_{mn} + \underline{\frac{\partial \bar{x}^k}{\partial x^\ell} \frac{\partial^2 x^\ell}{\partial \bar{x}^i \partial \bar{x}^j}}.$$

These differ from the correct laws by the underlined terms. But

$$\bar{u}_{i,j} = \frac{\partial \bar{u}_i}{\partial \bar{x}^j} - \bar{\Gamma}^k_{ij} \bar{u}_k = \frac{\partial x^k}{\partial \bar{x}^i} \frac{\partial x^\ell}{\partial \bar{x}^j} \left(\frac{\partial u_k}{\partial x^\ell} - \Gamma^m_{k\ell} u_m \right) = \frac{\partial x^k}{\partial \bar{x}^i} \frac{\partial x^\ell}{\partial \bar{x}^j} u_{k,\ell}.$$

7. These relations are obviously true in a Cartesian coordinate system and hence in all coordinate systems, since the equations involve only the components of tensors.

8. Set $f^i = 0$ in (8.27): $\dfrac{d^2 x^i}{ds^2} + \Gamma^i_{jk} \dfrac{dx^j}{ds} \dfrac{dx^k}{ds} = 0.$

9. In (7.39b) replace $\dfrac{\partial v_i}{\partial x_j}$ by $v_{i,j}$, raise the index on v_j, and replace $\dfrac{\partial^2 v_i}{\partial x_j \partial x_j}$ by $g^{jk} v_{i,jk}$ to obtain $\rho \dfrac{\partial v_i}{\partial t} + \rho v_{i,j} v^j = \rho g_i - p_{,i} + \mu g^{jk} v_{i,jk}$.

Problems 8.2

1. A single-valued inverse of the transformation (8.1) exists where $\mathbf{g}_1 \times \mathbf{g}_2 \cdot \mathbf{g}_3 \neq 0$. In orthogonal coordinates, $\mathbf{g}_1 \times \mathbf{g}_2 \cdot \mathbf{g}_3 = h_1 h_2 h_3$.

Cylindrical (8.36) : $\mathbf{g}_1 \times \mathbf{g}_2 \cdot \mathbf{g}_3 = \rho$, so for $x_1^2 + x_2^2 + x_3^2 > 0$,

$$\rho = \sqrt{x_1^2 + x_2^2}, \quad \theta = \tan^{-1}\frac{x_1}{x_2}, \qquad z = x_3.$$

Spherical (8.40) : $\mathbf{g}_1 \times \mathbf{g}_2 \cdot \mathbf{g}_3 = r^2 \sin\phi$, so for $x_1^2 + x_2^2 > 0$,

$$r = \sqrt{x_1^2 + x_2^2 + x_3^2}, \quad \phi = \tan^{-1}\frac{\sqrt{x_1^2 + x_2^2}}{x_3}, \quad \theta = \tan^{-1}\frac{x_2}{x_1}.$$

2. The physical components of the velocity vector in polar coordinates are

$$v_\rho = \frac{d\rho}{dt} = -\frac{v}{\sqrt{2}}, \quad v_\theta = \rho\frac{d\theta}{dt} = \frac{v}{\sqrt{2}}.$$

Integrating and applying the initial conditions

$$\rho(0) = \frac{\ell}{\sqrt{2}}, \quad \theta(0) = 0,$$

yields

$$\rho = \frac{\ell - vt}{\sqrt{2}}, \quad t = \frac{\ell(1 - e^{-\theta})}{v}.$$

Thus this ant moves along the logarithmic spiral $\rho = \dfrac{\ell e^{-\theta}}{\sqrt{2}}$ with ever increasing angular speed $\dfrac{d\theta}{dt} = \dfrac{ve^{\theta}}{\ell}$.

3. If $T = T(r)$, Laplace's equation in spherical coordinates is

$$\frac{1}{r^2}\frac{d}{dr}\left(r^2\frac{dT}{dr}\right) = 0.$$

This implies $T = \dfrac{C_1}{r} + C_2$. The boundary conditions $T(a_1) = T_1$ and $T(a_2) = T_2$ lead to

$$T = \frac{T_1\left(\dfrac{1}{r} - \dfrac{1}{a_2}\right) + T_2\left(\dfrac{1}{a_1} - \dfrac{1}{r}\right)}{\dfrac{1}{a_1} - \dfrac{1}{a_2}}.$$

4. The nonvanishing components are

$$u_\theta = \frac{Mz\rho}{\mu J}, \qquad T_{\theta z} = \frac{M\rho}{J}.$$

5. If $p = p(r, t)$, the wave equation in spherical coordinates becomes

$$\frac{\partial^2 p}{\partial t^2} = \frac{c^2}{r^2}\frac{\partial}{\partial r}\left(r^2 \frac{\partial p}{\partial r}\right).$$

This can be rewritten as

$$\frac{\partial^2 (rp)}{\partial t^2} = c^2 \frac{\partial^2 (rp)}{\partial r^2},$$

which is the one-dimensional wave equation for rp. The most general solution is (see Problem 8 in Section 7.2) $rp = f(r - ct) + g(r + ct)$. Throwing away the inward traveling wave, we obtain

$$p = \frac{f(r - ct)}{r}.$$

6. Use (8.27) to obtain Newton's law in spherical coordinates:

$$\frac{d^2 r}{dt^2} - r\left(\frac{d\phi}{dt}\right)^2 - r\sin^2\phi\left(\frac{d\theta}{dt}\right)^2 = -\frac{GM}{r^2},$$

$$\frac{d^2\phi}{dt^2} + \frac{2}{r}\frac{dr}{dt}\frac{d\phi}{dt} - \sin\phi\cos\phi\left(\frac{d\theta}{dt}\right)^2 = 0,$$

$$\frac{d^2\theta}{dt^2} + \frac{2}{r}\frac{dr}{dt}\frac{d\theta}{dt} + 2\cot\phi\frac{d\phi}{dt}\frac{d\theta}{dt} = 0.$$

The second equation is satisfied by $\phi = \frac{\pi}{2}$ (corresponding to a choice of the vertical coordinate axis through the angular momentum vector $\mathbf{h}$). The last equation then becomes

$$\frac{d}{dt}\left(r^2\frac{d\theta}{dt}\right) = 0,$$

and the solution is

$$\frac{d\theta}{dt} = \frac{h}{r^2}$$

or (7.58). The first equation becomes

$$\frac{d^2 r}{dt^2} - \frac{h^2}{r^3} = -\frac{GM}{r^2},$$

substituting $u = \dfrac{h^2}{GMr}$ reduces this to

$$\frac{d^2 u}{d\theta^2} + u = 1,$$

and the solution is $u = 1 + e\cos\theta$ or (7.57).

7. $\boldsymbol{\mathcal{E}} = \dfrac{2\cos\phi\mathbf{e}_r + \sin\phi\mathbf{e}_\phi}{r^3}$.

$(dr\mathbf{e}_r + rd\phi\mathbf{e}_\phi) \times \boldsymbol{\mathcal{E}} = 0$ leads to $\dfrac{dr}{r} = 2\cot\phi d\phi$. Integrate to obtain field lines $r = C\sin^2\phi$; see Fig. 7.16.

8. By symmetry $\boldsymbol{\mathcal{E}} = \mathcal{E}(\rho)\mathbf{e}_\rho$. Apply (7.61b) to an imaginary cylindrical surface coaxial with a length ℓ of the line: $2\pi\rho\ell\epsilon\mathcal{E} = \lambda\ell$. Hence

$$\boldsymbol{\mathcal{E}} = \frac{\lambda\mathbf{e}_\rho}{2\pi\epsilon\rho}.$$

9.
$$\mathbf{B} = \frac{\mu}{4\pi}\int_{-\infty}^{\infty}\frac{Idz\mathbf{e}_z \times (\rho\mathbf{e}_r - z\mathbf{e}_z)}{(\rho^2+z^2)^{3/2}} = \frac{\mu I\rho\mathbf{e}_\theta}{4\pi}\int_{-\infty}^{\infty}\frac{dz}{(\rho^2+z^2)^{3/2}}$$
$$= \frac{\mu I\mathbf{e}_\theta}{4\pi\rho}\left[\frac{z}{(\rho^2+z^2)^{1/2}}\right]_{z=-\infty}^{z=+\infty} = \frac{\mu I\mathbf{e}_\theta}{2\pi\rho}.$$

10. The (x, y) coordinate curves in a plane z = constant are confocal parabolas with a common axis.

$$h_1 = h_2 = \sqrt{x^2+y^2}, \qquad h_3 = 1,$$
$$\mathbf{e}_x = \frac{x\mathbf{e}_1 + y\mathbf{e}_2}{\sqrt{x^2+y^2}}, \qquad \mathbf{e}_y = \frac{-y\mathbf{e}_1 + x\mathbf{e}_2}{\sqrt{x^2+y^2}}, \qquad \mathbf{e}_z = \mathbf{e}_3.$$

Problems 8.3

1. At the origin of a geodesic coordinate system on the surface, $\boldsymbol{\nabla}\times\mathbf{u} = \epsilon_{\alpha\beta}\dfrac{\partial u^\beta}{\partial x^\alpha}\mathbf{n}$. The generalization of this to arbitrary surface coordinates is $\boldsymbol{\nabla}\times\mathbf{u} = e^{\alpha\beta}u_{\beta,\alpha}\mathbf{n}$. Use (8.49) to obtain

$$\begin{aligned}\mathbf{T}_{,\alpha} &= \left(T^{\beta\gamma}\mathbf{g}_\beta\mathbf{g}_\gamma\right)_{,\alpha} = T^{\beta\gamma}_{,\alpha}\mathbf{g}_\beta\mathbf{g}_\gamma + T^{\beta\gamma}\mathbf{g}_{\beta,\alpha}\mathbf{g}_\gamma + T^{\beta\gamma}\mathbf{g}_\beta\mathbf{g}_{\gamma,\alpha}\\ &= T^{\beta\gamma}{}_{,\alpha}\mathbf{g}_\beta\mathbf{g}_\gamma + K_{\alpha\beta}T^{\beta\gamma}\mathbf{n}\mathbf{g}_\gamma + K_{\alpha\gamma}T^{\beta\gamma}\mathbf{g}_\beta\mathbf{n}.\end{aligned}$$

2. Apply (6.38) to surface vectors $\mathbf{u}(x^1, x^2) = u_\alpha\mathbf{g}^\alpha$, using the expression for $\boldsymbol{\nabla}\times\mathbf{u}$ in the previous problem, to obtain the surface Stokes' theorem $\oint_C u_\alpha dx^\alpha = \iint_S e^{\alpha\beta}u_{\beta,\alpha}dS$. Then let $u_\alpha = e_{\gamma\alpha}v^\gamma$ to obtain the surface divergence theorem $\oint_C e_{\alpha\beta}v^\alpha dx^\beta = \iint_S v^\alpha_{,\alpha}dS$.

3. Surface area, geodesic curvature.

4. $R_{1212} = -R_{2112} = -R_{1221} = R_{2121}$.

5. The Gaussian curvature of a sphere of radius a is $\dfrac{1}{a^2} \neq 0$, so a sphere is not a developable surface.

6. Take the limit of (8.72) as $\rho \to 0$ with $\dfrac{d\rho}{d\ell} \to 1$, to obtain $K_{\ell\ell} = K_{\theta\theta} = \dfrac{d^2x_3}{d\ell^2}$.

7. Using (8.10), (8.20), (8.28), (8.66), we can reduce (8.57)–(8.58) to the form

$$\frac{\partial K_{11}}{\partial x^2} - \frac{1}{h_1}\frac{\partial h_1}{\partial x_2}K_{11} = \frac{h_1}{(h_2)^2}\frac{\partial h_1}{\partial x^2}K_{22},$$

$$\frac{\partial K_{22}}{\partial x^1} - \frac{1}{h_2}\frac{\partial h_2}{\partial x^1}K_{22} = \frac{h_2}{(h_1)^2}\frac{\partial h_2}{\partial x^1}K_{11},$$

$$K_G = -\frac{1}{h_1h_2}\left[\frac{\partial}{\partial x^1}\left(\frac{1}{h_1}\frac{\partial h_2}{\partial x_1}\right) + \frac{\partial}{\partial x^2}\left(\frac{1}{h_2}\frac{\partial h_1}{\partial x^2}\right)\right].$$

Introducing physical components via (4.30), and using (8.51), we obtain

$$\frac{\partial(h_1\bar{K}_{11})}{\partial x^2} = \bar{K}_{22}\frac{\partial h_1}{\partial x^2}, \quad \frac{\partial(h_2\bar{K}_{22})}{\partial x^1} = \bar{K}_{11}\frac{\partial h_2}{\partial x^1}, \quad K_G = \bar{K}_{11}\bar{K}_{22}.$$

For a surface of revolution, using (8.67) and (8.69), we obtain the identities

$$\frac{\partial K_{\ell\ell}}{\partial \theta} = 0, \qquad \frac{d^2x_3}{d\ell^2} = \left(\frac{d\rho}{d\ell}\frac{d^2x_3}{d\ell^2} - \frac{d^2\rho}{d\ell^2}\frac{dx^3}{d\ell}\right)\frac{d\rho}{d\ell},$$

$$K_G = \left(\frac{d\rho}{d\ell}\frac{d^2x_3}{d\ell^2} - \frac{d^2\rho}{d\ell^2}\frac{dx_3}{d\ell}\right)\left(\frac{1}{\rho}\frac{dx_3}{d\ell}\right) = -\frac{1}{\rho}\frac{d^2\rho}{d\ell^2}.$$

8. Use the same method as for Problem 1 in Section 5.4 to verify formulas, noting that along the geodesic $\dfrac{d\bar{e}_3}{ds}\cdot\bar{e}_2 = -\tau_g$.

9. The vector $\dfrac{d\bar{e}_1}{ds}$ for a great circle points to the center of the sphere and hence is normal to the spherical surface, so from (8.61) $\kappa_g = 0$. Similarly, the vector $\dfrac{d\bar{e}_1}{ds}$ for a circular helix points towards the axis of the cylinder and hence is normal to the cylindrical surface (see the example in Section 5.4).

10. $g_{11}(x^1,0) = \dfrac{\partial \mathbf{r}}{\partial x^1}(x^1,0)\cdot\dfrac{\partial \mathbf{r}}{\partial x^1}(x^1,0) = 1$ because x^1 is arc length along the x^1 coordinate curve; similarly $g_{22}(0,x^2) = 1$, $g_{12}(x^1,x^2) = 0$ because the coordinate system is orthogonal. Thus

$$\frac{\partial g_{11}}{\partial x^1}(x^1,0) = \frac{\partial g_{22}}{\partial x^2}(0,x^2) = \frac{\partial g_{12}}{\partial x^1}(x^1,x^2) = \frac{\partial g_{12}}{\partial x^2}(x^1,x^2) = 0.$$

Along the x^1 coordinate curve $\dfrac{dx^1}{ds} = 1$ and $\dfrac{dx^2}{ds} = 0$, so from (8.64) $\Gamma^1_{11}(x^1,0) = \Gamma^2_{11}(x^1,0) = 0$, similarly $\Gamma^2_{22}(0,x^2) = \Gamma^1_{22}(0,x^2) = 0$. From (8.28) this implies $\dfrac{\partial g_{11}}{\partial x^2}(x^1,0) = \dfrac{\partial g_{22}}{\partial x^1}(0,x^2) = 0$ and thus $\Gamma^1_{12}(x^1,0) = \Gamma^2_{12}(0,x^2) = 0$.

11. These relations are obviously true in a geodesic coordinate system and hence in all coordinate systems, since the equations involve only the components of surface tensors.

12. We must find the values of $e^\alpha = \dfrac{dx^\alpha}{ds}$ which maximize or minimize $\kappa_n = K_{\alpha\beta}e^\alpha e^\beta$. Since $h = g_{\alpha\beta}e^\alpha e^\beta - 1 = 0$, the extremum directions satisfy (see Problem 11 in Section 6.1) $\dfrac{\partial \kappa_n}{\partial e^\alpha} = \lambda \dfrac{\partial h}{\partial e^\alpha}$ or $\kappa_{\alpha\beta}e^\beta = \lambda g_{\alpha\beta}e^\beta = \kappa_n e_\alpha$. This is the equation for the eigenvectors of $\mathbf{K}$ (analogous to 3.20).

13. Since $ds^2 = g_{\alpha\beta}dx^\alpha dx^\beta = dx_1^2 + dx_2^2 + dx_3^2 = (a^2\cos^2\delta + b^2\sin^2\delta)d\delta^2$ $+a^2\sin^2\delta d\theta^2$, $g_{11} = a^2\cos^2\delta + b^2\sin^2\delta$, $g_{12} = 0$, $g_{22} = a^2\sin^2\delta$. Using $\rho = a\sin\delta$, $x_3 = b\cos\delta$, and $d\ell = (a^2\cos^2\delta + b^2\sin^2\delta)^{1/2}d\delta$ in (8.72), we obtain

$$K_{\ell\ell} = \frac{-ab}{(a^2\cos^2\delta + b^2\sin^2\delta)^{3/2}}, \quad K_{\ell\theta} = 0, \quad K_{\theta\theta} = \frac{-b}{a(a^2\cos^2\delta + b^2\sin^2\delta)^{1/2}}.$$

Probems 8.4

1. Consider an elemental section of the membrane having two sides of lengths $(\Delta x^1, \Delta x^2)$ coincident with the axes of a geodesic coordinate system. Require the sum of the moments in the direction normal to the surface caused by the stresses acting on the element to be zero (see Fig. 3.9): $T_{21}\Delta x^1\Delta x^2 - T_{12}\Delta x^1\Delta x^2 = 0$. Thus $T_{21} = T_{12}$.

2. For a sphere of radius a, from problem 13 in Section 8.3, $K_{\ell\ell} = K_{\theta\theta}$ $= -\dfrac{1}{a}$. Remembering (8.50) and (8.76), and noting that there are two air-water interfaces, we obtain $\Delta p = \dfrac{4T}{a} = 5600$ Pa ≈ 0.06 atm.

3. For $\mathbf{r}(\rho,\theta) = \rho\cos\theta\mathbf{e}_1 + \rho\sin\theta\mathbf{e}_2 + b\theta\mathbf{e}_3$, (8.44)–(8.46) and (8.48) imply $K_{11} = K_{22} = g_{12} = 0$, so $K_M = \frac{1}{2}(K_1^1 + K_2^2) = 0$.

4. From (8.72) and (8.81) with $C = 0$: $T_{\ell\ell} = -\dfrac{p}{2K_{\theta\theta}}$, $T_{\theta\theta} = -\dfrac{p}{K_{\theta\theta}}\left(1 - \dfrac{K_{\ell\ell}}{2K_{\theta\theta}}\right)$.

5. From (8.80) and the previous problem, the nonvanishing components are

$$\text{Spherical shell}: T_{\ell\ell} = T_{\theta\theta} = \frac{pa}{2}, \qquad u_3 = \frac{pa^2(1-\nu)}{2wY}.$$

$$\text{Cylindrical shell}: T_{\theta\theta} = pa, \qquad u_\ell = \frac{-\nu pa}{wY}\ell, \qquad u_3 = \frac{pa^2}{wY}$$

6. Guess that the displacement field in the $(x_1\text{-}x_2)$ system is $\mathbf{u} = \dfrac{ux_2}{L}\mathbf{e}_1$, and that the tension rays are parallel straight lines inclined at an angle α to the x_1 axis (see Fig. 8.11). It follows that the displacement components in the rotated (x^1, x^2) system are $u_1 = \dfrac{u}{L}(x^1\sin\alpha\cos\alpha + x^2\cos^2\alpha)$ and u_2 $= -\dfrac{u}{L}(x^1\sin^2\alpha + x^2\sin\alpha\cos\alpha)$. Then (8.86) imply $\alpha = \dfrac{\pi}{4}$ and $T_{11} = \dfrac{wYu}{2L}$.

7. By symmetry, the tension rays are the meridians $\theta =$ constant, $T_{\ell\ell}$ is independent of θ, and $u_\theta = 0$. Since for a sphere $h_2 = \rho = a \sin \dfrac{\ell}{a}$, from (8.85) the membrane stress in a tension ray is $T_{\ell\ell} = \dfrac{C}{h_2} = \dfrac{C}{\sin \frac{\ell}{a}}$, where C is a constant. Equations (8.86) become $\dfrac{du_\ell}{d\ell} = \dfrac{T_{\ell\ell}}{wY} = \dfrac{C}{wY \sin \frac{\ell}{a}}$. Integrating and applying $u_\ell\left(\dfrac{\pi a}{2}\right) = 0$ and $u_\ell(\ell_1) = -u$ yields

$$u_\ell = -\frac{u \ln\left(\cot \dfrac{\ell}{2a}\right)}{\ln\left(\cot \dfrac{\ell_1}{2a}\right)}, \qquad C = \frac{wYu}{a \ln\left(\cot \dfrac{\ell_1}{2a}\right)}.$$

From (8.84), the pressure on the membrane is $p = \dfrac{T_{\ell\ell}}{a}$.

Problems 9.1

1. m^n.

2. $|\mathbf{u}| = \sqrt{|u_0^2 - u_i u_i|}$. If $u_0 = -v_0$ and $u_1 = -v_1$ with other components zero, $|\mathbf{u} + \mathbf{v}| \leq |\mathbf{u}| + |\mathbf{v}|$. If $u_0 = u_1 = v_0$ with other components zero, $|\mathbf{u} + \mathbf{v}| > |\mathbf{u}| + |\mathbf{v}|$.

3. Check (9.5) to show that (9.6) represents a Lorentz transformation:

$$\eta_{\alpha\beta}\Lambda_0^\alpha\Lambda_0^\beta = \frac{1}{1 - \dfrac{v^2}{c^2}} - \frac{v^i v_i}{c^2 - v^2} = 1$$

$$\eta_{\alpha\beta}\Lambda_0^\alpha\Lambda_i^\beta = -\frac{v_i}{c\left[1 - \dfrac{v^2}{c_2}\right]} + \frac{1}{c\sqrt{1 - \dfrac{v^2}{c^2}}}\left[v_i + \left(\frac{1}{\sqrt{1 - \dfrac{v^2}{c^2}}} - 1\right)\frac{v_i v^j v_j}{v^2}\right] = 0,$$

$$\eta_{\alpha\beta}\Lambda_i^\alpha\Lambda_j^\beta = \frac{v_i v_j}{c^2 - v^2} - \left[\delta_{ik} + \left(\frac{1}{\sqrt{1 - \dfrac{v^2}{c^2}}} - 1\right)\frac{v_i v_k}{v^2}\right]$$

$$\times \left[\delta_{jk} + \left(\frac{1}{\sqrt{1 - \dfrac{v^2}{c^2}}} - 1\right)\frac{v_j v_k}{v^2}\right] = -\delta_{ij} = \eta_{ij}.$$

If particle has velocity components

$$\frac{dx_1}{dt} = v, \qquad \frac{dx_2}{dt} = \frac{dx_3}{dt} = 0,$$

then from (9.8), $\dfrac{d\bar{x}_i}{d\bar{t}} = 0.$

4. $\dfrac{10^{-6}}{\sqrt{1-(0.99)^2}} = 7 \times 10^{-6}\,\text{s}.$

5. (9.1) is the chain rule: $\dfrac{d\bar{x}^\alpha}{d\tau} = \dfrac{\partial \bar{x}^\alpha}{\partial x^\beta}\dfrac{dx^\beta}{d\tau}$

6. $\left(\dfrac{dx_1}{dt}\right)^2 + \left(\dfrac{dx_2}{dt}\right)^2 + \left(\dfrac{dx_3}{dt}\right)^2 =$

$$= \frac{\left(v + \dfrac{d\bar{x}_1}{d\bar{t}}\right)^2 + \left(1 - \dfrac{v^2}{c^2}\right)\left[\left(\dfrac{d\bar{x}_2}{d\bar{t}}\right)^2 + \left(\dfrac{d\bar{x}_3}{d\bar{t}}\right)^2\right]}{\left(1 + \dfrac{v}{c^2}\dfrac{d\bar{x}_1}{d\bar{t}}\right)^2}$$

$$= \frac{\left(v + \dfrac{d\bar{x}_1}{d\bar{t}}\right)^2 + \left(1 - \dfrac{v^2}{c^2}\right)\left[c^2 - \left(\dfrac{d\bar{x}_1}{d\bar{t}}\right)^2\right]}{\left(1 + \dfrac{v}{c^2}\dfrac{d\bar{x}_1}{d\bar{t}}\right)^2} = c^2.$$

7. Multiply spatial part by $\dfrac{dx_i}{d\tau}$ to obtain

$$\frac{m}{2}\frac{d}{d\tau}\left(\frac{dx^i}{d\tau}\frac{dx_i}{d\tau}\right) = \frac{m}{2}\frac{d}{d\tau}\left(\frac{v^2}{1 - \dfrac{v^2}{c^2}}\right)$$

$$= \frac{1}{\sqrt{1 - \dfrac{v^2}{c^2}}}\frac{d}{d\tau}\left(\frac{mc^2}{\sqrt{1 - \dfrac{v^2}{c^2}}}\right) = f^i\frac{dx_i}{d\tau}$$

or

$$\frac{d}{d\tau}\left(\frac{mc}{\sqrt{1 - \dfrac{v^2}{c^2}}}\right) = f^0.$$

8. (9.12) reduces to

$$\frac{d}{dt}\left(\frac{mv}{\sqrt{1-\frac{v^2}{c^2}}}\right)=F.$$

Integrating yields

$$x_1=\frac{mc^2}{F}\left(\sqrt{1+\frac{F^2t^2}{m^2c^2}}-1\right).$$

Note that $x_1 \to \frac{Ft^2}{2m}$ as $t \to 0$ (the Newtonian result), and that $x_1 \to ct$ as $t \to \infty$ (the particle speed approaches c).

9. From (9.9) and (9.14),

$$\frac{d}{d\tau}\left(m\frac{dx^\alpha}{d\tau}\right)=qF_\beta{}^\alpha\frac{dx^\beta}{d\tau}.$$

The spatial components of this equation imply

$$\frac{d}{dt}\left(\frac{m}{\sqrt{1-\frac{v^2}{c^2}}}\frac{dx_1}{dt}\right)=q\left(\mathcal{E}_1+\frac{dx_2}{dt}\mathcal{B}_3-\frac{dx_3}{dt}\mathcal{B}_2\right),\ \text{etc.}$$

10. From the preceding problem

$$m\frac{d}{dt}\left(\frac{1}{\sqrt{1-\frac{v^2}{c^2}}}\right)\mathbf{v}+\frac{m}{\sqrt{1-\frac{v^2}{c^2}}}\frac{d\mathbf{v}}{dt}=q\mathbf{v}\times\boldsymbol{\mathcal{B}}.$$

Since $\mathbf{v}\times\boldsymbol{\mathcal{B}}$ is perpendicular to $\mathbf{v}$, $\frac{dv}{dt}=0$ and the particle moves in a circle with constant speed v and centripetal acceleration κv^2. Then

$$\frac{m\kappa v^2}{\sqrt{1-\frac{v^2}{c^2}}}=qv\mathcal{B}\quad\text{with}\quad v=\frac{w}{\kappa}$$

leads to

$$\frac{w}{2\pi}=\frac{q\mathcal{B}\sqrt{1-\frac{v^2}{c^2}}}{2\pi m}.$$

11. $\frac{\partial^2F^{\alpha\beta}}{c^2\partial t^2}-\frac{\partial^2F^{\alpha\beta}}{\partial x_i\partial x_i}=0$

are equivalent to the Cartesian components of the wave equations (7.93) and (7.94).

12. See Problem 13 in Section 7.5: $\Box^2 A^\alpha = \mu J^\alpha$.

13. Let (ρ, θ, z) be a fixed coordinate system with z measured along the axis of the wire, and let $(\rho, \theta, \bar{z})$ be a cylindrical coordinate system moving with the wire. From problem 8 in Section 8.2,

$$\boldsymbol{\mathcal{E}} = \frac{\lambda \mathbf{e}_\rho}{2\pi\epsilon\rho} \quad \text{and} \quad \boldsymbol{B} = 0.$$

From the transformation laws (9.15)–(9.17),

$$\bar{\boldsymbol{\mathcal{E}}} = \frac{\lambda \mathbf{e}_\rho}{2\pi\epsilon\sqrt{1 - \dfrac{v^2}{c^2}}\,\rho} \quad \text{and} \quad \bar{\boldsymbol{B}} = \frac{-v\lambda\mu \mathbf{e}_\theta}{2\pi\sqrt{1 - \dfrac{v^2}{c^2}}\,\rho}.$$

From (9.18) and (9.19),

$$\bar{\lambda} = \frac{\lambda}{\sqrt{1 - \dfrac{v^2}{c^2}}} \quad \text{and} \quad \bar{I} = \frac{-v\lambda}{\sqrt{1 - \dfrac{v^2}{c^2}}}.$$

Thus

$$\bar{\boldsymbol{\mathcal{E}}} = \frac{\bar{\lambda} \mathbf{e}_\rho}{2\pi\epsilon\rho} \quad \text{and} \quad \bar{\boldsymbol{B}} = \frac{\mu \bar{I} \mathbf{e}_\theta}{2\pi\rho}$$

(in agreement with Problem 9 in Section 8.2).

Problems 9.2

1. With the use of the chain rule (9.9) becomes

$$\frac{d}{d\tau}\left(m\frac{\partial x^\alpha}{\partial \bar{x}^\beta}\frac{d\bar{x}^\beta}{d\tau}\right) = f^\alpha.$$

Expand and multiply by $\dfrac{\partial \bar{x}^\delta}{\partial x^\alpha}$ to obtain

$$\frac{d}{d\tau}\left(m\frac{d\bar{x}^\delta}{d\tau}\right) + m\frac{\partial \bar{x}^\delta}{\partial x^\alpha}\frac{\partial^2 x^\alpha}{\partial \bar{x}^\beta \partial \bar{x}^\gamma}\frac{d\bar{x}^\beta}{d\tau}\frac{d\bar{x}^\gamma}{d\tau} = \bar{f}^\delta.$$

Now

$$\bar{g}_{\alpha\beta}(\bar{O}) = \frac{\partial x^\lambda}{\partial \bar{x}^\alpha}(\bar{O})\frac{\partial x^\delta}{\partial \bar{x}^\beta}(\bar{O})\eta_{\lambda\delta},$$

$$\frac{\partial \bar{g}_{\alpha\beta}}{\partial \bar{x}^\gamma}(\bar{O}) = \frac{\partial^2 x^\lambda}{\partial \bar{x}^\alpha \partial \bar{x}^\gamma}(\bar{O})\frac{\partial x^\delta}{\partial \bar{x}^\beta}(\bar{O})\eta_{\lambda\delta} + \frac{\partial x^\lambda}{\partial \bar{x}^\alpha}(\bar{O})\frac{\partial^2 x^\delta}{\partial \bar{x}^\beta \partial \bar{x}^\gamma}(\bar{O})\eta_{\lambda\delta},$$

$$\bar{g}^{\alpha\beta}(\bar{O}) = \frac{\partial \bar{x}^\alpha}{\partial x^\lambda}(O)\frac{\partial \bar{x}^\beta}{\partial x^\delta}(O)\eta^{\lambda\delta},$$

so

$$\bar{\Gamma}^{\alpha}_{\beta\gamma}(\bar{O}) = \frac{1}{2}\bar{g}^{\alpha\delta}(\bar{O})\left(\frac{\partial \bar{g}_{\beta\delta}}{\partial \bar{x}^{\gamma}}(\bar{O}) + \frac{\partial \bar{g}_{\gamma\delta}}{\partial \bar{x}^{\beta}}(\bar{O}) - \frac{\partial \bar{g}_{\beta\gamma}}{\partial \bar{x}^{\delta}}(\bar{O})\right)$$

$$= \frac{\partial \bar{x}^{\alpha}}{\partial x^{\lambda}}(O)\frac{\partial^2 x^{\lambda}}{\partial \bar{x}^{\beta}\partial \bar{x}^{\gamma}}(\bar{O}),$$

in agreement with (8.7). Thus (9.21) is valid at $\bar{O}$.

2. With use of the chain rule (9.13) becomes

$$\frac{\partial}{\partial x^{\beta}}\left(\bar{F}^{\gamma\delta}\frac{\partial x^{\alpha}}{\partial \bar{x}^{\gamma}}\frac{\partial x^{\beta}}{\partial \bar{x}^{\delta}}\right) = \mu J^{\alpha}.$$

Expand and multiply by $\dfrac{\partial \bar{x}^{\lambda}}{\partial x^{\alpha}}$ to obtain

$$\frac{\partial \bar{F}^{\lambda\delta}}{\partial \bar{x}^{\delta}} + \bar{F}^{\gamma\delta}\frac{\partial^2 x^{\alpha}}{\partial \bar{x}^{\gamma}\partial \bar{x}^{\delta}}\frac{\partial \bar{x}^{\lambda}}{\partial x^{\alpha}} + \bar{F}^{\lambda\delta}\frac{\partial^2 x^{\beta}}{\partial \bar{x}^{\delta}\bar{x}^{\alpha}}\frac{\partial \bar{x}^{\alpha}}{\partial x^{\beta}} = \mu \bar{J}^{\lambda}.$$

From (8.7),

$$\frac{\partial \bar{F}^{\lambda\delta}}{\partial \bar{x}^{\delta}}(\bar{O}) + \bar{\Gamma}^{\lambda}_{\gamma\delta}(\bar{O})\bar{F}^{\gamma\delta}(\bar{O}) + \bar{\Gamma}^{\delta}_{\gamma\delta}(\bar{O})\bar{F}^{\lambda\gamma}(\bar{O}) = \mu \bar{J}^{\lambda}(\bar{O}).$$

Thus the first set of (9.23) is valid at $\bar{O}$. Use a similar argument to obtain the second set of (9.23).

3. From (4.3) and (5.16), $\bar{x}^i = R_j{}^i(t)x^j - \bar{y}^i(t)$. Since

$$\Delta\tau = \sqrt{1 - \frac{v^2}{c^2}}\Delta t \approx \Delta t,$$

the spatial part of (9.21) reduces to

$$m\frac{d^2\bar{x}^i}{dt^2} + m\bar{\Gamma}^i_{\beta\gamma}\frac{d\bar{x}^{\beta}}{dt}\frac{d\bar{x}^{\gamma}}{dt} = \bar{f}^i.$$

From (8.7),

$$\bar{\Gamma}^{\alpha}_{\beta\gamma} = \frac{\partial \bar{x}^{\alpha}}{\partial x^{\lambda}}\frac{\partial^2 x^{\lambda}}{\partial \bar{x}^{\beta}\partial \bar{x}^{\gamma}}.$$

Calculating the partial derivatives and using (5.6b) we obtain

$$\bar{\Gamma}^i_{jk} = 0, \qquad \bar{\Gamma}^i_{0j} = -\frac{1}{c}\varepsilon_{ijk}\bar{\omega}^k,$$

$$\bar{\Gamma}^i_{00} = -\frac{1}{c^2}\varepsilon_{ijk}\frac{d\bar{\omega}^k}{dt}(\bar{x}^j + \bar{y}^j) - \frac{1}{c^2}\bar{\omega}_j\bar{\omega}^j(\bar{x}^i + \bar{y}^i) + \frac{1}{c^2}\bar{\omega}_i\bar{\omega}_j(\bar{x}^j + \bar{y}^j)$$

$$+\frac{2}{c}\varepsilon_{ijk}\bar{\omega}^j\frac{d\bar{y}^k}{dt} + \frac{1}{c^2}\frac{d^2\bar{y}^i}{dt^2}.$$

Thus

$$m\frac{d^2\bar{x}^i}{dt^2} = \bar{f}^i + m\bar{\omega}_j\bar{\omega}^j(\bar{x}^i + \bar{y}^i) - m\bar{\omega}_i\bar{\omega}_j(\bar{x}^j + \bar{y}^j)$$

$$-2m\varepsilon_{ijk}\bar{\omega}^j\left(\frac{d\bar{x}^k}{dt} + \frac{d\bar{y}^k}{dt}\right) - m\varepsilon_{ijk}\frac{d\bar{\omega}^j}{dt}(\bar{x}^k + \bar{y}^k) - m\frac{d^2\bar{y}^i}{dt^2}.$$

This agrees with the answer to Problem 6 in Section 5.3.

4. (9.23) with the Christoffel symbols of the preceding problem.

5. Smallness of $h_{\alpha\beta}$ implies

$$R_{\alpha\beta} \approx \frac{\partial\Gamma^\gamma_{\gamma\alpha}}{\partial x^\beta} - \frac{\partial\Gamma^\gamma_{\alpha\beta}}{\partial x^\gamma} \quad \text{and} \quad \Gamma^\alpha_{\beta\gamma} \approx \frac{\eta^{\alpha\delta}}{2}\left(\frac{\partial h_{\delta\gamma}}{\partial x^\beta} + \frac{\partial h_{\delta\beta}}{\partial x^\gamma} - \frac{\partial h_{\beta\gamma}}{\partial x^\delta}\right)$$

so

$$R_{\alpha\beta} \approx \frac{1}{2}\left(\Box^2 h_{\alpha\beta} - \frac{\partial^2 h^\delta_\beta}{\partial x^\delta \partial x^\alpha} - \frac{\partial^2 h^\delta_\alpha}{\partial x^\delta \partial x^\beta} + \frac{\partial^2 h^\delta_\delta}{\partial x^\alpha \partial x^\beta}\right).$$

Harmonic condition implies

$$\frac{\partial h^\delta_\alpha}{\partial x^\delta} \approx \frac{1}{2}\frac{\partial h^\delta_\delta}{\partial x^\alpha}, \qquad \text{so} \qquad R_{\alpha\beta} \approx \frac{1}{2}\Box^2 h_{\alpha\beta}.$$

Einstein equations (9.24) imply $\Box^2 h_{\alpha\beta} = 0$, so metric components of a weak field satisfy the wave equation (see Problem 11 in Section 9.1).

6. If we neglect $\dfrac{dx^i}{d\tau}$ terms, (9.22) become

$$\frac{d^2t}{d\tau^2} + c\Gamma^0_{00}\left(\frac{dt}{d\tau}\right)^2 = 0 \quad \text{and} \quad \frac{d^2x^i}{d\tau^2} + c^2\Gamma^i_{00}\left(\frac{dt}{d\tau}\right)^2 = 0.$$

But

$$\Gamma^0_{00} = 0 \quad \text{and} \quad \Gamma^1_{00} = -\frac{1}{2}g^{ij}\frac{\partial g_{00}}{\partial x^j} \approx -\frac{1}{2}\eta^{ij}\frac{\partial h_{00}}{\partial x^j},$$

so

$$\frac{dt}{d\tau} = \text{constant} \quad \text{and} \quad \frac{d^2x^i}{dt^2} = \frac{c^2}{2}\eta^{ij}\frac{\partial h_{00}}{\partial x^j}.$$

Thus, with $\Phi = \dfrac{c^2}{2}h_{00}$, $\dfrac{d^2\mathbf{r}}{dt^2} = -\nabla\Phi$, in agreement with (7.51). From the preceding problem $R_{00} \approx \frac{1}{2}\Box^2 h_{00} = -\frac{1}{2}\dfrac{\partial^2 h_{00}}{\partial x^i \partial x^i}$. Thus, from (9.24), $\nabla^2\Phi = 0$, in agreement with Newtonian theory.

7. Solve (9.36) to obtain

$$u_1 = 1 + \frac{e^2}{2} - \frac{e^2}{6}\cos 2y + c_1\cos y + c_2\sin y.$$

Initial conditions

$$r(0) = a(1-e), \qquad \frac{dr}{d\theta}(0) = 0 \qquad \text{imply} \qquad c_1 = -1 - \frac{e^2}{3}, \qquad c_2 = 0.$$

Thus the complete first-order solution is

$$r = \frac{a(1-e^2)}{1 + e\cos[(1-\varepsilon)\theta] + \varepsilon\left[1 + \frac{e^2}{2} - \frac{e^2}{6}\cos 2\theta - \left(1 + \frac{e^2}{3}\right)\cos\theta\right] + \cdots}.$$

8. $\dfrac{6\pi GM}{a(1-e^2)c^2} = 5 \times 10^{-7}\ \dfrac{\text{radians}}{\text{revolution}} = 0.104\ \dfrac{\text{seconds}}{\text{revolution}},$

$$0.104\ \frac{\text{seconds}}{\text{revolution}} \times 415\ \frac{\text{revolutions}}{\text{century}} = 43\ \frac{\text{seconds}}{\text{century}}.$$

9. From (9.27), the time change $\Delta\tau_1$ recorded by a clock on the earth's surface at radius r_1 and colatitude ϕ is related to the time change Δt recorded by a clock at rest and unaffected by the gravitational field by

$$\Delta\tau_1 = \sqrt{1 - \frac{2GM}{c^2 r_1} - \frac{r_1^2 \sin^2\phi}{c^2}\left(\frac{d\theta}{dt}\right)^2}\,\Delta t.$$

Expanding the radical and neglecting small terms such as those arising from the Earth's angular speed $\dfrac{d\theta}{dt}$ yields

$$\Delta\tau_1 \approx \left(1 - \frac{GM}{c^2 r_1}\right)\Delta t.$$

The time change $\Delta\tau_2$ recorded by a satellite clock traveling in a circular orbit of radius r_2 and speed $v = \sqrt{\dfrac{GM}{r_2}}$ is

$$\Delta\tau_2 = \sqrt{1 - \frac{2GM}{c^2 r_2} - \frac{v^2}{c^2}}\Delta t \approx \left(1 - \frac{3GM}{2c^2 r_2}\right)\Delta t.$$

Thus

$$\frac{\Delta\tau_2}{\Delta\tau_1} \approx 1 + \frac{GM}{c^2 r_1} - \frac{3GM}{2c^2 r_2}.$$

For a geosynchronous satellite, this gives

$$\frac{\Delta\tau_2}{\Delta\tau_1} = 1 + 5.4 \times 10^{-10}\,\text{s}.$$

INDEX

Q

R

S

V

W

Y

Z

Printed in the United States
by Baker & Taylor Publisher Services